Modern Power Systems

Modern Power Systems

Edited by **Linda Morand**

C WILLFORD PRESS

New York

Published by Willford Press,
118-35 Queens Blvd., Suite 400,
Forest Hills, NY 11375, USA
www.willfordpress.com

Modern Power Systems
Edited by Linda Morand

International Standard Book Number: 978-1-68285-135-7 (Hardback)

Printed in the United States of America.

Contents

Preface

Energy efficiency and sustainable technologies are becoming a necessity with the passage of time. This has increased the scope of modern technologies to a large extent. This book elucidates the concepts and innovative models around prospective developments with respect to the modern power systems. The topics included herein like smart grid technologies, sustainable electric power, clean energy, etc. are of utmost significance and bound to provide incredible insights to readers about the construction and working of modern power systems, their efficiency and impact on the environment. This book is most suitable for graduate and post graduate engineering students and research scholars.

The researches compiled throughout the book are authentic and of high quality, combining several disciplines and from very diverse regions from around the world. Drawing on the contributions of many researchers from diverse countries, the book's objective is to provide the readers with the latest achievements in the area of research. This book will surely be a source of knowledge to all interested and researching the field.

In the end, I would like to express my deep sense of gratitude to all the authors for meeting the set deadlines in completing and submitting their research chapters. I would also like to thank the publisher for the support offered to us throughout the course of the book. Finally, I extend my sincere thanks to my family for being a constant source of inspiration and encouragement.

Editor

Improved thermal design methodology for wind power converters

Fei YANG (✉), Liang GUO

Abstract This paper presents an improved thermal design methodology for wind power converters. It combines analysis and experimental thermal design tools, including heat transfer correlations, flow network modeling (FNM), computational fluid dynamics (CFD), and experimental measurement techniques. Moreover, a systemic product development process is introduced and an effective combination between the product development process and the thermal design methodology is achieved. The draft CFD modeling at the initial design stage is done. Furthermore, it uses the detailed CFD modeling and experimental measurement techniques to provide a higher degree of accuracy at latter design stages. The key advantage of the improved methodology is its emphasis on the use of varied design tools, each of which is actively applied at its optimal point in the proposed product development process. Thus, during the earlier stages of the product development process, the thermal risk is systematically reduced, and long-term reliability of products is maintained in a higher degree.

Keywords Flow network modeling (FNM), Computational fluid dynamics (CFD), Prototype, Simulation, Experimental measurement

1 Introduction

The ultimate goal of the thermal design is not the prediction of component temperatures, but rather the reduction of risk, which is thermally associated to the product [1]. This risk, inherent to today megawatt level wind power converters, especially the power cabinet involving insulated gate bipolar transistor (IGBT) modules, is manifested in compromised designs that cannot satisfy projected schedules due to unforeseen thermal and/or reliability issues. Therefore, thermal engineers should use temperature and airflow predictions to uncover the potential risk areas and develop feasible solutions as early as possible in the product design process. Many methods are helpful during this process, including heat transfer correlations, flow network modeling (FNM), computational fluid dynamics (CFD), and experimental measurement techniques.

This paper begins with an overview of the thermal design methodology and product development process. Direct application of an improved methodology to the thermal design for a megawatt level wind power converter is discussed. Numerical modeling and empirical results are presented and compared, followed by a discussion of method improvement in the future product design.

Biber and Belady [2] first introduced a thermal design methodology which had later evolved into the "enhanced product design cycle" method [3]. According to this methodology, the development cycle comprises three distinct phases: concept development, detailed design and hardware test. During each phase, the most practical and efficient thermal design tools are employed.

1.1 Concept development

The concept development phase is the initial stage in the product design cycle. This phase is characterized by rapidly changing product layouts and requirements, as representatives from all disciplines meeting to discuss requirements and to brainstorm. At this stage, the responsibilities of thermal designers include:

1) Completing an original mechanical layout capable of supporting the power dissipation of units;

F. YANG, L. GUO, NARI Technology Development Co., Ltd., Nanjing 210061, China
(✉) e-mail: yangfei1@sgepri.sgcc.com.cn

2) Completing the design and selection of thermal related parts (fans, heat sinks, main power components);
3) Identifying the areas with thermal risks and proposing necessary design requirements to reduce these risks.

Based upon their ease of use, quick solution times and input data which are limitedly required (usually geometry and fluid data only), the common tools used in this phase are generalized correlations for heat transfer and fluid flow, and FNM techniques [3–5].

Though FNM techniques are easy ones, thermal engineers are still required for comprehensive fluid mechanics foundation and adequate thermal design experience in order to build two-dimensional (2D) flow paths in the system. Some specialized data, such as heat transfer coefficients, interface thermal resistances and system flow resistances, are selected and calculated by thermal engineers. For engineers with limited professional background, it is difficult to build this kind of model. Moreover, FNM cannot solve technical problems of three-dimensional (3D) flow paths, or additional modeling may be required if such modeling exists.

1.2 Detailed design

Once the initial layout is confirmed, the detailed design work begins. Here, thermal analysis is oriented in further details, while results are more refined. This is the preparation for actual prototype sample building. The thermal designers work with other engineers together to determine the product design details and identify the areas with thermal risks within the product. Experimental measurements of thermal related parts may be required as inputs for models, or accurate information from vendors is needed. After this phase, thermal engineers must have enough confidence to proceed into the prototype sample building stage.

Tools commonly used in the electronics industry for the detailed thermal design are CFD and finite element analysis (FEA) solvers. Typically, the 3D modeling or existing simplified 3D computer aided design (CAD) models may be required. CFD solvers marked for the electronics industry, such as Fluent [6] and Flotherm [7], incorporate the ability to perform heat transfer calculations and fluid flow solutions. Users should be able to deal with various details in the CFD modeling, including system boundary condition input, power component modeling and model grid subdivision [8–11], etc. In addition, users need to analyze simulation results, dig deeply into critical areas, and offer design suggestions for actual products.

1.3 Hardware test

Once the prototype product is available, the hardware test phase begins. The designers' goal at this stage is to experimentally measure critical components and risk factors in the product to verify the design. Additionally, measurements are compared with estimates to calibrate or fine-tune earlier models. These comparisons are used to determine the accuracy of initial predictions and to assist designers to develop the thermal intuition.

While the most commonly used tool for temperature measurement is thermocouple, other tools, such as infrared imaging technique and optical radiation meter, are also available. Detailed information can be found in [12, 13].

The air speed measurement is also valuable during this phase. Hot-wire anemometer is the normal tool in the electric device field. It can capture precise and repeatable flow speed, but needs clean testing condition. The impeller anemometer is another easily operated tool for the main flow channel, but it is unsuited for breeze testing.

2 Product development process

A systemic product development process defines the key process steps in the design and development of the entire product for customers. A successful process should meet the following goals [14]:

1) Design and development schedules should be optimized for cost, quality and delivery, to meet customers' needs or to exceed customers' expectation.
2) The design should meet all of the customers' specification and internal requirements; any deviation should be confirmed and approved by customers with written feedback.
3) The product should be designed to meet the company's internal requirements for manufacturing ability, quality and reliability.

Based on the wind power converter design experience and referenced development process from some international companies, the proposed systemic product development process is shown in Fig. 1.

Referring to Fig. 1, six major development stages are defined as follows: request for quotation (RFQ), prototype, engineering verification test (EVT), design verification test (DVT), pilot, and mass production stages. Regarding the proposed product development process, the design team can follow or skip any of stages in the development cycle.

At the RFQ and prototype stages, the design team provides technical proposals, detailed quotations and prototype samples, which are the design foundation for the total product development process. Thus, this is where thermal engineers provide the most important input for product design. At latter stages, the thermal designers'

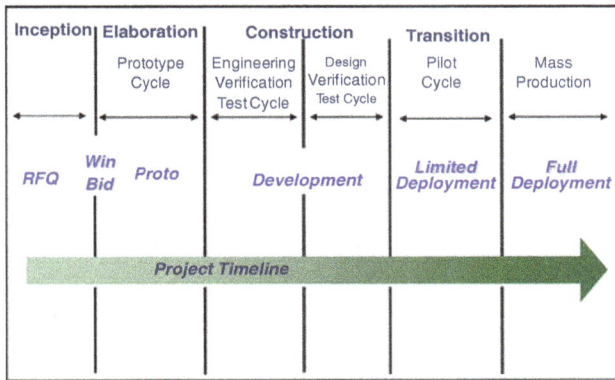

Fig. 1 Product development process

responsibility turns into experiment measurement and simulation model optimization.

3 Power cabinet design: improved methodology

The initial "enhanced product design cycle" method is shown in Fig. 2.

The improved methodology integrates the "enhanced product design cycle" method into the proposed product development process of the thermal design for wind power converters. It introduces the draft CFD modeling at the concept stage, and focuses on the detailed CFD modeling at the design stage. In this way, each step is validated. Thus thermal designers can test their intuition and assumptions in the earlier work.

3.1 Concept development phase

Corresponding to the concept development, there is an RFQ stage in product development process. At the RFQ stage, the bill of materials (BOM) and product technical proposals

are main outputs of the design team. The design team includes electronic, mechanical and thermal engineers.

Since the wind power converter is to maintain "a long life" of 20 years, the thermal performance and the reliability of products are of the utmost importance. Especially, as a key section in the converter design, the power cabinet design is the primary element at the beginning.

The architecture design of the power cabinet begins with IGBT power module selection (to meet the customers' specification), system cabinet size and price consideration. After several days' work, one power module is selected finally.

IGBT modules are the pressure contact ones without base plates. The direct bonded copper (DBC) substrate is not soldered to the base plate. Instead, it is pressed directly to the heat sink. The elimination of a base plate assures the high temperature cycling capability (no solder fatigue caused by the thermal expansion coefficient mismatch) and a low thermal resistance [15].

Six IGBT modules are tired upon a power cabinet. The lower group with three modules (Modules A, B, C) just like a sandwich for the motor side, which is air-forcedly cooled by a radial fan. The upper group with three modules (Modules E, F, G) is for the grid side, which uses another radial fan as well. The power cabinet with two main groups is designed by using generalized empirical correlations, geometric data from mechanical layout drawings and vendors' specifications. The power cabinet layout is modeled by using a commercial FNM solver, called the MacroFlow, as shown in Fig. 3.

Airflow estimates from the flow network models are used to predict the airflow rate through each power module region, as well as to estimate the surface and junction temperatures of each power module heat sink. The

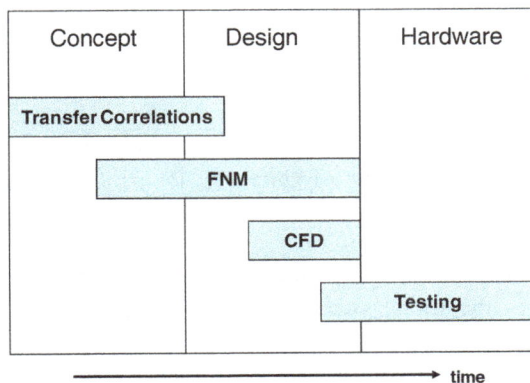

Fig. 2 Enhanced product design cycle [3]

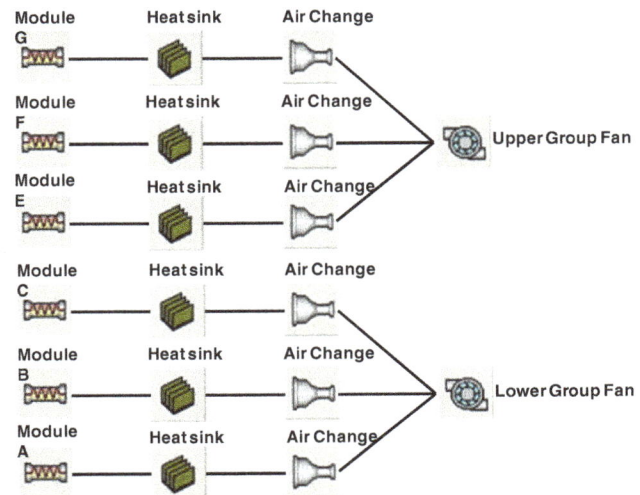

Fig. 3 FNM used in the modeling of a power cabinet

maximum junction temperature of an IGBT module is 150 °C in the data sheet. Considering the general derating in the converter field, it is determined to be 105 °C. Worst-case IGBT module temperature is estimated with the use of extrusion heat sink and the thermal interface data is described by (1) [16–19]. The thermal resistance Θ, is defined by (2).

$$T_j = (T_a)_{\max} + (\Theta_{jc} + \Theta_{ch} + \Theta_{ha})P_{\max} \tag{1}$$

$$\Theta = \Delta T/P \tag{2}$$

where T_j is the junction temperature (°C); T_a is the ambient temperature (°C); Θ_{jc} is the thermal resistance between the junction and the case (°C/W); Θ_{ch} is the thermal resistance between the case and the heat sink (°C/W); Θ_{ha} is the thermal resistance between the heat sink and the ambient (°C/W); $P_{\max}$ is the maximum power loss (W); ΔT is the temperature difference (°C); and P is the power loss (W).

In this manner, the minimum required airflow through each power module is determined. Then the total air volume at the working point of every radial fan is gained simultaneously. Based on the estimation above, two radial fans are chosen initially.

Then, thermal engineers could provide the main BOM on the thermal design aspect. However, at the RFQ stage, it is critical to ensure the accuracy of commercial quotation. The big error range using a FNM tool is a potential risk because of difficulties on estimation of system resistances and airflow distribution. Here, introducing the draft CFD model in advance is a good option. Then, thermal designers could verify the fan selection and airflow distribution with ease, and ensure that each power model will get enough air volume, while reducing the thermal risk because of improper fan selection at the RFQ stage.

The draft CFD modeling is accomplished by using a commercial CFD software package, called the Flotherm. The CFD modeling includes two groups: the motor-side modules and the gird-side modules.

The IGBT modules are directly derived from a specific 3D model on website. Regarding two flow channels, their sizes and specific geometry are not designed at the system level. Instead, two simple cubic channels are modeled to substitute the actual structure. Additionally, the CFD model is not generally solved for heat radiation effects because heat transfer and convection are main ways in a power cabinet. In this manner, the simplified system-level CFD model is constructed and solved. The system-level flow distribution section is shown in Fig. 4.

Referring to the simulation data of draft CFD modeling, the flow distribution in each power module is average. It is proved that two chosen radial fans could meet the thermal design requirement at the RFQ stage.

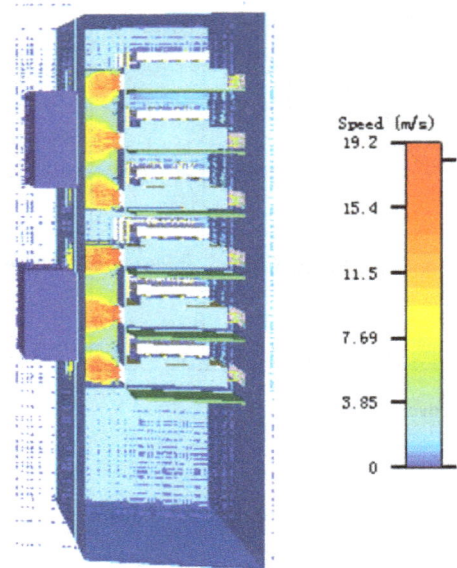

Fig. 4 Flow distribution section in the draft CFD modeling

3.2 Detailed design phase

Corresponding to the detailed design, there is a design period of prototype stage in the product development process. Once the commercial contract is signed and the conceptual design is approved by the customers, the thermal designers begin to concentrate on the areas with thermal risks identified in the initial modeling. The emphasis is placed on developing the draft CFD model at the RFQ stage. In order to do this, more detailed modeling techniques are employed. Furthermore, the result of detailed CFD simulation instructs the prototype product preparation.

Compared with the draft CFD modeling, the detailed CFD modeling begins to set up with an actual 3D structure in order to simulate the airflow distribution and the junction temperature of each power module. Besides, accurate power loss data, detailed flow channel structures and fan positions are determined. The system-level flow and the temperature distribution section are shown in Fig. 5.

3.3 Hardware test phase

Corresponding to the hardware test, there is a test period of prototype stage in the product development process. Once a prototype product is available, the thermal designers shift their focus once again. The emphasis is moved to temperature collection and airflow data measurement in the actual system. The prototype product is shown in Fig. 6.

The system is fully instrumented with thermocouples in order to obtain the thermal profile of power modules. Each

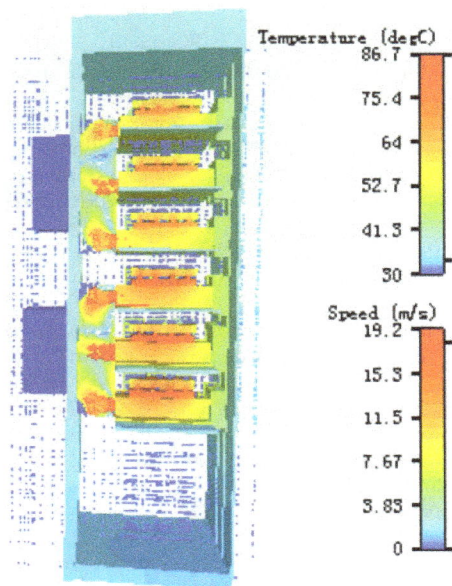

Fig. 5 Flow and temperature distribution section in the detailed CFD modeling

inlet of power module heat sinks is tested by an impeller anemometer. The testing conditions are 2,000 m altitude, 30 °C ambient and maximum power using (1). Extrapolated results are compared with component specifications for design verification.

Fig. 6 Prototype sample of a power cabinet

4 Results

4.1 Airflow

Volumetric airflow predictions of FNM, draft CFD modeling, detailed CFD modeling, and measured values are compared in Table 1.

Airflow predictions are very close by using the draft and the detailed CFD modeling because of the simple structure of two main flow channels. The similar flow channels in that two phase results in the close airflow predictions. Moreover, it validates the significance to introduce the draft CFD modeling at the RFQ stage.

Airflow predictions based upon FNM and CFD modeling are lower than the measured airflow values in the prototype system. There are some reasons for these deviations. First, the 2D FNM technique cannot completely represent the 3D flow present there. Airflow estimates taken from the flow network models are functional requirement for the system fan selection. Considering the system airflow redundancy and commercial quotation, the thermal engineer usually chooses the higher level fans at the RFQ stage. That explains the reason why CFD predictions are bigger than FNM results. Secondly, it is difficult to build a precise CFD modeling because of the complex airflow distribution in the actual system. In general experience, the simulation results based on the software Flotherm are conservative. That is the possible reason why CFD predictions are less than experimental results.

The increasing tendency of airflow data from FNM predictions to measured values are seen in Table 1. It reflects the redundancy design idea at earlier stages, and relatively reduces the thermal design risk.

4.2 Temperature

Junction temperature estimates are compared with measured values in Table 2.

As shown in Table 2, junction temperature predictions reflect the same rules with airflow: measured values are

Table 1 Volumetric airflow predictions

Power module section	Airflow predictions (m³/h)			Measured airflow values (m³/h)
	FNM	Draft CFD	Detailed CFD	
Module A	270.4	363.7	364.1	451.4
Module B	274.3	358.2	363.6	460.9
Module C	270.2	358.4	361.0	427.7
Module E	204.5	327.6	317.9	403.9
Module F	210.6	326.4	315.8	413.4
Module G	203.6	325.3	312.5	403.9

Table 2 Component junction temperature at 30 °C

Power module section	Junction temperature (°C)		
	FNM	Detailed CFD	Measured
Module A	98.4	85.6	75.42
Module B	95.2	86.9	69.94
Module C	97.2	86.2	69.55
Module E	96.4	74.6	80.53
Module F	93.5	74.4	77.59
Module G	95.4	72.4	79.52

better than temperature predictions based upon FNM and CFD. The volumetric airflow is the key factor to affect the junction temperature of a IGBT module. The decreasing tendency of temperatures from FNM predictions to measured values indicates the redundancy design idea again.

Although there are discrepancies between the measured temperatures and the predicted values, the goal of the design is accomplished: all measured temperatures are lower than 105 °C. As a result, the product has been successfully released, thus it requires no changes to the system.

5 Conclusions

A systemic thermal design methodology is presented and applied to the design of the power cabinet in a megawatt level converter. The improved methodology combines the product development process tightly, and supports the document approval at the RFQ stage, while reducing the systematic risk through application of available thermal design tools and techniques at earlier stages. For the design of a power cabinet, the methodology has key advantages:

1) Decreasing the commercial and design risk at the RFQ stage through comprehensive application of FNM and draft CFD modeling;
2) Utilizing the commendable combination between the thermal design methodology and the product development process;
3) Applying various thermal design tools to reduce the system thermal risk and maintain the long-term reliability of products.

Finally, through enormous frequent applications in different cabinets and experience accumulation, the improved thermal methodology will play an important role in the future wind power converter design.

References

[1] Belady C, Minichiello A (2003) Effective thermal design for electronic systems. Electron Cool Mag 9(2):16–21
[2] Biber C, Belady C (1997) Pressure drop prediction for heat sinks: What is the best method? In: Proceedings of the Pacific Rim/ASME international electronic packaging technical conference (INTERpack'97), Kohala Coast, HI, USA, 15–19 Jun 1997
[3] Belady C (1999) Improving productivity in electronics packaging with flow network modeling (FNM). Electron Cool Mag 5(1):36–40
[4] Ellison G (1984) Thermal computations for electronic equipment. Van Nostrand Reinhold Company, New York
[5] Minichiello A (2000) Flow network modeling: A case study in expedient system prototyping. In: Proceedings of the 7th intersociety conference on thermal and thermomechanical phenomena in electronic systems (ITHERM'00), Vol 1, Las Vegas, NV, USA, 23–26 May 2000, pp 70–77
[6] Fluent corporate website.
[7] Flotherm corporate website.
[8] Patel CD, Belady CL (1997) Modeling and metrology in high performance heat sink design. In: Proceedings of the 47th electronic components and technology conference (ISPS'97), San Jose, CA, USA, 18–21 May 1997, pp 296–302
[9] He Y (2008) Application of flotherm in thermal analysis on electronic equipments. Electron Qual 1:39–40 (in Chinese)
[10] Tiersten HF, Sham TL, Lwo BJ, et al (1993) A global–local procedure for the heat conduction analysis of multichip modules. In: Proceedings of the 1993 ASME international electronics packaging conference, Vol 1, Binghamton, NY, USA, 29 Sept–2 Oct 1993, pp 103–118
[11] Oda J, Sakamoto J (1998) Applications of FEM for multiple laminated structure in electronic packaging. Finite Elem Anal Des 30(1/2):147–162
[12] Moffat RJ, Ortega A (1997) Experimental methods in air cooling of electronics. In: Proceedings of the 13th annual IEEE semiconductor thermal measurement and management symposium, Austin, TX, USA, 28–30 Jan 1997, pp 15–16
[13] Ortega A, Moffat R (2000) Experimental methods in air cooling of electronics. In: Proceedings of the 7th intersociety conference on thermal and thermomechanical phenomena in electronic systems (ITHERM'00), Vol 1, Las Vegas, NV, USA, 23–26 May 2000
[14] Emerson corporate website.

[15] Ye H, Lin M, Basaran C (2002) Failure modes and FEM analysis of power electronic packaging. Finite Elem Anal Des 38(7):601–612
[16] Hanreich G, Nicolics J (2001) Measuring the natural convective heat transfer coefficient at the surface of electronic components. In: Proceedings of the 18th instrumentation and measurement technology conference (IMTC'01), Vol 2, Budapest, Hungary, 21–23 May 2001, pp 1045–1050
[17] Aghazadeh M, Mallik D (1990) Thermal characteristics of single and multi-layer high performance PQFP packages. IEEE Trans Compon Hybrid Manuf Technol 13(4):151–153
[18] Hassselman DPH, Donaldson KY, Barlow FD (2000) Interfacial thermal resistance and temperature dependence of three

adhesives for electronic packaging. IEEE Trans Compon Pack Technol 23(4):175–179

[19] Zhang X, Wong EH (2002) Thermo-mechanical finite element analysis in a multichip build up substrate based package design. Microelectron Reliab 44(4):611–619

Author Biographies

Fei YANG was born in China, in 1975. He received the B.E. degree in Mechanical Manufacture and Technology from Southwest Jiaotong University in 1998, and the M.E. degree in Engineering Mechanics from Dalian University of Technology in 2005. He worked for a number of years on telecom devices in Huawei Technologies Co., Ltd., and power supply units in Emerson Telecom Power, U.S. In 2012, he joined NARI Technology Development Co., Ltd. as an engineer. His main research interests are focused on thermal design, environmental suitability and structural design.

Liang GUO was born in China, in 1982. He received the B.E. degree and the M.E. degree in Power Electronics from Nanjing University of Aeronautics and Astronautics in 2005 and in 2008, respectively. He is currently an engineer in NARI Technology Development Co., Ltd. His research interests include power electronic converters applied to renewable energy sources, power systems and energy efficiency.

2

Demand response: a strategy to address residential air-conditioning peak load in Australia

Robert SMITH (✉), **Ke MENG, Zhaoyang DONG,**
Robert SIMPSON

Abstract Rapid growth in electricity network peak demand is increasing pressure for new investment which may be used for only a few hours a year. Residential air-conditioning is widely believed to be the prime cause of the rise in peak demand but, in the absence of detailed residential demand research, there is no bottom-up empirical evidence to support this supposition or to estimate its impact. This paper first examines the developments in network peak demand, at a national, network distribution, and local distribution feeder level to show recent trends in peak demand. Secondly, this paper applies analytics to the half-hourly consumption data of a sample of Ausgrid's interval metered customers, combined with local weather data, to develop an algorithm which can recognize air-conditioner use and can identify consumption patterns and peak load. This estimate is then compared to system peaks to determine residential air-conditioning's impact on overall demand. Finally, this paper considers the future impacts of air-conditioning load on peak demand as penetration rates reaches saturation levels and new minimum energy performance standards take effect reducing new units peak impacts.

Keywords Air conditioner, Demand response, Peak load

R. SMITH, R. SIMPSON, Ausgrid, Sydney, NSW 2000,
Australia
(✉) e-mail: rsmith@ausgrid.com.au
R. SIMPSON
e-mail: robert.simpson@ausgrid.com.au
K. MENG, Z. DONG, Center for Intelligent Electricity Networks
(CIEN), The University of Newcastle, Newcastle, NSW,
Australia
K. MENG
e-mail: ke.meng@newcastle.edu.au
Z. DONG
e-mail: joe.dong@newcastle.edu.au

1 Introduction

Electricity peak demand is increasing rapidly in Australia in the past decade, driven by the connection of new businesses and homes, and also by a rise in the use of air conditioners. The residential air-conditioning load is widely believed to be a significant portion of electric utility peak load, which is a major driver of investment for electricity network and generating facilities. This rapid uptake has major ramifications for national electricity infrastructures and requires large investment to cope with short peaks in load. The federal government energy white paper [1] characterizes this problem as a customer' purchase of a $1,500 air-conditioner driving $7,000 of extra augmentation capital expenditure across the electricity grid. Therefore, if without detailed residential air-conditioning load research, there will be no bottom-up empirical evidence to estimate its impacts and to reduce growth-related infrastructure investment.

Many Australian utilities have identified the enormous challenges of the increasing peak demand to the operation of future power grids [2]. They are devoting increasing resource to address the peak demand problems, and are promoting industry and academia to conduct joint research to improve transition of latest research results into industrial applications. Shaving or shifting the peak demand is one of the most direct ways to address the air-conditioning load issues. A lot of interests have been directed towards shifting peak load to off-peak periods. Four key initiatives are listed as follows [3].

1) Keep the customers updating with varying prices and let them decide when to modify or curtail their energy use;
2) Direct load control of major contributors of peak demand (air conditioning, pool pumps, water heaters, and etc);

3) Small-scale distributed generations (solar, wind, electric vehicles, energy storage, and etc);

4) Improve energy efficiency standards (buildings thermal performance, electric appliance efficiency, and etc).

Large-scale deployment of these programs has the potential to reduce the need for expensive peak generations, to provide significant bill savings for customers, to enhance electric system reliability, and to increase the penetrations of intermittent renewable energy [4]. However, at the current stage, most of the activities and initiatives aiming at reducing peak load focused on the commercial and industrial sectors where are logistically easier to be implemented than in the residential sector [5]. But it is reported that the peak demand from residential air conditioners will grow more rapidly in the near future. Consequently, the key element for further development is to design a program which can help to address the peak demand of residential air-conditioning units directly. While progress has been made in recent years, more effort is needed in the future.

This paper is organized as follows - after introduction section, the developments in network peak demand, at a national, network distribution, and local distribution feeder level is examined to show recent trends in peak demand. After that, half-hourly metered customer load data from Ausgrid, combined with local weather data is used to develop an algorithm which can predict air-conditioner existence and identify consumption patterns and peak demand, followed by the determination of residential air-conditioning's impacts on the whole power network. Finally, future impacts of air-conditioning load on peak demand as penetration rates reaches saturation levels are discussed and new minimum energy performance standards take effect reducing new units peak impacts are proposed. Conclusions and further developments are provided in the last section.

2 Customer demand and temperature

In this section, the developments in network peak demand are studied to show recent trends in peak load. Moreover, the load patterns and the temperature profiles are studied together so that the average phase relationship between the demand and the temperature in four seasons can be observed respectively.

2.1 National demand

Summary of national demand in different years is shown in Table 1.

Table 1 Summary of demand of Australian NEM (mw)

Year	Maximum	Mean	Median	Minimum
2009	35,433.42	23,408.38	23,561.79	15,881.00
2010	33,752.42	23,329.49	23,543.23	15,455.58
2011	34,887.65	22,924.67	23,166.51	15,144.41

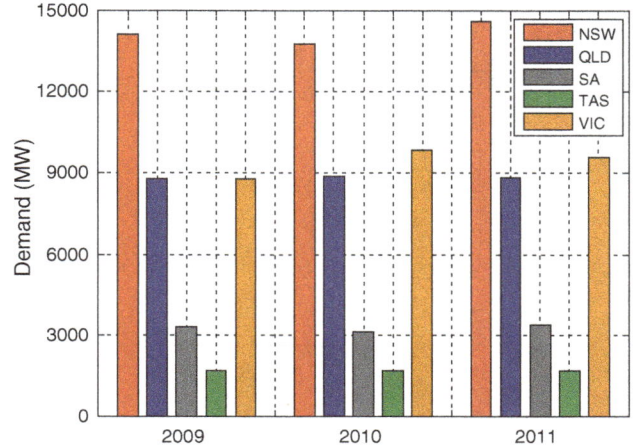

Fig. 1 Regional peak demand in Australian NEM

Regional peak demand in Australian national electricity market (NEM) is shown in Fig. 1. The Australian NEM is composed of the states of New South Wales (NSW), Queensland (QLD), South Australia (SA), Tasmania (TAS), and Victoria (VIC) [6]. The average electricity demand in the east Australian states of NSW, QLD, and VIC are much higher than that of the other two states. In the NEM, demand for electricity has been falling in recent years; however the rise in peak demand has been even more pronounced [7]. Our study shows that a significant portion of the reduction in electricity demand in the NEM is attributed to a shift away from grid-sourced electricity.

2.2 Household demand

In the past decade, residential load has undergone substantial changes with higher volatility and greater unpredictability. In the following study, the half-hourly end-user metered load data from 180 of Ausgrid's customers is analysed. The demand data covers one year, from 01 May 2010 to 30 April 2011. The comparisons of the half-hourly average customer household load of four seasons are shown in Fig. 2. The four seasons in Australia are winter (Jun.–Aug.), spring (Sep.–Nov.), summer (Dec.–Feb.), and autumn (Mar.–May).

The figures above illustrate the different level of averaged customer load of four different seasons. It is easy to find the household demand in winter is higher than in the

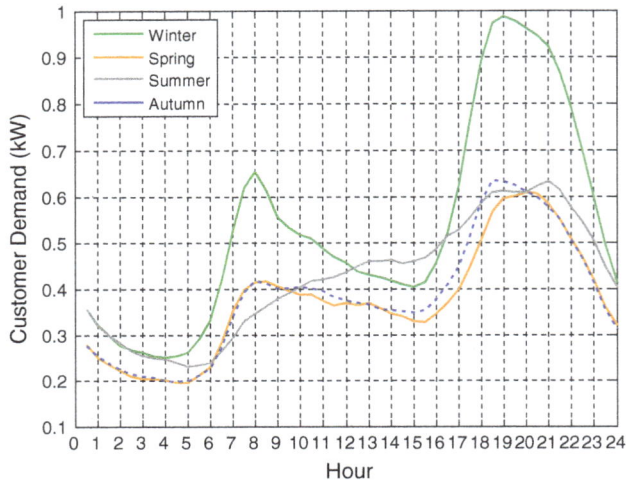

Fig. 2 Half-hourly average customer household load of different seasons

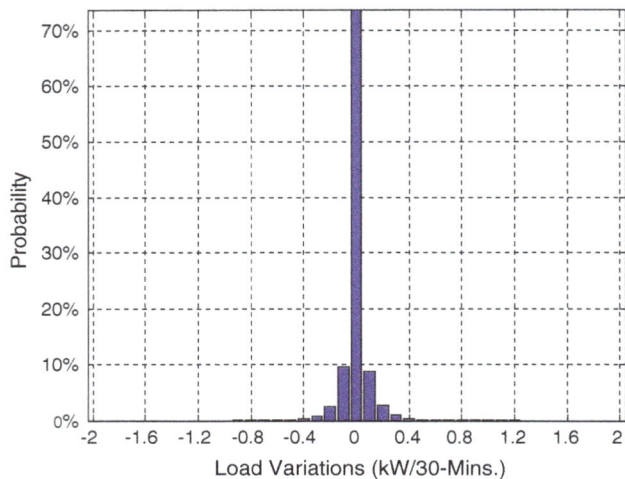

Fig. 3 Distribution of load variations of one customer

Table 2 Statistics of customer load variations (kW)

Customer ID	Mean	Std.	Max. J.U.	Max. J.D.
1	−6.56E−06	9.48E−02	1.236	0.935
2	4.28E−06	2.42E−01	2.637	2.055
3	1.06E−05	2.64E−01	2.511	2.130
4	−1.71E−07	2.14E−01	1.796	1.402
5	1.82E−05	2.09E−01	2.379	1.745
6	−1.20E−05	1.31E−01	1.465	1.144
7	−9.93E−06	1.55E−01	1.719	1.653
8	−2.11E−06	4.77E−01	4.095	3.422
9	2.28E−07	2.61E−01	1.523	1.768
10	−1.25E−05	1.62E−01	2.077	2.008
11	−3.03E−06	3.40E−01	3.960	3.662
12	7.64E−05	2.88E−01	2.416	2.083
13	5.71E−08	1.02E−01	2.294	1.559
14	6.28E−07	2.28E−01	1.864	1.510
15	1.37E−06	1.26E−01	1.365	1.290
16	2.45E−06	2.03E−01	1.765	1.962
17	−8.56E−07	2.77E−01	2.350	2.085
18	4.77E−05	2.89E−01	2.615	2.064
19	1.49E−05	4.13E−01	4.592	4.328
20	2.28E−06	1.97E−01	2.008	2.515
21	−5.71E−07	1.58E−01	2.054	1.721
22	−6.28E−06	1.37E−01	1.338	1.323
23	−7.47E−05	4.76E−01	3.715	3.130
24	−6.85E−06	2.17E−01	2.417	1.712
25	3.11E−05	6.13E−01	6.125	4.142

Note: "J.U." means "Jump Up", "J.D." means "Jump Down"

other seasons during the studied period. It is noticeable that the daily demand in winter is between 0.2 and 1.0 kW, with two peaks in the morning and in the evening, respectively. The lowest point is at around 04:00 in the early morning. On the contrary, in summer, the daily load curve is relatively flat, which climbs from early morning to late evening gradually, and then drops rapidly to the morning level again. The demand in spring and autumn show basically the same patterns as it is in winter, but only in much lower levels. Generally, in Australia, the equipments that drive summer peak demand include air conditioners, pool pumps, and etc. In winter, electricity needed to operate heating systems significantly contributes to the peak demand. The basic load is driven by devices that are used year-round, like fridges, electric water heaters, dishwashers, clothes washers and dryers, hot tubs and waterbed

heaters. In Fig. 3, the results show that the half-hourly variations of all household demand are distributed according to normal distributions, with same mean values but with different standard deviations. In Table 2, the statistic results of 25 customers are provided.

2.3 Temperature

Half-hourly average temperature of different seasons is shown in Fig. 4. From the result above, we can find that temperature changes across one day, normally with a high period of temperature in the early afternoon followed by a drop until the early morning. The peak temperature occurs at approximately 13:00–15:00 in the afternoon, which is different as the peak demand at 07:00–09:00 and 18:00–21:00. Moreover, there are significant variations in average temperature between four seasons, with higher temperature expected in summer compared to in spring, autumn, and winter. In addition, the duration of high

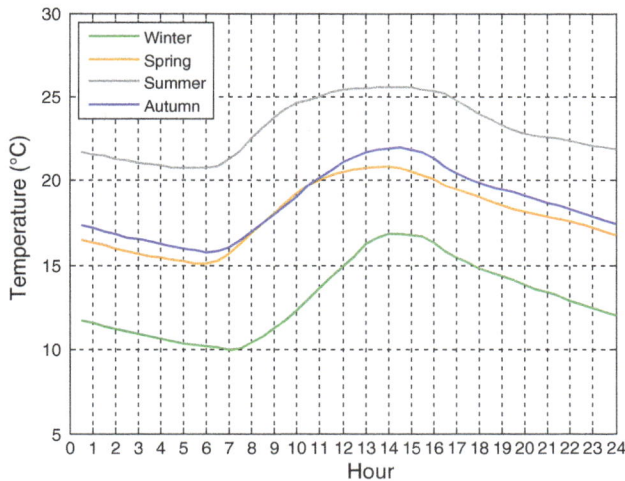

Fig. 4 Half-hourly average temperature of different seasons

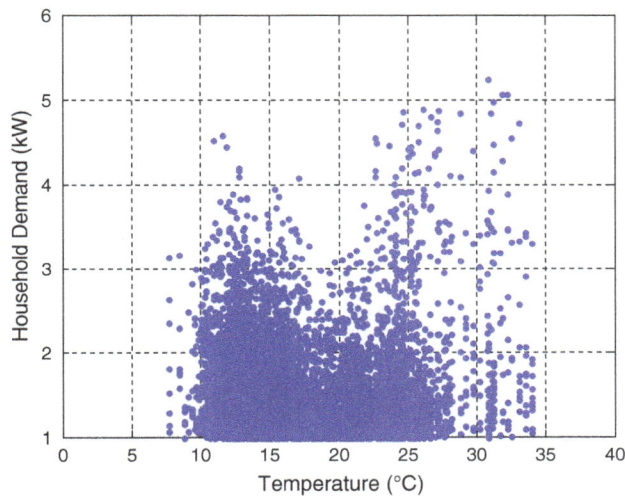

Fig. 5 Temperature vs. load of one customer

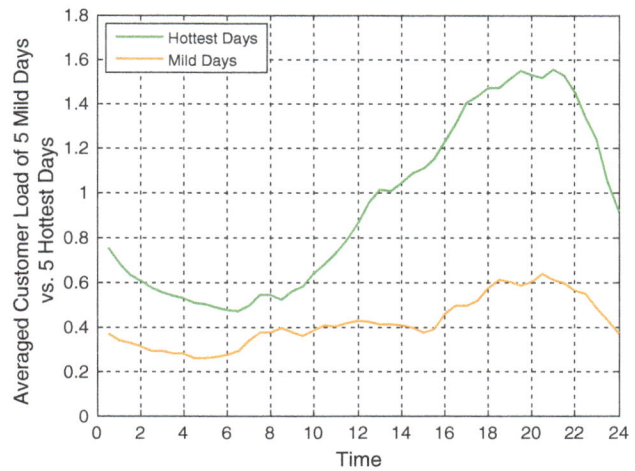

Fig. 6 Averaged customer load of five mild days vs. five hottest days

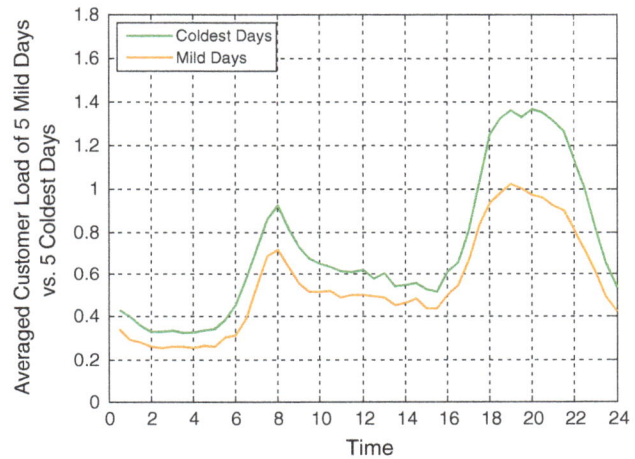

Fig. 7 Averaged customer load of five mild days vs. five coldest days

temperature periods change between four seasons, with peaks in spring and summer lasting longer.

2.4 Temperature vs. demand

The average household demand vs. average temperature is shown in Fig. 5. Each point represents an observation of temperature versus demand in that time period for a given working day. The striking feature of the graph is the nearly v-shaped structure. The joint of the "v" in peak periods is around 18–20 °C. As temperatures increase above this level, demand increases; similarly, at temperatures below this level, loads also increase. Therefore, outdoor weather conditions are crucial in determining residential energy consumption for household air-conditioning appliances.

In order to show the weather impacts more clearly, all the household demand from Monday to Friday is aggregated to show the difference between five mild summer days and five hottest days in Fig. 6, five mild winter days and five coldest days in Fig. 7. The average demand curves are peakier for hot and cold days due to increased air conditioning load.

3 Air conditioner use pattern study

In this section, half-hourly metered customer load data from Ausgrid, combined with local weather data is used to develop an expert system which can predict air-conditioners existence and identify consumption patterns. And then air-conditioning load is extracted from household load data in order to evaluate the impacts of household air conditioners.

3.1 Air conditioner existence prediction

An expert system is developed to identify the existence of air conditioner from the household load data. The expert system contains a knowledge base containing accumulated experience and a set of rules, which are expressed with natural language rules "IF … THEN …" [8]. This expert system could be enhanced with additions to the knowledge base or to the set of rules. With the expert system, the customer load data is firstly analyzed for jump up/down that larger than 0.5 kW load. The followings are some expert knowledge used to predict the existence of air conditioners.

1) Larger than median value of half-hourly load in this month [0.5–1.0 kW (+1), 1.0 kW–1.5 kW (+2), >1.5 kW (+3)];
2) Load jump up or jump down 0.5–1.5 kW [0.5–1.0 kW (+1), 1.0 kW–1.5 kW (+2), >1.5 kW (+3)];
3) Find some patterns which happen at the same time every day or most of the days in this month (e.g. pool pump);
4) Do not turn on or turn off air conditioners frequently, normally keep on/off for at least 1 h (e.g. electric water heater) and do not use for over 24 h;

Table 3 Survey vs. prediction results

Survey		Prediction	
"W"	"WO"	"W"	"WO"
108	72	103	77
Misclassified customer		13	
Prediction accuracy		92.78%	

Note: "W" means "WITH", "WO" means "WITHOUT"

5) Normally turn on air conditioners in hot summer days or cold winter days.

Table 3 shows the survey vs. prediction results.

The promising prediction performance based on the demand data illustrates the efficiency of the proposed expert system. The recognized air conditioner ON/OFF state of one customer is provided in Fig. 8. It shows that the air conditioner is normally turned on during hot summer days. Although the power ratings of air-conditioning appliances for each house are not known, based on prediction results above, the power drawn by air conditioners at each house can be estimated.

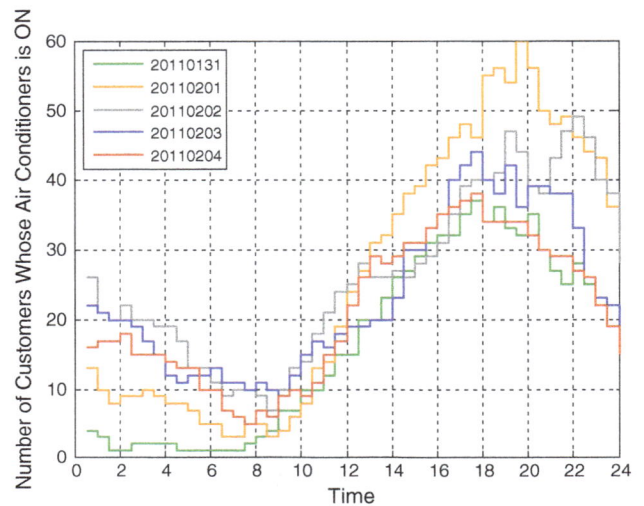

Fig. 9 Number of air conditioners is ON in hot summer days

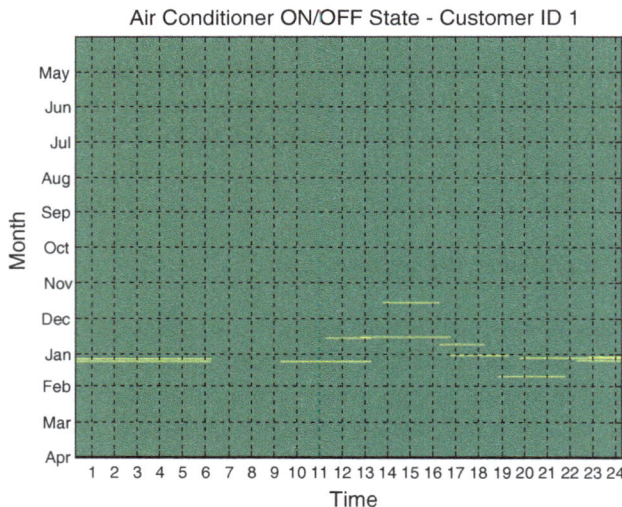

Fig. 8 Air conditioner ON/OFF states recognition of one customer

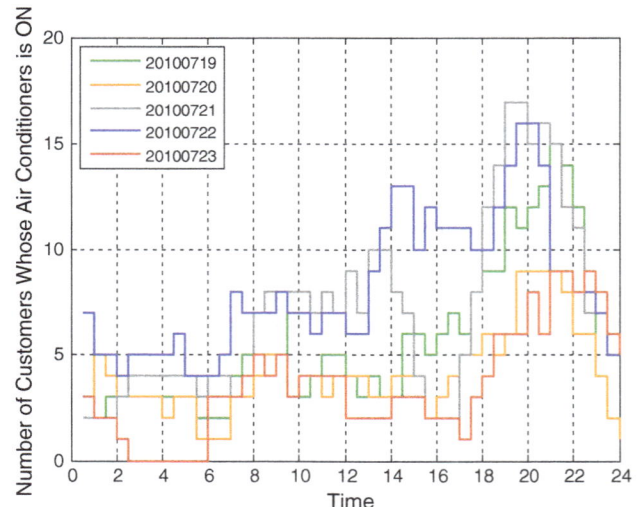

Fig. 10 Number of air conditioners is ON in cold winter days

Table 4 Air conditioner on ratio in summer and winter peaks

Season	Date	ON ratio/%	Average time/h
Summer	31/01/2011	62.04	11.67
Summer	01/02/2011	83.33	14.02
Summer	02/02/2011	75.93	15.44
Summer	03/02/2011	73.15	14.20
Summer	04/02/2011	65.74	14.30
Winter	19/07/2010	2.78	1.00
Winter	20/07/2010	2.78	1.00
Winter	21/07/2010	1.85	1.00
Winter	22/07/2010	6.48	1.00
Winter	23/07/2010	2.78	1.00

3.2 Impacts of residential air-conditioning load

Another factor which is of interest is how many of air conditioning units are turned on at peaks. In the studied area, the number of air conditioners turned on during hot summer days and cold winter days are presented in Figs. 9 and 10. Air conditioner on ratio in summer and winter peaks is shown in Table 4. The study shows that more than 60% air conditioners are turned on during peak summer days, and this figure reaches 83% on extreme hot day. A lot of customers run their air conditioners over 12 h on peak days, but in winter, the average operation time is only 1 h. It is reasonable because the weather at summer peak is extremely hot in Australia, so air conditioners continuously operate for longer time.

The summer heat and winter cold pushes an increase in the penetration of air-conditioning load, therefore, change both the magnitude and composition of the system demand. Along with the higher load come several problems, the extensive usage of household air-conditioner is leading to capacity straining load peaks, increasing the costs of electricity supply and volumes of greenhouse gases emissions. Moreover, if the air-conditioning load reaches certain portion of feeder load, then it is very likely to cause security and reliability problem [9, 10]. The recent load peaks have resulted in supply disruptions, and on some occasions, high spot price on the NEM [11]. Accordingly, initiatives to curtail/shift peak demand have significant future potential value in terms of reducing growth-related infrastructure investment and improving system reliability. Fortunately, demand response programs provide an effective solution to the residential air-conditioning load problem. Demand response refers to "changes in electric usage by end-use customers from their normal consumption patterns in response to changes in the price of electricity over time, or to incentive payments designed to induce lower electricity use at times of high wholesale market prices or when system reliability is jeopardized" [12]. Turning on and off air conditioning appliances doesn't change indoor temperature immediately. As a result, air conditioning appliances become important and effective objects in demand response programs. Nowadays, increasing number of Australian utilities is moving towards supporting wide scale introduction of interval metering and design of demand response programs to improve the efficiency of the NEM. In the next section, a number of barriers blocking the further applications of demand response programs will be discussed.

4 Barriers of demand response

Demand response programs can effectively adjust the imbalance between supply and demand, prevent overinvestment and ensure safe and reliable operation of power network, and can promote fairer marketing by introducing new competition mechanisms [13]. However, limits on the number of customers called for demand response programs reflect the deficiency of current demand response technologies and low level of customer acceptance. Some of the major barriers are summarized as follows.

1) Lack of advanced metering infrastructure;
   At current stage, a great majority of residential customers in Australia are billed for electricity using the traditional meters, which can only record the accumulated electricity consumption. Customers are billed according to the difference between the previous reading and the latest one. In the future, real-time information, such as price, will be delivered to customers more effectively through smart meters. Control signals become more widely accessible through communication networks. All these technologies will facilitate the application of more intelligent and customer-friendly demand response programs, allowing customers to identify and target discretionary loads that can be curtailed or shifted [4]. A roll-out of smart meters and a shift to time-of-use pricing structures along with in-home energy displays provides the best way to reduce air-conditioning peak load.

2) Lack of interoperability and open standards.
   Although the increasing number of air conditioner suppliers ensures that their products are compatible with different demand response programs, there is no open standard for all brands of air-conditioning appliances. Furthermore, a lot of installed air-conditioning units cannot be enhanced to provide demand response function. Some elements are needed to be incorporated into these air conditioners specifically to facilitate participation in different demand response programs.

To some extent, it may raise the overall implementation cost and reduce the potential benefits of these programs correspondingly. Governments and policy makers should encourage demand response programs with generous subsidies and with regulatory support, which will help to reduce the costs and risks to programs sponsors and to air-conditioning unit users. Furthermore, programs sponsors will be more willing to develop and offer air conditioner demand response program and end-users will have more options.

3) Lack of effective demand response programs.

A series of demand response strategies involving air conditioning systems are provided in [13]. Generally speaking, each strategy has its own merits and drawbacks. Many attempts try to merge some of the individual implementations together into a new strategy, so that it can overcome individual disadvantages and benefit from each others' advantages. The simplest but most useful strategy is to increase the temperature setting point for an entire facility, thereby reducing the electric load. It can be implemented by changing set points in one step or several steps, or by increasing over time. In the future, one or multiple strategies can be selected and applied together or in sequence to realize demand response according to specific characters of air-conditioning appliances.

4) Lack of customer awareness and participation.

Creating customer awareness and increasing active participation is the most significant factors in demand response programs. A lack of understanding of initiatives has negative effects on customer awareness and participation in demand response programs. This situation is likely to change if participations are driven by consumer demand due to increased awareness, and if programs are supported by governments and utilities. For example, time-of-use pricing is necessary to motivate consumers to act on their own but is not the only way to motivate consumers to participate in demand response programs [14]. In order to recruit customers to participate in these programs, it is necessary to offer certain financial benefit. Utilities could offer a flat or graduated incentive in return for the authority to consumer's appliances up to certain number of hours each year.

5 Conclusion

Due to the rapid growth in residential air-conditioning unit, the development of demand response programs to directly address the air-conditioning peak load effects is becoming increasingly urgent. In this paper, the residential load data from Ausgrid combined with local weather data is used to develop an expert system, which can predict air-conditioner existence and identify usage patterns. The promising prediction performance illustrates that this expert system can be used as an accessorial tool by demand response sponsors. Due to many barriers, the current development of demand management measures targeting air conditioners is relatively slow. Governments and policy makers should work together to encourage demand response programs with generous subsidies and with regulatory support. In the near future, the widespread demand response programs will play a critical role in shifting air-conditioning peak load.

Acknowledgement The authors would like to thank Mr Eric Pozorski from Ausgrid for his valuable inputs to this work. This work was supported in part by an ARC Grant LP110200957.

References

[1] The federal government energy white paper. Australian Government, Department of Resources, Energy and Tourism, Canberra, Canberra, Australia, 2012

[2] Demand management and planning program. Final report. TransGrid, Sydney, Australia, 2008

[3] Analysis of initiatives to lower peak demand. Final report. Energy Supply Association of Australia, Canberra, Australia, 2012

[4] Goldman C, Reid M, Levy R et al (2010) Coordination of energy efficiency and demand response. LBNL-3044E. Ernest Orlando Lawrence Berkeley National Laboratory, Berkeley, CA, USA

[5] A national demand management strategy for small airconditioners: The role of the national appliance and equipment energy efficiency program (NAEEEP). National Appliance and Equipment Energy Efficiency Committee and the Australian Greenhouse Office, Canberra, Australia, 2004

[6] Meng K, Dong ZY, Wong KP (2009) Self-adaptive RBF neural network for short-term electricity price forecasting. IET Gener Transm Dis 3(4):325–335

[7] Chen X, Dong ZY, Meng K et al (2012) Electricity price forecast with extreme learning machine and bootstrapping. IEEE Trans Power Syst 27(4):2055–2062

[8] Wong KP (1989) Expert system for protection current transformer design specification preparation. IEE Proc Gener Transm Dis 136(6):391–400

[9] Demand response spinning reserve demonstration. LBNL-62761. Ernest Orlando Lawrence Berkeley National Laboratory, Berkeley, CA, USA, 2007

[10] Demand response spinning reserve demonstration–Phase 2 findings from the summer of 2008. LBNL-2490E, Ernest Orlando Lawrence Berkeley National Laboratory, Berkeley, CA, USA, 2009

[11] Australian Energy Market Operator (AEMO). Accessed 18 Jun 2012

[12] Benefits of demand response in electricity markets and recommendations for achieving them. US Department of Energy, Washington, DC, USA, 2006

[13] Han J, Piette MA (2008) Solutions for summer electric power shortages: Demand response and its applications in air conditioning and refrigerating systems. Refrig Air Cond Electr Power Mach 29(1):1–4 (in Chinese)

[14] Power of choice—giving consumers options in the way they use electricity. Directions Paper. Australian Energy Market Commission, Sydney, Australia, 2012

Author Biographies

Robert SMITH is a senior engineer with Ausgrid, Sydney, Australia. His research interest includes energy management and grid applications.

Ke MENG (M'10) obtained Ph.D. from the University of Queensland, Australia in 2009. He is currently with the Centre for Intelligent Electricity Networks (CIEN), The University of Newcastle, Australia. His research interest includes pattern recognition, power system stability analysis, wind power, and energy storage.

Zhaoyang DONG (M'99–SM'06) obtained his Ph.D. degree from the University of Sydney, Australia in 1999. He is now a Professor and Head of School of Electrical and Information Engineering, The University of Sydney. He is previously Ausgrid Chair and Director of the Centre for Intelligent Electricity Networks (CIEN), the University of Newcastle, Australia. His research interest includes smart grid, power system planning, power system security, load modeling, renewable energy systems, electricity market, and computational intelligence and its application in power engineering. He is an editor of IEEE Transactions on Smart Grid, and IEEE Power Engineering Letters.

Robert SIMPSON is a manager with Ausgrid, Sydney, Australia. His research interest includes energy management and grid applications.

3

Power transmission risk assessment considering component condition

Lei GUO (✉), Qiwei QIU, Jian LIU,
Yu ZHOU, Linglei JIANG

Abstract This paper proposes a new method for power transmission risk assessment considering historical failure statistics of transmission systems and operation failure risks of system components. Component failure risks are integrated into the new method based on operational condition assessment of components using the support vector data description (SVDD) approach. The traditional outage probability model of transmission lines has been modified to build a new framework for power transmission system risk assessment. The proposed SVDD approach can provide a suitable mechanism to map component assessment grades to failure risks based on probabilistic behaviors of power system failures. Under the new method, both up-to-date component failure risks and traditional system risk indices can be processed with the proposed outage model. As a result, component failure probabilities are not only related to historical statistic data but also operational data of components, and derived risk indices can reflect current operational conditions of components. In simulation studies, the SVDD approach is employed to evaluate component conditions and link such conditions to failure rates using up-to-date component operational data, including both on-line and off-line data of components. The IEEE 24-bus RTS-1979 system is used to demonstrate that component operational conditions can greatly affect the overall transmission system failure risks.

Keywords Risk assessment, Component failure risk, Outage probability, Condition assessment, Support vector data description

1 Introduction

With the continuous increase of energy demand, the accurate risk assessment of power systems is of great importance, since risks are increased when a power system is operated close to its stability limits due to distributed generation and market competition. With regard to power system assessment, higher risks lead to lower reliability, and vice versa. The probabilistic behaviors of power system failures are the root origin of risks [1], and an effective risk assessment model can provide quantitative risk indices to represent system reliability. Traditionally, only historical failure statistics are employed in power system risk assessment, however the overall system risk is also related to component operational conditions. When component failure risks change, the overall system risk varies accordingly. Incorporating component risks into the power system risk assessment can improve the accuracy and rationality of risk evaluations.

In the past decade, considerable efforts have been devoted to probabilistic risk assessment of power transmission systems and substation configurations. A widely

L. GUO, J. LIU, Y. ZHOU, North Subsection of State Grid Corporation of China, Beijing 100053, China
(✉) e-mail: stones_wall@163.com
J. LIU
e-mail: liu.jian@nc.sgcc.com.cn
Y. ZHOU
e-mail: zhou.yu@nc.sgcc.com.cn
Q. QIU, Shanghai Electric Power Company,
Shanghai 201400, China
e-mail: q2water@163.com
L. JIANG, Maintenance Company of Shanghai Electric Power Company, Shanghai 200000, China
e-mail: ryandavidsharp@gmail.com

used framework for power system risk assessment was reported in [1, 2], in which the approach, objective, application and economic cost were discussed in detail [3]. However, in this traditional framework, failure risks of components, such as transformers and circuit breakers, were not considered. Generally, the risk assessment of components in substations was performed separately [4–6]. As a result, there is a lack of a mechanism to convert component operational conditions into failure risks in the traditional framework. In [4], a risk assessment model of a combinative system in a transmission network and substations was proposed. Compared with the traditional framework, in which system risks of transmission networks and substation configurations are assessed separately, the method presented in [4] can evaluate system risks considering both transmission networks and substations by assessing new load curtailments at load points for each failure state. As an improvement, substations are no longer treated as a transmission node and substation configurations, and individual components, such as breakers and transformers, are linked to system risks by analyzing statistical data of substation components. However, component failure data are still based on historical statistics. Consequently, the impact of online component operational conditions cannot be integrated in risk evaluations.

A multi-objective risk assessment framework was presented in [7], and probabilistic indices for assessing real-time power system security levels were derived. However, operation risks of components were still not considered. A failure probability model was developed based on the evidential reasoning (ER) theory for overhead lines in [8], which can accurately reflect the impact of surroundings on failure probabilities. However, component outage rates were set as a fixed value, which was not linked to operational conditions of components. Based on the ER theory and the functional group decomposition principle, a contingency identification method for components was presented in [9].

However, in that research, component conditions, such as operational conditions and monitoring data, were not considered, and components were just treated as part of transmission lines. But actually, each component has its own failure risk, which is influenced by its operational condition. In practice, component condition assessment is usually conducted by experts or trained on-site engineers. As operational conditions could be affected by faults or environment, such as loading conditions and temperatures [5, 6], the failure probability of components is not fixed. Thus, the outage probability of transmission lines changes accordingly. As the component failure probability changes, the results of risk assessment are not fixed values as those of traditional risk assessment models [1], which should be determined by both operational conditions and historical data.

The support vector data description (SVDD) approach is developed for classification and evaluation with machine learning, which can be employed to aggregate diagnosis information [10, 11]. In particular, regarding the probabilistic and uncertain behaviors of component failures, SVDD is a suitable solution for presenting evaluation of various failure conditions. Based on the outputs of SVDD component evaluation, system operators can obtain overall evaluations of studied components, which can be classified into different condition levels accordingly. The SVDD approach is capable of providing the most recent condition for components in power transmission systems. The objective of this paper is to develop a new risk assessment method for power transmission systems, in which component conditions are considered based on on-line and off-line data. The method comprises of three parts: component evaluation, index transition and system risk evaluation. The proposed method employs SVDD for component risk assessment and the Monte Carlo (MC) simulation [1] for system state selections.

2 SVDD approach to component condition assessment

The SVDD approach is an one-class classification data description method which proposed by Tax [10]. By training with a set of certain samples, the distribution of target class can be obtained by SVDD, so the outliers can be divided. The SVDD approach can provide well distribution area and can be used in condition detection, fault diagnosis and multi-classification, etc. [12–14].

By applying SVDD approach to mechanical condition monitoring and fault diagnosis, machine conditions can be monitored only by using normal condition signals instead of abnormal condition signals. With the method, the machine set conditions (normal or abnormal) can be described by using quantitative indices, and the scientific decision-making basis for equipment management and predictive maintenance can be offered. The method is used to evaluate the condition of the key equipment in power transmission lines, and it correctly evaluates an abnormal condition of the equipment in time and contributes to a successful diagnosis of the incipient fault of a bolt crack.

As a data set containing N data objects: $\{x_i, i = 1, 2, ..., N\}$, the basic concept of SVDD is trying to find a sphere with minimum volume, containing all (or most of) the data objects [10]. This is very sensitive to the most outlying object in the target data set. When one or a few very remote objects are in the training set, a very large sphere is obtained which will not represent the data very well. Therefore, [15] considered some data points outside the sphere and introduced slack variable $\xi_i (\xi_i \geq 0, i = 1, 2, ..., n)$. Of the sphere, described by center a and radius R, the radius is minimized as follows:

$$\min_{R} F(R, a) = R^2 + C \sum_{i=1}^{n} \xi_i$$

$$\text{s.t.} \begin{cases} [\varphi(x_i) - a][\varphi(x_i) - a]^{\mathrm{T}} \leq R^2 + \xi_i \\ \xi_i \geq 0, \quad i = 1, 2, \ldots, n \end{cases} \quad (1)$$

where the variable C gives the trade-off between the simplicity (or volume of the sphere) and the number of errors (number of target objects rejected); the function φ is a nonlinear mapping function used for mapping objects into the high dimensional.

The dual form of (1) is written as:

$$\max_{\alpha_i} L = \sum_{i=1}^{n} \alpha_i K(x_i \cdot x_j)$$

$$\text{s.t.} \begin{cases} \sum_{i=1}^{n} \alpha_i = 1 \quad i = 1, 2, \ldots, n \\ 0 \leq \alpha_i \leq C \quad i = 1, 2, \ldots, n \end{cases} \quad (2)$$

where $K(x_i \cdot x_j)$ is the kernel function which satisfies Mercer's theorem:

$$K(x_i \cdot x_j) = [\varphi(x_i) \cdot \varphi(x_j)]. \quad (3)$$

The kernel function implicitly maps the objects x_i into some feature space and when a suitable feature space is chosen, a better and more tight description can be obtained. No explicit mapping is required, the problem is expressed completely in terms of $K(x_i \cdot x_j)$.

Finally, parameter α_i can be obtained, and x_i satisfies $\alpha_i > 0$, called the support vector. From the basic concept and definition of SVDD [11], the equation is obtained as follows:

$$R^2 = K(x_k \cdot x_k) - 2 \sum_{i=1} \alpha_i K(x_i \cdot x_k)$$

$$+ \sum_{i,j} \alpha_i \alpha_j K(x_i \cdot x_j) \quad \forall x_k \in \text{SV} < C, \quad (4)$$

where R^2 is the distance to the center of the sphere (a); SV means support vector.

Based on (4), each support vector can provide the value of R^2. For the test sample z, assuming that:

$$D(a, z) = K(z \cdot z) - 2 \sum_{i=1} \alpha_i K(z \cdot x_i)$$

$$+ \sum_{i,j} \alpha_i \alpha_j K(x_i \cdot x_j) \quad i, j = 1, 2, \ldots, n. \quad (5)$$

If $D(a, z) \leq R^2$, then z is considered as the target, otherwise is considered as the outlier.

For illustration purposes, we define evaluation level as the SVDD result of each component. It corresponds to the relationships between component condition characteristic values and failure probability. For example, based on a transformer dissolved gas analysis (DGA) value, the SVDD evaluation level of this component is graded as 'good', 'normal', 'poor' or 'serious'. Therefore, each level equals an area of failure rate (from 0 to 1), afterwards, the component condition is considered in power transmission risk assessment.

3 Power transmission system risk evaluation considering component risks

In practice, power system risk assessment is concerned with two aspects: i.e., system adequacy and system security [1]. System adequacy mainly relates to the existence of sufficient facilities within a system to satisfy consumer load demands and system operational constraints, while system security relates to the ability of a system to respond to dynamic and transient disturbances arising within the system. Thus, security is associated with the response of a system to perturbations. As most of the risk assessments

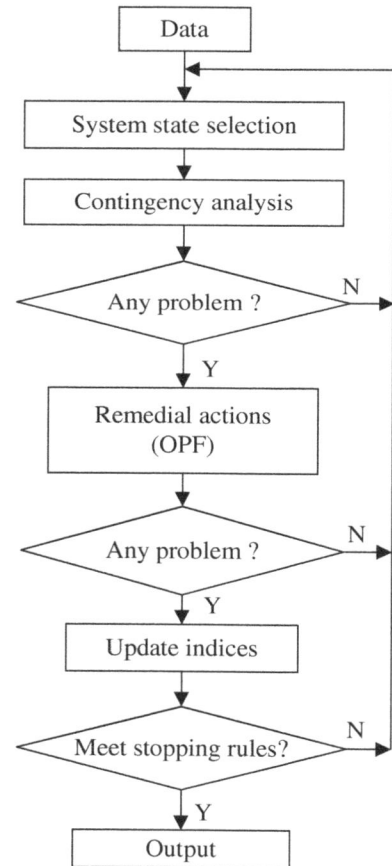

Fig. 1 Procedure of power transmission risk assessment

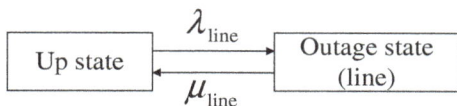

Fig. 2 State space diagram of a two-state repairable forced outage

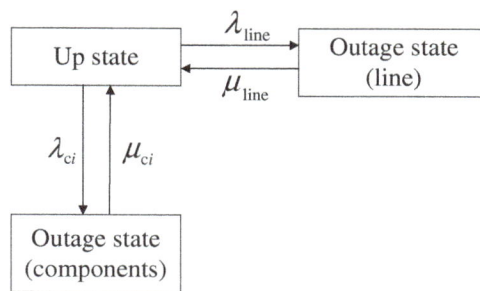

Fig. 3 State space diagram considering both line and component outages

carried out by utilities are in the domain of adequacy assessment [1], in this paper the system adequacy analysis is set as the risk assessment objective.

The basic procedure of power transmission system risk assessment is shown in Fig. 1 [1]. Firstly, a system sate is selected based on historical failure statistics. Then, the contingency analysis and optimal power flow (OPF) method are performed to identify whether a selected state causes any problems. Finally, risk indices are calculated. In this paper, procedures of outage modeling in system state selections have been modified using the updated outage model considering component failure risks.

3.1 Traditional component failure models

Traditionally, for power transmission system risk evaluation, only failures of transmission components are considered, whereas generating units are assumed to be 100% reliable. Key transmission components include overhead lines, cables, transformers, capacitors, and reactors. Generally, these components are represented by a two-state (up and down) model. Figure 2 shows the diagram of a basic two-state repairable forced outage, which can be used to describe a typical steady up-down-up cycle process.

The average unavailability of a transmission line in a long-term process is defined as follows [1]:

$$P_{\text{line}} = \frac{\lambda_{\text{line}}}{\lambda_{\text{line}} + \mu_{\text{line}}}, \tag{6}$$

$$\begin{cases} \lambda_{\text{line}} = \dfrac{8,760}{MTTF_{\text{line}}} \\ \mu_{\text{line}} = \dfrac{8,760}{MTTR_{\text{line}}} \end{cases}, \tag{7}$$

where P_{line} is the outage probability of lines; λ_{line} and μ_{line} are the line failure and repair rates (1/year), respectively; $MTTR_{\text{line}}$ and $MTTF_{\text{line}}$ are the mean time to repair (MTTR) (hours per year) and mean time to failure (MTTF) (hours/year) of lines, respectively.

The historical data are recorded over a one year period and subsequently the failure and repair rates can be derived based on MTTF and MTTR, respectively.

3.2 Outage model integrating component failure risk

A new outage model has been developed based on both historical failure statistics (IEEE RTS-79) [16] and component failure risks. In the traditional outage model, the forced outage probability of transmission lines is denoted by P_{line}. Based on (1), the planned outage model and Markov equations [1], the state space diagram considering both the line and component risks is shown in Fig. 3.

Applying the Markov method based on the state space diagram, the outage probabilities can be obtained as follows:

$$P_{ci} = \frac{\lambda_{ci}\mu_{\text{line}}}{\lambda_{ci}\mu_{\text{line}} + \lambda_{\text{line}}\mu_{ci} + \mu_{\text{line}}\mu_{ci}}, \tag{8}$$

$$P_{\text{up}} = \frac{\mu_{\text{line}}\mu_{ci}}{\lambda_{ci}\mu_{\text{line}} + \lambda_{\text{line}}\mu_{ci} + \mu_{\text{line}}\mu_{ci}}, \tag{9}$$

$$P_{\text{line}} = \frac{\lambda_{\text{line}}\mu_{ci}}{\lambda_{ci}\mu_{\text{line}} + \lambda_{\text{line}}\mu_{ci} + \mu_{\text{line}}\mu_{ci}}, \tag{10}$$

$$\begin{cases} \lambda_{ci} = \dfrac{8,760}{MTTF_{ci}} \\ \mu_{ci} = \dfrac{8,760}{MTTR_{ci}} \end{cases}, \tag{11}$$

where P_{up}, P_{line} and P_{ci} are the probabilities of the up state, the historical line outage state and the failure risk-based outage state of the ith component; λ_{line} and λ_{ci} are the transition rates of historical line outages and component failure risk-based outage states; μ_{line} and μ_{ci} are the recovery (repair) rates of the historical line outage and the component failure risk-based outage state (repairs/year); $MTTR_{ci}$ and $MTTF_{ci}$ are the MTTR and MTTF of component, respectively.

In Section 3, the condition assessment of components based on SVDD approach is used as an example to illustrate the procedures for calculating component failure risks. As discussed previously, the overall assessment of a component can be expressed using SVDD approach as a set of evaluation grades, and then component failure rates can be derived in association with the SVDD evaluation levels.

There are normally a number of components in the same transmission line or bus, and the line or bus fails when one of the main components fails. As a result, the maximum component failure risk value is selected to represent the overall failure probability of components in the same line. Considering failure risks of components, the transmission line outage probability integrating both historical failure statistics and component failure risks can be expressed in the following (13):

$$P_c = \max(P_{c1}, P_{c2}, \ldots, P_{cn}), \tag{12}$$

$$P_L = P_{line} + P_c = \frac{\lambda_{line}\mu_{ci} + \lambda_{ci}\mu_{line}}{\lambda_{ci}\mu_{line} + \lambda_{line}\mu_{ci} + \mu_{line}\mu_{ci}}, \tag{13}$$

where P_L is the overall outage probability of transmission lines; P_c is the maximum component failure risk among n components in the same transmission line; P_{cn} is the outage probability of the nth component in the same line.

As shown in (13), it is assumed that the ith component has the maximum failure risk.

3.3 Load curve models and contingency analysis

In this paper, for the state enumeration or state sampling method (nonsequential MC simulation), a nonchronological load duration curve is utilized. A single load curve is considered and loads at all buses are scaled proportionally to follow the shape of the given load curve. A multiple-step model is established to represent the load duration curve [1]. Regarding the contingency analysis on adequacy risk assessment, the capacity balance between the generation and the load demand is important. As a result, the DC power-flow-based contingency analysis is employed in this study, because it provides fast and sufficiently accurate real power flows following line outages for risk assessment, in which a large number of outage events are considered.

3.4 Optimization models for load curtailment

When an outage causes system problems, a special OPF model is used to reschedule generations and alleviate constraint violations. At the same time, load curtailment needs to be avoided if possible or the total load curtailment is required to be minimized if unavoidable. The objective function of an OPF model is to minimize the total load curtailment, whereas load curtailment at buses is the solution of the OPF model. The risk indices are then calculated based on load curtailments in selected system outage states and their probabilities of occurrence. To reduce the computational burden, the DC power-flow-based OPF model is usually employed in the adequacy risk assessment [1]. It can be expressed in the following equations:

$$\min \sum_{i \in ND} C_i, \tag{14}$$

$$\text{s.t.} \begin{cases} T(S) = A(S)(PG - PD + C) \\ \sum_{i \in NG} PG_i + \sum_{i \in ND} C_i = \sum_{i \in ND} PD_i \\ PG_i^{min} \leq PG_i \leq PG_i^{max} \quad i \in NG \\ 0 \leq C_i \leq PD_i \quad i \in ND \\ |T_k(S)| \leq T_k^{max} \quad k \in L \end{cases} \tag{15}$$

where i is the bus number; C_i is the load curtailment at the ith bus; $T(S)$ is the real power flow vector in the outage state; $A(S)$ is the relation matrix between real power flows and power injections in the outage state S; PG and PD are the generation output and load power vectors, respectively; C is the load curtailment vector; PG_i, PD_i, C_i and $T_k(S)$ are the elements of PG, PD, C and $T(S)$, respectively; the subscript "min", "max" are the limits, respectively; NG, ND and L are the sets of generation buses, load buses and branch circuits in a system.

The objective of the model is to minimize the total load curtailment while satisfying the power balance, DC power flow relationships and limits on line flows and generation outputs.

3.5 Risk indices

There are various risk indices, which are used for quantifying system risks. In practice, loss-of-load probability (LOLP) and expected demand not supplied (EDNS) are two most popular indices, which are employed in this research. LOLP indicates the probability of load loss caused by element capacity shortage (1/year). It can be expressed in the following equation:

$$L_{LOLP} = \sum_{x \in X} I_f(x)P(x), \tag{16}$$

where $P(x)$ is the probability of system state x; $I_f(x)$ is a two valued function of system state x. If x

Table 1 Evaluation level corresponding to failure rate

Evaluation level	Failure rate (1/year)
Good	0–0.2
Normal	0.2–0.8
Poor	0.8–1.0
Serious	Outage

Table 2 Results of SVDD classification

Actual/evaluation	Good	Normal	Poor	Serious	Total
Good	39	7	4	0	50
Normal	5	42	2	1	50
Poor	0	1	26	3	30
Serious	0	2	1	27	30
Total	44	51	35		130

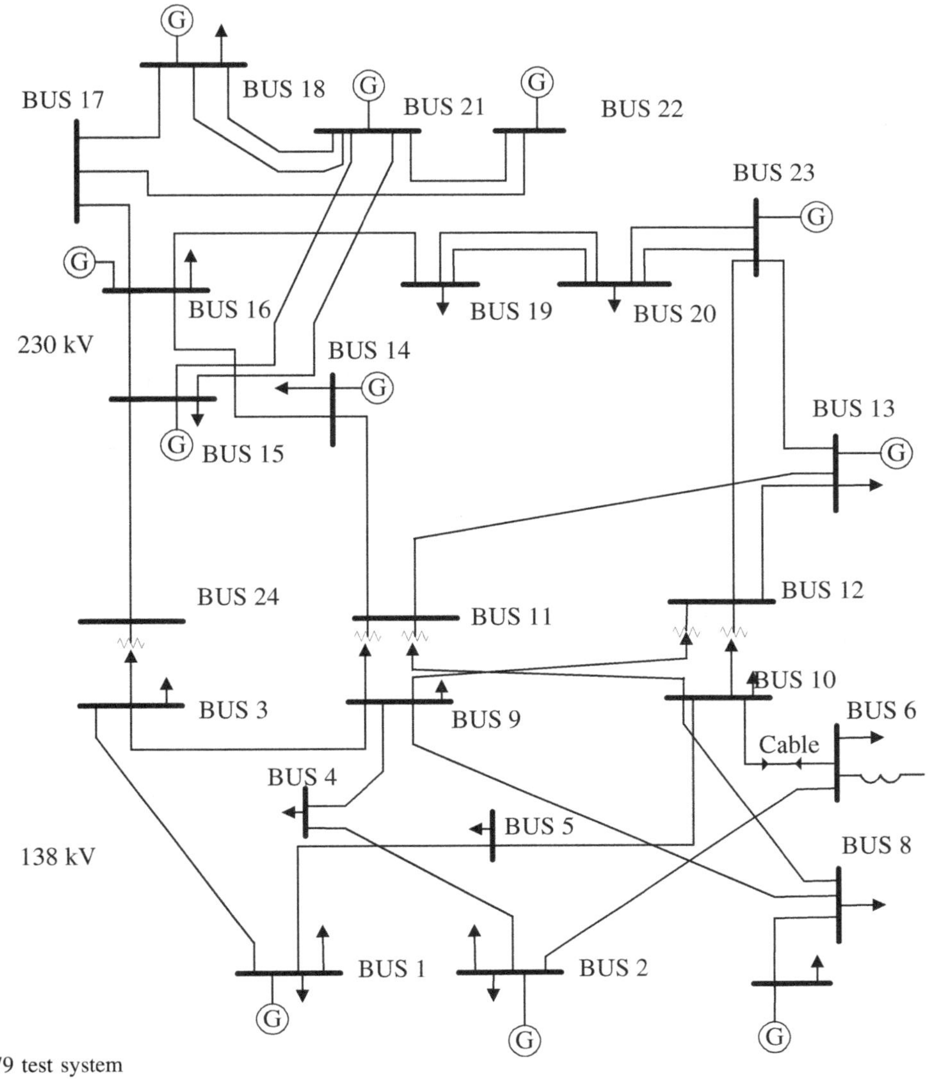

Fig. 4 IEEE RTS-79 test system

indicates the failure state, then I_f is equal to 1, otherwise 0.

EDNS denotes the average shortage of power supply per year (MW/year).

$$L_{EDNS} = \sum_{x \in X} I_f(x) L_C(x) P(x), \tag{17}$$

where $L_C(x)$ represents the minimum load loss for recovery in the outage state x.

4 Case study

4.1 Component failure risk mapping

The condition assessment of transformers is used as an example to illustrate the procedures for calculating component failure risks. For other components, the procedures

can be performed in a similar manner. As discussed previously, the overall assessment of a component can be expressed as a set of evaluation grades, then component failure rates can be derived in association with the SVDD evaluation levels. For illustration purposes, Table 1 lists the corresponding relationships between evaluation levels and failure rates. For example, based on historical statistics or operation experience, the failure rate of a component is 0.14 per year, and then the SVDD evaluation level of this component is graded as 'good'.

Table 1 only gives reference values for illustration purposes, and in practice, this table may be modified based on operation situations and historical statistics analysis. It is defined that the SVDD evaluation grade, that is 'serious', 'poor', 'normal' or 'good', with the maximum value is treated as the final evaluation grade of a component. Using this mapping table, the failure rate λ_c can be derived by the proposed SVDD approach. Likewise, the repair rates μ_c can

Table 3 Case I: LOLP and ENDS results at different sampling frequencies

Sampling times	LOLP (%)	LOLP considering component	EDNS (MW/ year)	EDNS considering component
10,000	8.72	9.38	14.29	14.42
20,000	8.69	9.25	14.62	14.39

Table 4 Case II: LOLP and ENDS results at different sampling frequencies

Sampling times	LOLP considering component	EDNS considering component
10,000	12.01	16.68
20,000	10.65	15.28

be defined similarly. In the meantime, λ_{line} and μ_{line} can be derived from historical statistic data, and finally the outage probability of components can be calculated by using (8).

4.2 SVDD-based component failure risk evaluation

In this paper, 260 sets of transformer monitoring data sample are adopted. These samples are acquired from the DGA monitoring of transformers. Thereinto, 130 sets are used for training (50 sets are in good condition, 50 sets are normal, 30 set are poor and 30 sets are serious) and others are used for testing. By adopting the SVDD approach, the evaluation results are listed in Table 2.

In the evaluation results, the total accuracy is about 84%. The accuracy of 'good', 'normal', 'poor' and 'serious' are 78%, 84%, 87% and 90%, respectively.

Based on the mapping relations listed in Table 1, the failure rate of a component can be generated randomly between the areas. However, the mapping relation is only for illustrating reference values, those values can be defined based on the practice or experience.

4.3 System risk assessment: Case I

In this case study, the IEEE RTS-79 system is employed as the test system [15] as shown in Fig. 4. In the IEEE RTS-79 test system, the load model gives hourly loads for one year on a per unit basis, expressed in chronological fashion so that daily, weekly and seasonal patterns can be modeled. The generating system contains 32 units, ranging from 12 to 400 MW. The transmission system contains 24 load/generation buses connected by 38 lines or autotransformers at two voltage levels, i.e. 138 kV and 230 kV. The transmission system includes cables, lines on a common right of way, and lines on a common tower. The transmission system data include the line length, impedance, ratings, and reliability data.

In MC simulations different sampling frequencies lead to different convergences, therefore the sampling frequencies of the MC simulation are set with different values. As a result, the derived risk index values at different sampling frequencies are presented in Table 3.

It is assumed that all the transformers in the system are operated under a 'normal' condition, and the component

failure rates are set as a random value between 0.2 and 0.8 randomly according to Table 1.

Compared with the results without considering component risks the LOLP and EDNS have rarely increased.

4.4 System risk assessment: Case II

In Case II, the component failure rates are set higher than that of Case I and all system components are assumed to be operated under a 'poor' condition. The component failure risks are generated between 0.8 and 1.0 mapping to the 'poor' level for each line. Different sampling frequencies are also compared, as shown in Table 4.

In Case II, it is clear that when evaluation levels of all transformers are changed from 'normal' to a worse level like 'poor', LOLP increases significantly. Compared with these of Case I, LOLP considering component risks increases nearly by 15%, while ENDS considering components is raised by 6%.

4.5 System risk assessment: Case III

In this case, only the component between BUS 3 and BUS 24 is set as outage, while other components in the system are considered as 'normal'. According to Table 1, the probability of the outage component is 100%, and failure rates of other components are between 0 and 0.2. Different sampling frequencies are applied and its results concerning LOLP and ENDS are listed in Table 5. Compared with Case II, LOLP and EDNS are 8 and 3% more than those of Case I, which indicates that when a transformer is working under a 'serious' condition, the overall system risk increases. However, the risks are less than those in Case II, which indicates that when failure risks of several components change from 'normal' to a 'poor' or 'serious' grade, the risk values

Table 5 Case III: LOLP and ENDS results at different sampling frequencies

Sampling times	LOLP considering component	EDNS considering component
10,000	9.81	14.70
20,000	9.92	14.82

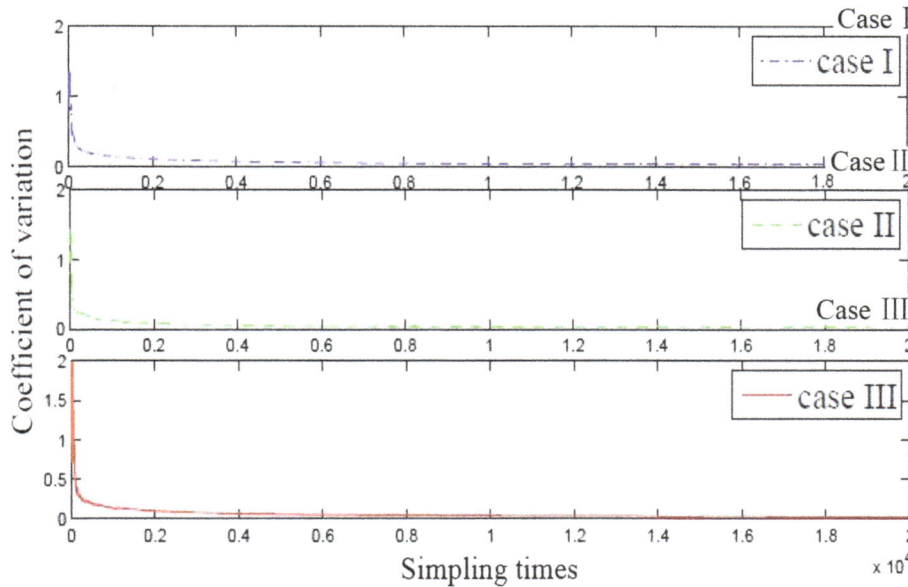

Fig. 5 Convergence variation of EDNS

increase much more than the situation that only one component is in the 'serious' grade.

The variation convergence curves of EDNS in three cases are illustrated in Fig. 5, which shows that when the sampling frequency is over 20,000 times, the variation convergence is relatively small under 0.2. It means that, for three cases the derived coefficients are reliable when the sampling frequency is over 20,000 times.

5 Conclusion

A new method for transmission system risk assessment considering component monitoring data is proposed. The proposed SVDD-based approach can provide a suitable mechanism to map component evaluation grades to failure risks based on the probabilistic behaviors of power system failures. Using the new method, both up-to-date component condition status and traditional system risk indices can be processed with the developed outage model. In this study, transformer DGA data have been used to calculate component failure risks. The simulation results indicate that transmission system risks are affected not only by component operational conditions, but also by historical statistics data. In case studies, the implementation procedures of component risk evaluation using SVDD and system risk assessment are demonstrated.

References

[1] Li WY (2005) Risk assessment of power systems: models, methods, and applications. Wiley, Hoboken
[2] Billinton R, Li W (1994) Reliability assessment of electric power systems using Monte Carlo methods. Plenum, New York
[3] Ghajar RF, Billinton R (2006) Economic costs of power interruptions: a consistent model and methodology. Int J Electr Power Energy Syst 28(1):29–35
[4] Li WY, Lu JP (2005) Risk evaluation of combinative transmission network and substation configurations and its application in substation planning. IEEE Trans Power Syst 20(2): 1144–1150
[5] Tang WH, Spurgeon K, Wu QH et al (2004) An evidential reasoning approach to transformer condition assessments. IEEE Trans Power Deliv 19(4):1696–1703
[6] Shintemirov A, Tang WH, Wu QH (2010) Transformer winding condition assessment using frequency response analysis and evidential reasoning. IET Electr Power Appl 4(3):198–212
[7] Xiao F, McCalley JD (2009) Power system risk assessment and control in a multiobjective framework. IEEE Trans Power Syst 24(1):78–85
[8] Zhang GH, Duan MY, Zhang JH et al (2009) Power system risk assessment based on the evidence theory and utility theory. Automat Electr Power Syst 33(23):1–12 (in Chinese)
[9] Song Y, Wang CS (2008) N-K contingency identification method under double failure incident based on evidence theory and functional group decomposition. Proc Chin Soc Electr Eng 28(28):47–53 (in Chinese)
[10] Tax DMJ (2004) Support vector data description. Mach Learn 54(1):45–66
[11] Tax DMJ, Duin RPW (2000) Data description in subspaces. In: Proceedings of the 15th international conference on pattern recognition (ICPR'00), vol 2, Barcelona, Spain, 3–7 Sept 2000, pp 672–675
[12] Zhui XK, Yang DG (2007) Multi class support vector domain description for pattern recognition based on a measure of expansibility. Acta Electron Sin 37(3):464–469 (in Chinese)

[13] Tang MZ, Wang YB, Yang HC (2011) Modified support vector data description for fault diagnosis. Control Decis 26(7):967–972 (in Chinese)

[14] Li LJ, Han J, Hao W et al (2005) Condition evaluation for mechanical equipment by means of support vector data description. J Mech Sci Technol 24(12):1426–1429 (in Chinese)

[15] Tax DMJ, Duin RPW (1999) Support vector domain description. Pattern Recogn Lett 20(11–13):1191–1199

[16] IEEE RTS Task Force of APM Subcommittee (1979) IEEE reliability test system. IEEE Trans Power App Syst 98(6):2047–2054

Lei GUO was born in 1984. He obtained his Ph.D. in Electrical Engineering from Zhejiang University, Hangzhou, China, in 2012. He is currently working in the North Subsection of State Grid Corporation of China, Beijing, China. His research interests include power system stability, power grid planning, intelligent information processing technology and its application in power systems.

Qiwei QIU was born in 1978. He is currently working in the Shanghai Electric Power Company. His research interests include power transmission technology and management.

Jian LIU was born in 1979. He obtained his master degree in Tsinghua University, Beijing, China, in 2006. He is currently working in the North Subsection of State Grid Corporation of China, Beijing, China. His research interests include power system stability and power grid dispatching.

Yu ZHOU was born in 1983. He is currently working in the North Subsection of State Grid Corporation of China, Beijing, China. His research interests include power system stability and power grid dispatching.

Linglei JIANG was born in 1980. He is currently working in the Maintenance Company of Shanghai Electric Power Company. His research interests include power transmission technology and management.

4

R&D and application of voltage sourced converter based high voltage direct current engineering technology in China

Guangfu TANG (✉), Zhiyuan HE, Hui PANG

Abstract As a new generation of direct current (DC) transmission technology, voltage sourced converter (VSC) based high voltage direct current (HVDC) has been widely developed and applied all over the world. China has also carried out a deep technical research and engineering application in this area, and at present, it has been stepped into a fast growing period. This paper gives a general review over China's VSC based HVDC in terms of engineering technology, application and future development. It comprehensively analyzes the technical difficulties and future development orientation on the aspects of the main configurations of VSC based HVDC system, topological structures of converters, control and protection technologies, flexible DC cables, converter valve tests, etc. It introduces the applicable fields and current status of China's VSC based HVDC projects, and analyzes the application trends of VSC based HVDC projects both in China and all over the world according to the development characteristics and demands of future power grids.

Keywords Voltage sourced converter based high voltage direct current (VSC based HVDC), Two-level converter, Modular multi-level converter (MMC), Direct current grid (DC grid)

1 Introduction

The transmission technology has gone from DC to AC, AC to DC and then to coexist technological evolution. With the electrical and electronic technological improvement, as a new generation of DC transmission technology, flexible DC may help to solve many difficulties that the current AC/DC transmission technologies face, and provide a new solution [1, 2] for innovation of transmission methods and establishment of future grid.

The earliest concept of voltage sourced converter (VSC) based high voltage direct current (HVDC) transmission was proposed by Boon-Teck et al. from Canada McGill University [3, 4] in 1990. The operating principle of this concept is that the AC active power and reactive power may be controlled by switching on and off the VSC or changing the voltage phase angle and amplitude so as to effectively overcome some existing defects. This concept was formally named as "VSC-HVDC" in the International Conference on Large HV Electric Systems (CIGRE) and America Electrical and Electronic Project in 2004, also as HVDC Light [5], HVDC Plus and HVDC MaxSine by the international companies of ABB, Siemens and Alstom respectively. It has also been named as high voltage direct current flexible (HVDC Flexible) in China.

Early VSC based HVDC (referred to as HVDC Flexible in this paper) adopts two-level or three-level converter technology, but it always has several defects, such as high harmonic contents and high switching losses, etc. With the continuous growing demand for high voltage level and power transmission capacity, these defects are represented more and more obviously, which become a bottleneck for application of the two-level or three-level technology. Therefore, the future of the two-level or three-level technology may be only used for low power transmission or on some special occasions such as offshore platform power supply or variable speed drives, etc. In 2001, R. Marquart and A. Lesnicar from Bundeswehr Munich University of Germany jointly proposed a modular multi-level converter (MMC) topology [6, 7]. The proposal and application of the MMC technology is an important milestone in the history of technical development for HVDC Flexible engineering. The occurrence of such technology improves the operational

G. TANG, Z. HE, H. PANG, State Grid Smart Grid Research Institute, Beijing 102200, China
(✉) e-mail: gftang@sgri.sgcc.com.cn

(a) **Symmetric monopole system**

(b) **Asymmetric monopole system**

(c) **Symmetric bipole system**

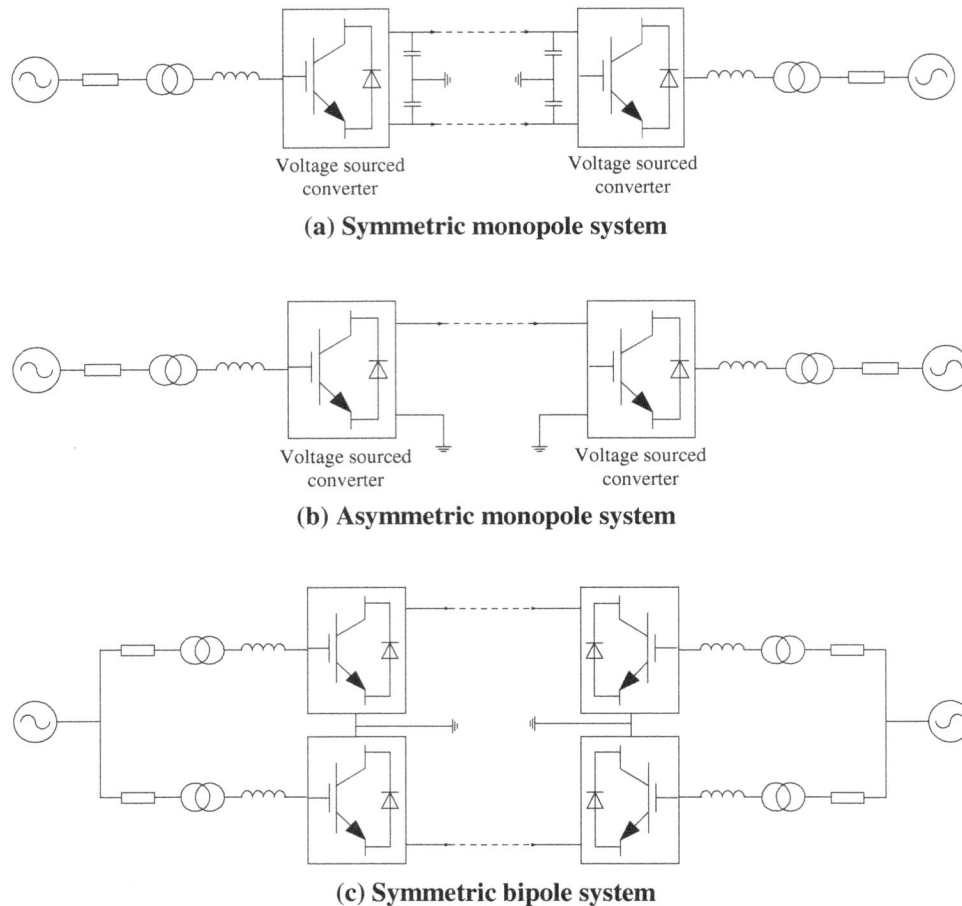

Fig. 1 Typical two-terminal HVDC Flexible system

benefits of HVDC Flexible projects, greatly promotes the development and application of the technology.

Based on three aspects, i.e. engineering technology, application and future development, this paper firstly analyzes the challenges that the technical development of HVDC Flexible projects currently faces, future development orientation of relevant technology and expected targets. Secondly, it briefs China's HVDC Flexible projects and points out the technology and application characteristics. Finally, it analyzes the potential development of the HVDC Flexible projects and the project application prospect both in China and all over the world.

2 HVDC flexible technology

2.1 HVDC flexible system

The HVDC Flexible system with two-level or three-level converter configurations usually adopts the grounding point on the DC side, but on the AC side while the system with MMC configuration. Regardless the grounding

location a monopole- symmetric system is always used for the HVDC Flexible system. During normal operating conditions, no current passes through grounding path thus it is no need to set special grounding pole, but when the DC line or converter fails, the whole system will stop running. Furthermore, it may forms a monopole asymmetric structure via grounding path or metallic return, which is similar to a pole of a traditional HVDC transmission system.

In order to increase the power capacity and voltage level of the HVDC Flexible system, and to meet the requirements of super-high voltage and remote large power transmission, the converter of the monopole convertor station may also consist of several small converter units in series or/and parallel connections. As shown in Fig. 1, two asymmetric monopole systems may form a bipole symmetric system via series connections, which is similar to a traditional HVDC transmission system.

Transformers used for a bipole system are required to withstand the transformer DC bias voltage caused by the asymmetry of DC voltage. Unlike conventional DC transformers, the transformers are not required to withstand the

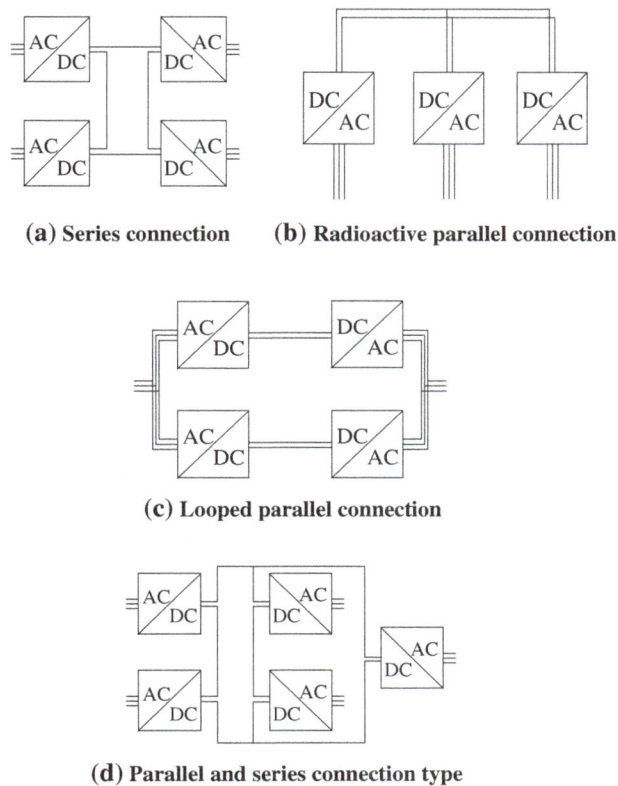

(a) Series connection   (b) Radioactive parallel connection

(c) Looped parallel connection

(d) Parallel and series connection type

Fig. 2 Typical connection diagrams of multi-terminal HVDC

harmonic components generated by the convertor stations. At present, in order to reduce the fault incidence on the DC side, the HVDC Flexible transmission projects usually use cables as the transmission lines, which is the main reason that the monopole structure is used in the HVDC Flexible transmission system. In this way, it is more reliable to use single converter and can also reduce the project costs.

The multi-terminal flexible DC transmission system is generally connected in parallel so as to ensure the converters are operated in the same DC voltage level. The parallel connected multi-terminal flexible DC networks can also be divided into two basic structures, i.e. star and ring types. Other complicated structures can be regarded as the extension and combination of these two structures. There are four kinds of topological structures as showed in Fig. 2.

Compared with the series connection type, the parallel connection type has less line losses, larger adjusting range, more accessible insulation coordination, more flexible extension mode and significant economic efficiency. Therefore, all existing operational multi-terminal DC transmission projects use the parallel connection mode.

2.2 HVDC flexible converter technology

Based on the equivalent characteristic of bridge arms, the HVDC Flexible converter techniques can be divided into controllable switch type and controllable power supply type. The converting bridge arm of the controllable switch type converter is equivalent to the controllable switch, controlling the turn-on and turn-off of the bridge arms by appropriate pulse width modulation techniques and transmitting the voltage from the DC side to the AC side. The energy storage capacitors of the controllable power supply type converter are–distributed in different bridge arms. Their converting bridge arms are equivalent to the controllable voltage sources which can indirectly change the output voltage on the AC side by changing the equivalent voltage of the bridge arms. Both equivalent circuits shown in Fig. 3 and Fig. 4 show the output waveforms at the AC outlets of the two topological converter valves [8–10].

The controllable switch type converter is represented typically by a two-level converter and its topological structure and operating control are relatively simple. But the switching frequency and losses of the converter are higher, and the harmonic contents on the AC and DC sides are larger resulting in multiple filter units required. Although a 3-level converter has relatively lower harmonic contents in the output voltage waveform, and its switching frequency, total harmonic level and losses have also been reduced, the converter has complicate topological structure, higher costs and lower reliable system. In addition, as each bridge arm of switch type voltage sourced converter is directly connected by a large number of switching devices in series, it needs to solve the static and dynamic voltage sharing and other problems caused by the turn-on and turn-off of the switching devices.

Modular multi-level (MMC) converters are the typical representative of controllable power supply type converters [11–13]. The equivalent output voltage is achieved by changing the number of the series connected sub-modules into the bridge arms. As shown in Fig. 5, this type of converters can also be divided into semi-bridge mode, full bridge mode, clamping twin module mode and other diversified forms according to the types adopted in sub-modules. In addition, the cascaded two-level converters (CTL) are cascaded by semi-bridge circuits, so they also belong to the controllable power supply type converters in nature [14].

When the sub-modules in the bridge arms exceed a certain number and the output voltage waveform of the converter is approximated to a sinusoidal step waveform, filtering devices are not required. Compared with the two-level converters, the MMC converters have the following outstanding advantages: (1) modularization design which can make the upgrade of voltage levels and capacities easier; (2) the switching frequency and stress of devices are reduced significantly; (3) the harmonics and total harmonic distortion in the output voltage are greatly reduced and filtering devices are not required on the AC side.

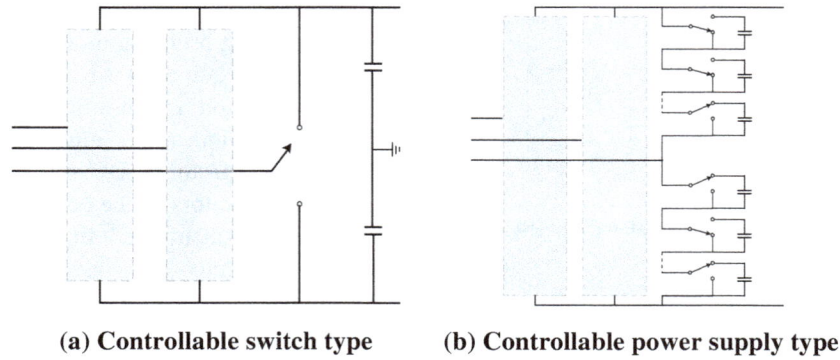

(a) Controllable switch type   (b) Controllable power supply type

Fig. 3 Equivalent circuits of switch-controllable converter and voltage-controllable converter

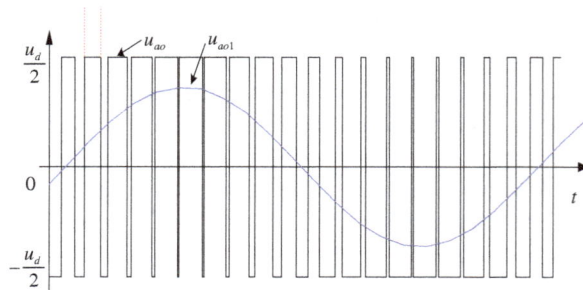

(a) Controllable switch type

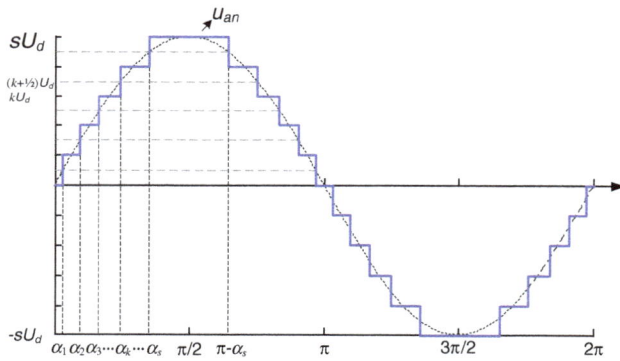

(b) Controllable power supply type

Fig. 4 Output waveform in converter AC side

Comparing with the two-level converters, the disadvantages of the MMCs are: (1) as there are a large number of sub-modules in series connection in each bridge arm, the valve control system needs to process a large amount of data in each period, therefore high requirements are needed for the control system; (2) distributed energy storage capacitor increases the balance control for the sub-module capacitor voltages [15]; (3) the energy is not evenly distributed in each bridge arm, which destroys the internal stability within the sub-modules and causes the distortion of the current waveform.

At present, the converter techniques in engineering application, no matter the two-level type or the semi-bridge MMC, all have a prominent problem, i.e. it is unable to achieve the isolation of AC and DC systems under DC fault conditions. However the full bridge and clamping twin sub-module MMCs can still support the AC voltage to achieve the inhibiting effect on the short-circuit current on the AC side when the DC voltage is rapidly reduced, because they can make the equivalent output voltage of the bridge arms become negative [16].

2.3 Control and protection of HVDC flexible

The flexible DC control and protection system, as the core to ensure normal operation of the system, is used to realize the control function of normal operation of the system and the protection function under faults. The control and protection system includes the convertor station level control and protection system and the converter valve level control and protection system.

Unlike the conventional DC transmission, the converter valve level control and protection system in HVDC Flexible is far more complicated. Especially, in the MMC HVDC Flexible system, the convertor station level controller (called pole controller or station controller for short) just undertakes a part of control and protection function. The control and protection for the valves relies more on the converter valve level controller. The functions include producing the control signals of converter valve sub-modules, handling and summarizing the data, achieving the protection of converter valves and so on, according to the signal requirements of convertor station level control (as shown in Fig. 6). Therefore, the HVDC Flexible control and protection system is usually required to achieve the high speed sync control in nanosecond level to meet the high real-time requirements of HVDC Flexible control system.

In addition to realizing the normal start and shut down of the system, the flexible HVDC convertor station level control system also includes the steady power control and

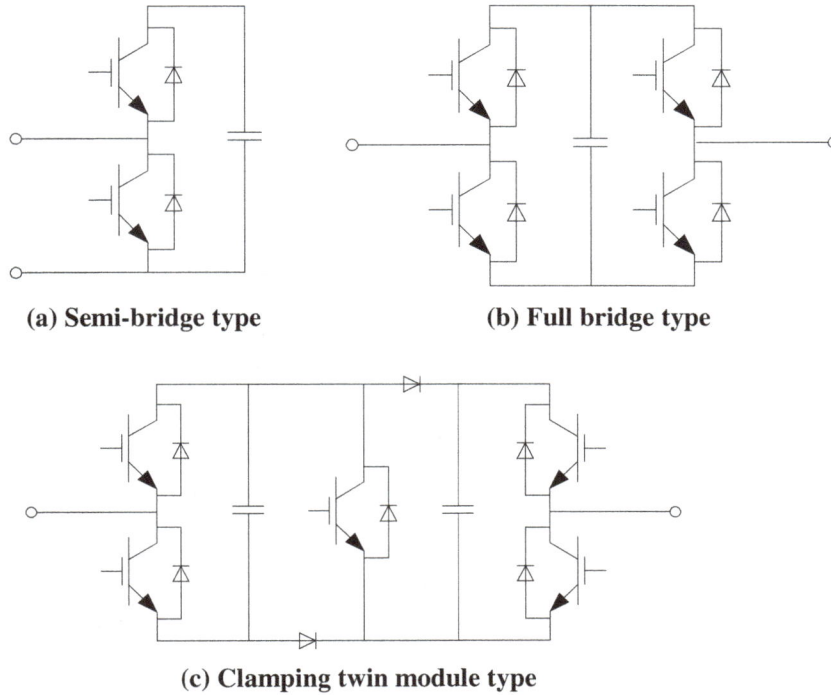

Fig. 5 Sub-module topologies of modular multi-level converter

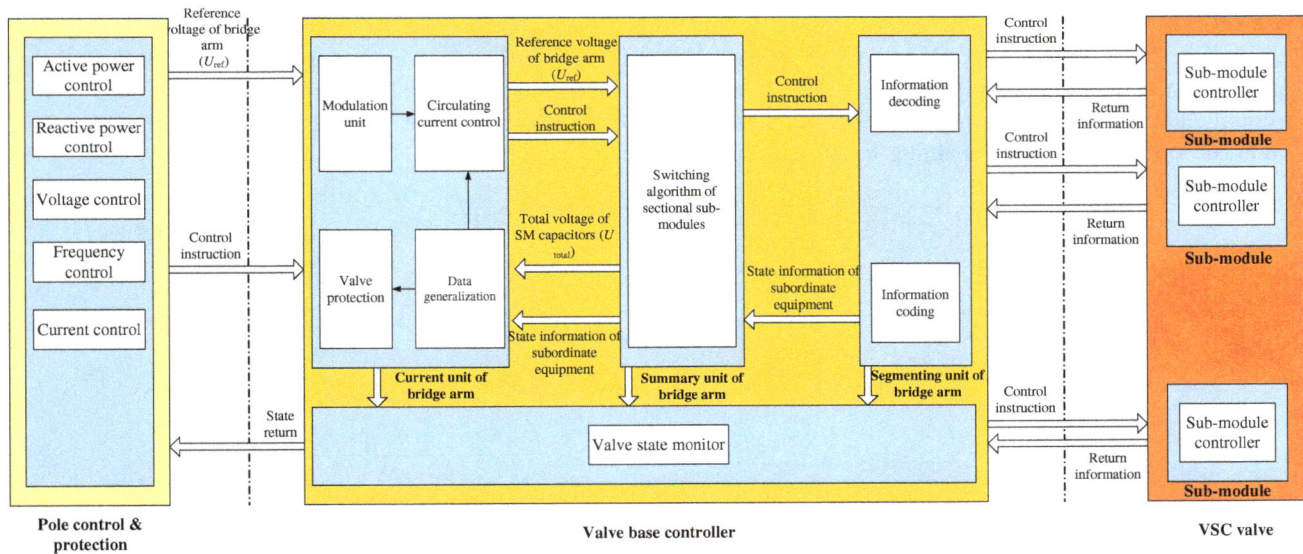

Fig. 6 Station & converter control system

adjustment. Its power controller includes an active power controller and a reactive power controller. The active power controller includes active power control and DC voltage control and the reactive power controller includes reactive power control and AC voltage control. Generally speaking, the normal operation of two-terminal HVDC Flexible system needs one station to control the DC voltage and the other terminal to control the active power, while the two-station reactive adjustments are mutually independent and can freely choose to control the reactive power or the AC voltage. On the control strategy, no matter the two-level or MMC technique is used, the equivalent mathematical model on the AC side is similar. Therefore, the same station level control strategy can be used. In numerous station level control strategies, the DC vector control strategy has become the mainstream control technique of the VSC for its high current responding speed and accurate current control effect [17–19].

The valve level control is the main difference of control system between MMC, HVDC Flexible, conventional HVDC and two-level HVDC Flexible. The valve base controller (VBC) in the HVDC Flexible is the middle connector to realize the station level control system and low-level sub-module control. It is used to realize the control, protection and monitoring of the valve arms and the communication between the station control system and converter valve. Meanwhile it is also used to realize the balance function of the sub-module capacitor voltage and circulation control function, which is the key to ensure the normal operation of the MMC HVDC Flexible system. Because the valve arm in the system with high voltage and large capacity is generally made up of hundreds of sub-modules, in order to ensure the voltage balance between each sub-module, VBC has a high requirement on the data disposal speed of the sub-modules, usually less than 100 μs. It is a huge challenge for the valve control design to achieve this high speed control balance technique of large-scale sub-modules. The unique circulation phenomenon of modular multi-level technique will cause the increase of the converter valve current stress and the loss level. If serious enough, it may make the system lose the balance and be unable to operate normally. For this reason, the design of circulating current control strategy is also a key factor in valve control.

The main function of HVDC Flexible protection system is to rapidly trip a fault or abnormally operating equipment in the system under the fault operating conditions, so as to ensure the safe operation of the remaining health system.

As shown in Fig. 7, the protection system can be roughly divided into AC-side protection, converter protection and DC-side protection.

2.4 Cable technology of HVDC flexible

It is difficult for the HVDC Flexible system to trip faults occurred on the DC side, so DC cables are usually used in the established HVDC Flexible projects to reduce the fault occurring rate.

Compared with AC cables, the conductors of the DC cables do not have skin effect and proximity effect, so even though they transmit large current, the complex segmentation conductor structure is not required for them. The electric field intensity of DC cables is distributed in direct proportion to the resistance coefficients of insulation which will change along with the temperature. If the load increases, the electric field intensity on the insulation surface will gradually increase. Therefore, the maximum load allowed by the DC cables should not make the electric field intensity on the insulation surface exceed its permissible value, i.e. not only the maximum working temperature of the cables shall be considered but also the temperature distribution of the insulating layer.

Compared with traditional DC cables, the DC cables used for the HVDC Flexible are not required to withstand the reversal of polarity of the voltage. Therefore, in some sense, the technical requirement to the cables for HVDC Flexible is lower than that of the traditional DC cables.

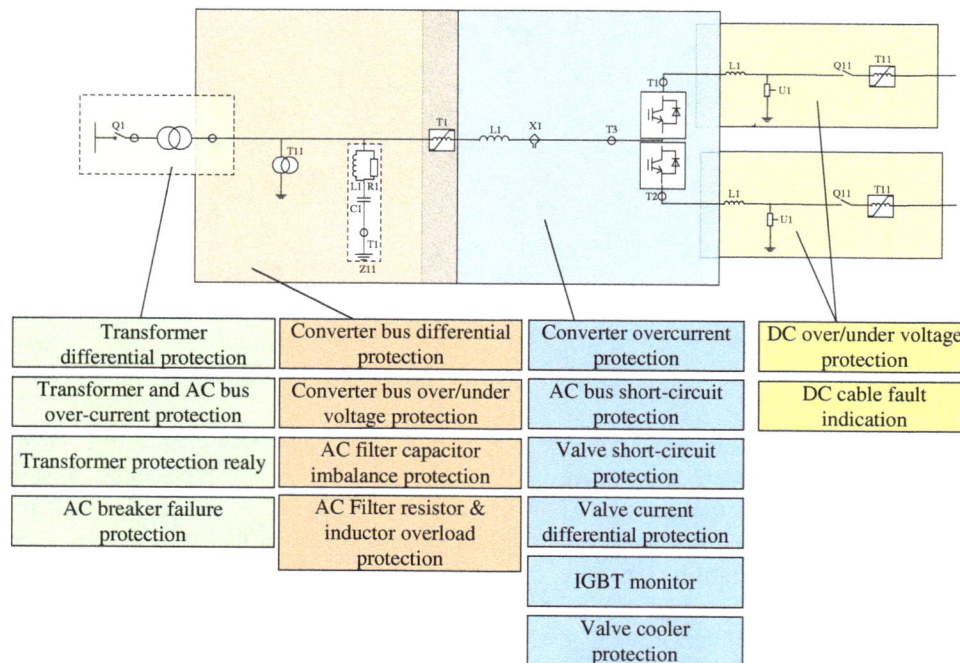

Fig. 7 Typical protection zoning of flexible HVDC system

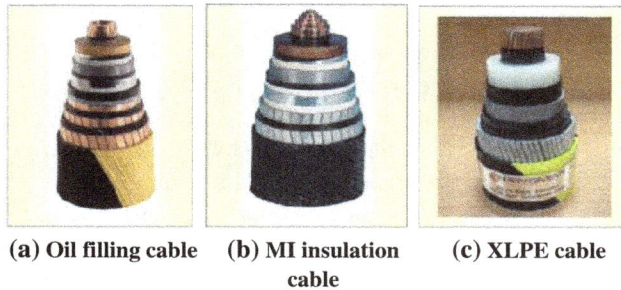

(a) Oil filling cable **(b) MI insulation cable** **(c) XLPE cable**

Fig. 8 Three types of DC cable technology

At present, based on the different insulation forms the cables used for HVDC Flexible are mainly divided into three different types, i.e. self-contained oil filling (SCOF) cable, mass-impregnated (MI) cable and cross-linked polyethylene (XLPE) cable, as shown in Fig. 8.

The self-contained oil filling cable uses a very mature technique and its rated voltage level can reach 800 kV. The cable is filled with low viscosity cable oil and tits insulating paper is made from brown paper of coniferous wood pulp. When the cable is damaged due to the external force resulting in oil leakage, it is unnecessary to stop the operation to deal with the leakage immediately. The cables can keep normal operation by adding oil form the oil compensating equipment. But form the environment point of view, the cable oil leakage will cause environmental pollution, especially the pollution caused by the seabed cables to the marine environment. The oil filling cable needs fuel tanks and other accessory equipment, so the workload of its operation and maintenance is large and the cost is also high.

The mass-impregnated (MI) cable is also using a very mature technique, which has been used in the DC transmission system for more than 100 years. The MI cable is suitable for up to 500 kV DC. At present, even though the longest MI cable route is 580 km for NorNed project built in 2008, its length for use is almost unlimited. Its operating maximum temperature is 55 °C and it is not suitable to operate in high temperature different conditions.

The XLPE insulated cable for HVDC Flexible is made of cross-linked polyethylene insulated material. Through ultra-clean and high purity technology or adding nanometer materials into the cross-linked AC cable insulation, the space charge problem of cross-linked DC cable has been solved. Due to the high softening point, the cable has small thermal deformation, high mechanical strength in high temperature and good thermal aging resistance of cross-linked polyethylene, thus its highest operating temperature can reach up to 90 °C, and its short-time allowable temperature can reach up to 250 °C. The XLPE insulated HVDC Flexible cable is extruded into new-style monopole cable by three lays of polymeric materials. The insulating

layer is extruded simultaneously by the conductor shielding layer, the insulating layer and the insulation shielding layer. The middle conductor is generally the single-core conductor made of aluminum or copper [20]. The highest parameters of current HVDC Flexible XLPE cable which can meet the project application requirements are 320 kV and 1,560 A. The flexible DC cables with a voltage level of above 500 kV are in developing now. China has currently completed the development of 160 kV flexible DC cables and already apply them into the practical engineering application; the 200 kV flexible DC cables have already passed the type tests and are in production and the 320 kV flexible DC cables have been started to be developed, but it still needs certain time to put them into practical application.

2.5 HVDC flexible test technology

The testing technology of HVDC Flexible mainly includes the testing technology of the converter valves and valve control equipment. Specifically to the converter valve testing technology, the CIGRE B4.48 working group in-charged by a Chinese expert explained and analyzed the stresses suffered by the converter valves in different working conditions in detail and made relevant testing suggestions in their report (447-Components Testing of VSC System for HVDC Applications). IEC 62501 has currently formulated relevant testing standards for converter valves. China has already had relevant capacities to carry out the HVDC Flexible converter valve type tests and completed the type tests for the 1,000 MW/±320 kV converter valves.

The HVDC Flexible converter valves are divided into switch type and controllable voltage source type and converters are voltage source type. Their basic operating principles are different from those for the conventional HVDC. Therefore, the transient state and steady state working conditions of HVDC Flexible converter valves are quite different from those of conventional HVDC converter valves. Due to the different stresses suffered by the flexible converter valves, the test items, methods and equipment for the originally conventional DC converter valves are almost not applicable. Therefore, it is necessary to undertake further researches on the operating principles of valves and the voltages, currents, heat, forces, other stresses and waveforms on the power electronic devices and components, so as to recommend relevant test items and equivalent testing methods.

In the steady state operation, the voltage and current stresses suffered by the HVDC Flexible converter valves are the super-impose of continuous DC and AC components. During transient, under the clamping action of sub-module capacitor voltage, there will be short-term capacitor discharge current in the converter valves and the discharge current is gradually reduced due to the protection actions.

The type test items of HVDC Flexible converter valves mainly include insulating type tests and operation type tests. The insulating type tests can also be divided into valve ground insulation tests and valve body insulation tests. The details are given in Table 1.

The valve control test technology is the important link to test the function and reliability of the valve control system. From the perspective of the HVDC Flexible valve base controller test system, as the trigger signals for all devices in a bridge arm in the conventional DC are identical, so the test can be carried out with a single equivalent scheme of thyristor devices. But as the trigger signals for each sub-modules in a bridge arm in the HVDC Flexible are different, the valve control system of HVDC Flexible needs to

Table 1 Main test items for HVDC Flexible converter

Test items	Contents
Type tests of insulation to ground for HVDC Flexible converter valves	AC voltage withstand test of valve support
	DC voltage withstand test of valve support
	Operation impulse test of valve support
	Lightning impulse test of valve support
Insulation type tests of HVDC Flexible converter valves	DC voltage withstand test
	AC & DC voltage withstand test
Operation tests of HVDC Flexible converter valves	Maximum operating load test
	Maximum transient overload operation test
	Minimum DC voltage test
	Valve short circuit current test
	Valve over-current turn-off test
	Valve electromagnetic interference test

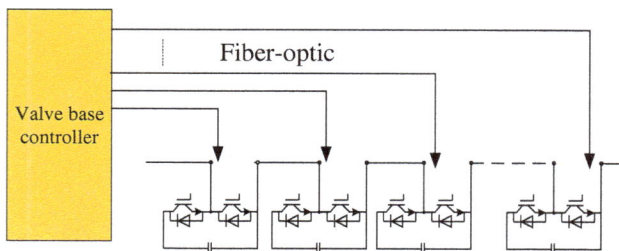

Fig. 9 Diagram of test system for valve base controller

provide different control command for each convertor sub-module. The original equivalent test methods for the valve system are not suitable to the HVDC Flexible converter valve control system (see Fig. 9).

The HVDC Flexible digital-analog hybrid simulation system constructed by dynamic simulation technique is currently the important technology means to study the MMC HVDC Flexible system. Dynamic modelling system can accurately simulate the dynamic behaviors of the HVDC Flexible converter valves, and provide real-time test function for the valve control system and pole control and protection system, as shown in Fig. 10.

At present, China has built a MMC HVDC Flexible real-time simulation system with up to 3,000 nodes. It can be used to undertake online tests and system simulation of the valve base control equipment of ± 320 kV voltage level and within 100 µs. The simulation systems for the projects of ± 500 kV and high, the multi-terminal HVDC Flexible and DC grid have been basically constructed and accomplished.

The digital real-time simulation system can finish the power grid modeling and realize the simulation of electromagnetic transients of the HVDC Flexible, such as start-up, shutdown, operation mode switching process, low frequency oscillation phenomena and faults. It can also be used to simulate the unlock and lock tests of the valve control system, the communication tests between the valve control and pole control equipment, the start-up/shutdown control tests of the converter valves, the valve fault simulation tests and so on. The digital real-time simulation is the necessary means to research and test the HVDC Flexible system. It can also be combined with the dynamic simulation tests to constitute the simulation test platform with more thorough functions and reduce the development cost and time of dynamic test system [21, 22].

3 Application status of HVDC Flexible projects in China

3.1 Application fields of HVDC Flexible projects

Based on HVDC Flexible technology characteristics, the constituted system is widely used in the fields of renewable energy source (RES) integration, island power supply, urban power supply, grid interconnection, etc.

The use of HVDC Flexible technology for the integration of wind power, solar energy and other RES with high power fluctuation can reduce the voltage fluctuation caused by the power fluctuation of RES and improve power quality. When a short circuit fault occurs in the AC system, the HVDC Flexible system can efficiently isolate the fault to ensure the stable operation of the RES. According to the calculation of CIGRE, the HVDC Flexible system is the best solution for an off-shore wind farm if the off-shore distance is over 60 km.

The use of HVDC Flexible technology in power supply to islands and offshore drilling platforms can give full play to its self-commutated technical advantages. Meanwhile, with respect to the AC circuits, DC circuits have advantages of lower investment and operating costs, no additional compensation equipment required for long distance power transmission and so on.

The use of HVDC Flexible technology in power supply to the urban center can not only quickly control the active power and reactive power to solve the voltage flicker and other power quality issues; but also provide system damping to improve the stability of the system and "black start function" when a serious fault occurs. In addition, the use of buried DC cables in HVDC Flexible will not cause alternation electromagnetic field and oil pollution, can achieve the capacity expansion and reformation of the urban power grid and can meet the urban central demand requirements and

environmental protection and energy saving requirements under the circumstances of no electromagnetic interference and no influence to the city appearance.

The use of HVDC Flexible technology to realize the power grid interconnection can not only complete the power exchange function between power grids, but also solve dynamic stability, black start of power grids and excessive short-circuit current and other issues in a large-scale power grid. All of these advantages of the HVDC Flexible rely on its quick and independent reactive power adjustment, "black start" capability, no short-circuit current indeed and other technical characteristics. Furthermore, the dimension of a HVDC Flexible convertor station is smaller than that of a conventional DC convertor station with the same capacity, thus the HVDC Flexible convertor station can be built close to the load center.

3.2 Application status of HVDC Flexible project in China

At the beginning of 2006, several Chinese institutes started to research the HVDC Flexible technology. They have obtained a series of achievements in basic theoretical research, key technologies, core equipment development, test capability building, engineering, system integration etc. The China's first HVDC Flexible demonstration project was commissioned into operation in Nanhui of Shanghai in July, 2011 [18, 23].

The modular multilevel converter (MMC) structure is used in Shanghai Nanhui HVDC Flexible demonstration project. It has a capacity of 20 MW, ±30 kV DC with a transmission length of approximately 8 km [23]. Nanhui wind power plant has been integrated in Shanghai Power Grid through this project (see Fig. 11). To verify the effect of the HVDC Flexible system in wind power integration, the manual shot-circuit tests were carried out, and the results indicated that this project can substantially improve the low voltage ride through capacity of wind power plants.

In order to meet the increased power demands of the economy development in the south of Dalian downtown,

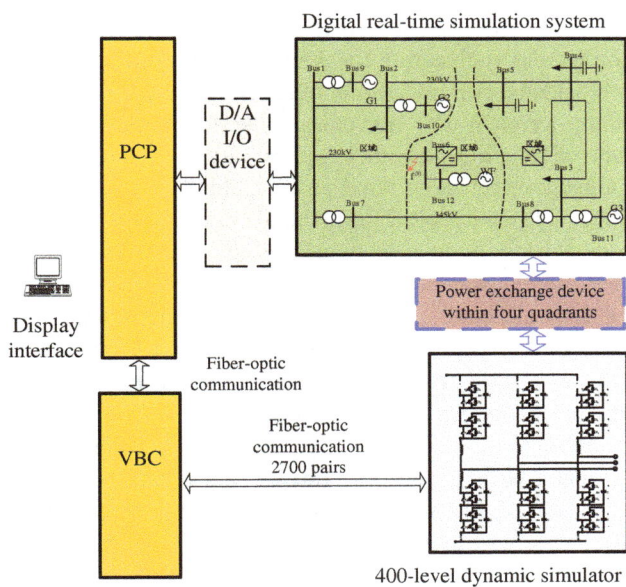

Fig. 10 Diagram of real-time hybrid simulation system

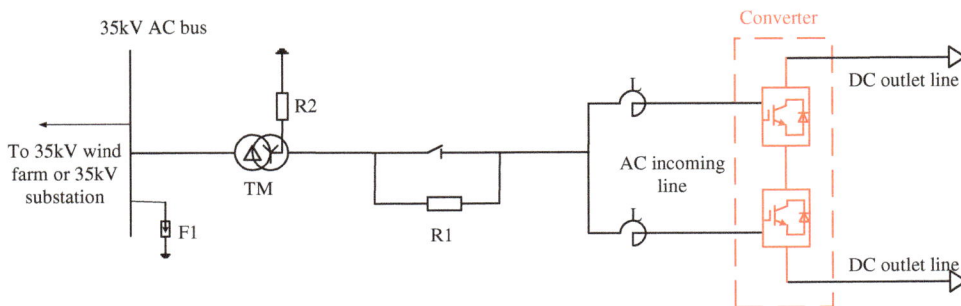

Fig. 11 Circuit diagram of Nanfeng station in Shanghai Nanhui project

avoiding the serious influence of natural disasters on the urban power supply and eliminating the potential safety hazards of power grids, the Station Grid Corporation of China (SGCC) started to construct a HVDC Flexible project which will to connect the north major network and the harbor east areas of the southern part downtown of Dalian city in 2012. This project has a rated capacity of 1,000 MW, a direct current (DC) voltage of ±320 kV [24], and a delivery transmission distance of about approximately 60 km. By end of December 2012, the world's first set of 1,000 MW/±320 kV converter valves and valve base controllers were developed relying on this project, and went through the witness tests of DNV KEMA (see Fig. 12). Meanwhile, relying on this project, China has built the world's largest dynamic model platform with 400 levels which effectively verified the validity of Dalian valve control system design and all kinds of functions. On the basis of engineering design, China has grasped a whole set of packaged design techniques on high-capacity HVDC Flexible system, convertor station construction techniques and system operation and maintenance techniques.

Fig. 12 1,000 MW/±320 kV HVDC Flexible converter valve

In order to improve the power distribution reliability and operation flexibility of Zhoushan Power Grid and consider the digestion integration and save of the rich wind power resources in Zhoushan islands, the Station Grid Corporation of China SGCC has planned to construct a 5-terminal HVDC Flexible project which is scheduled to be completed in 2014. This project contains 5 convertor stations with a total system capacity of 1,000 MW, and the largest convertor station will have a capacity of 400 MW and a direct current (DC) voltage class of ±200 kV (see Fig. 13). This project is the HVDC Flexible project [25] with the largest terminal number in the current world currently. It can meet the increasing load demands in Zhoushan region, becomes the second power source in the power supply of the northern islands and improves the power distribution reliability after being completed. It has the advantages of providing dynamic reactive power compensation ability and, improving the electric energy quality of Zhoushan Power Grid, reliving the grid connection problem of wind power fields in Zhoushan islands and improving the flexibility of the power grid dispatching and operation. The construction and implementation of this project can also provide a good reference in technology and engineering for the future power supply of offshore islands, grid integration of renewable energy sources, multi-terminal DC transmission system, even construction of DC grids and other applications.

In Guangdong Nan'ao wind power field, a three-terminal HVDC Flexible project [26] is under construction now. Because a large number of wind turbine generators are installed around Nan'ao Island, in order to achieve integrating the wind power, two wind farms are firstly planned to be connected into the two-terminal convertor station through a 110 kV transformer substation. The power energy will be transmitted to Shantou Power Grid after being collected by the HVDC Flexible system. Nan'ao project has a DC voltage of ±160 kV and a rated transmitting power of 200 MW. Its system diagram is as shown in Fig. 14.

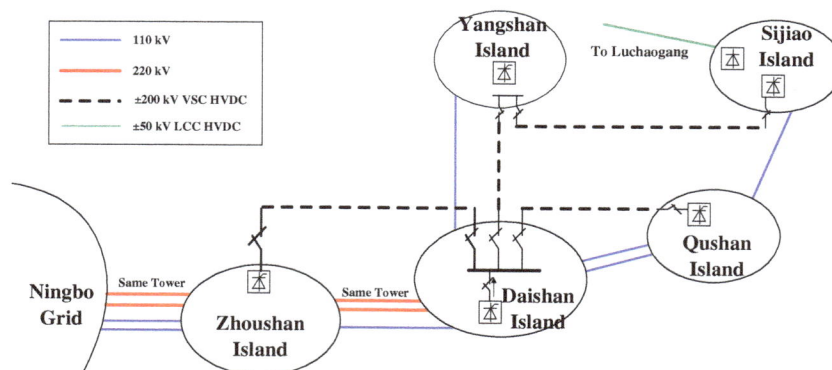

Fig. 13 Diagram of Zhoushan five-terminal HVDC Flexible project

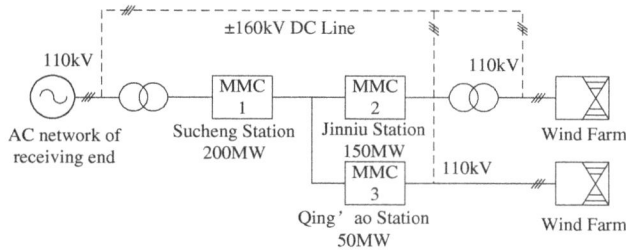

Fig. 14 Diagram of Nan'ao three-terminal HVDC Flexible project

4 Future development trend of HVDC flexible

The rapid progress of the HVDC Flexible technology promoted their applications in wind power integration, grid interconnection and other situations, while the development of market drives the improvement of the technique levels in return. From the current application demands, we can find that the main development directions of future HVDC Flexible technologies include: High-capacity HVDC Flexible technique, DC grid technique and overhead line HVDC Flexible technique.

4.1 High-capacity HVDC flexible technologies

Integration of large wind farms and urban loads, power grid interconnection and other application situations put forward higher requirements on the transmission capacity of HVDC Flexible. Currently, the world's largest HVDC Flexible project is 1,000 MW/±320 kV. In 3 to 5 years, the projects of 1,200 MW/±400 kV and higher will be implemented soon.

Technically, the improvement of voltage level and capacity of the current HVDC Flexible system is mainly restricted by the voltage level of the XLPE cables and the development of the existing IGBT devices. In previous projects, the use of single converter also limited the increase of the system capacity. Therefore, the capacity improvement of the future HVDC Flexible will mainly concentrate on the XLPE cables with higher rated voltages, new type high-capacity power electronic devices, new system topology applications and other aspects.

1) XLPE cable technology

On the choice of cables, comparing to XPLE cables, MI cables can be considered to be used in the higher capacity HVDC Flexible projects. This kind of cables has higher voltage levels, but the fabrication cost is relatively higher, which makes them difficultly be promoted in great range. Therefore, it will still need a breakthrough in XLPE cables in the future.

On the aspect of XLPE cables, the main difficulties faced currently are the charge distribution and

manufacturing technique of the cable insulating materials and the design and process of the cable connectors. The principal cable manufacturers in the world all have put a great development effort on this aspect. At present, the ±500 kV XLPE cables are at the test phase and are expected to be put into project application in the next 2 to 3 years. With the continuous promotion of the future projects, XLPE cables of higher voltage levels may also appear. It is expected that in 5 years, the voltage and capacity levels of XLPE DC cables will be promoted to 600 kV/2 GW, and in 10 years, the voltage and capacity levels of DC cables will reach more than 750 kV/3 GW.

2) Power electronic device technology

To improve the capacity of IGBT devices, it is required to solve the problems in the fabrication and encapsulation of IGBT chips due to the increase of voltage and current. It will be faced with quite high difficulty to substantially improve the production technique at the same time of ensuring the product reliability. At present, the main device manufactures in the world are carrying out research and development on the IGBT devices of high voltage and large current levels, which are expected to be put into commercial utilization in the next few years. Meanwhile, in the application process of the devices, a series of technological difficulties shall be solved, such as the driver design of new type IGBT, the overshoot suppression of current turn-off, the quick protection design and so on. These difficulties need to be solved by careful design and long-term tests for verification.

In respect of new type semiconductor devices, "large energy gap" semiconductor materials (such as silicon carbide (SiC), etc) are a kind of materials with the best development prospect in the future. That is why the power semiconductor devices have currently become one of the research hotspots in the worldwide. Compared with the existing devices, IGBT devices made of this kind of materials will be promoted several to tens of times on the aspects of voltage withstand level, through-current capacity, operating temperature and so on. And meanwhile, the losses will only be a fraction of that of the traditional silicon devices. Converters and systems constituted upon this kind of devices will directly make the existing engineering capacity improve several to tens of times, which will bring revolutionary change to the development of HVDC Flexible. But due to lots of problems existed in the quality and technique control and other aspects of SiC and equivalent materials currently, the process is just remained at the sample and product phase of small capacity. In the short term, they are unable to be really put into project application of large capacity. It is expected that SiC and equivalent devices will be put into a certain scale of demonstration application in the electric power system in 20 years.

3) Combined type system topology technique

Because cable and device development cycle is relevant long, if projects in the short term need to increase the capacity, two kinds of schemes can be used. One is in the form of converter combination; and the other is to use full bridge type converters constituted by sub-modules which makes the use of negative level output to increase AC voltage level, so as to upgrade the system capacity. But both kinds of schemes will increase the systematic costs.

The basic approaches of multiple converter combination include: series connection, parallel connection or series-parallel connection and so on, which are basically the same as the multiple converter scheme used in conventional DC transmission. The system structure of multiple converter combination can not only reduce the requirements of the circuital insulating level and improve the system reliability under the fault circumstances, but also facilitate the construction of the staging projects (such as Caprivi Link project). In addition, in the present HVDC Flexible projects, VSCs are used in both terminals of each system. But other forms of converter station structures also have high feasibility in special situations, such as the use of hybrid DC with current sourced converter (CSC) and VSC at each of the two terminals respectively, or the use of CSC and VSC for the two converters in a convertor station, etc [27–29]. Some schemes in these system structures have been used in several projects (such as Skagerrak4 project and GBX project and so on).

Based on the current technologies with the addition of the combined topology applications, the voltage level of the HVDC Flexible system can be directly improved to ±640 kV or above, in which single converter capacity can be promoted to 2,000 MW. However, if considering using the combination of series-parallel connection of converters, for example using eight 320 kV, 1 kA converter units in which every two units will be connected in parallel first then in series. In such combination the voltage level will reach 640 kV and the total capacity will reach 2,560 MW. If the series-parallel connection units are recombined into a double pole system, the system parameter will achieve 640 kV/5, 120 MW.

4.2 Multi-terminal DC and DC grid technologies

With the continuous development of renewable energy sources as well as the demands of existing grid upgrade and other aspects, the future development of HVDC Flexible will continue to focus on the networking and concentrated on transmission of wind power, local grid interconnection, electric power transmission to urban center load and other aspects. In many cases these applications need to realize the power supply of multiple power input and output points, thus the multi-terminal DC and even DC grid technologies are required to be used.

1) Concepts of multi-terminal DC and DC grid

Multi-terminal HVDC transmission (MTHVDC) is the primary stage of the DC grid development. It is a transmission system connected by more than three convertor stations through series, parallel or combination connection of parallel and series. It can realize multiple power input and output points. Because the HVDC Flexible has unique technical advantages in construction multi-terminal system, it will be rapidly developed in the MTHVDC system in the future.

The DC grid is equivalent to the extension of the MTHVDC. It is a stable AC/DC hybrid wide-area transmission network which has the intelligence of advanced energy management system. In the network, different clients, existing transmission networks, micro power grids and different power sources can get efficient management, optimization, monitoring and control, and real-time response to any problems. It can integrate multiple power sources and transmit and distribute the power energy with minimum losses and maximum efficiency in a larger range.

The most fundamental difference between DC grid and MTHVDC is that DC grid is a transmission system with "loops", there are multiple transmission circuits connected between each convertor stations and the whole system has redundancy and high reliability [30] (see Fig. 15). Therefore, the future development direction of MTHVDC shall be towards the development of network, i.e. DC based power transmission and distribution network.

2) Future development trend of DC grid

Specific to the development of DC grid, Europe has proposed a "Super Grid" to construct a new generation of transmission grid based on the HVDC Flexible technology

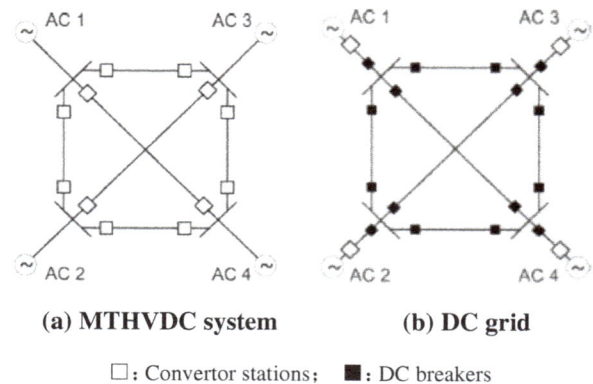

(a) **MTHVDC system**   (b) **DC grid**

□: Convertor stations； ■: DC breakers

Fig. 15 MTHVDC and HVDC grid

and to, establish a wide-area intelligent transmission network. The aims of this proposal are to realize the power fluctuation suppression of renewable distribution generations in a wide-area range and large-scale efficient integration of renewable energy sources, guarantee the safe and stable operation of the power grids, improve the power supply quality, promote the harmonious development of renewable energy sources and power system and so on [31–33]. China can also construct DC grids based on the large-scale offshore wind farms in the future.

3) DC grid technology and equipment development

It shall be noted that the DC grid is still at the beginning development stage. There are lots of critical issues needing to be solved, such as standardization of the DC grid and equipment, development of core equipment, control technology, wide-area measurement and fault detection technologies, protection technology, the safety and reliability assessment techniques of the DC grids, etc.

From the current technique development point of view, the following problems need to be solved emphatically before the DC grids are really put into application.

First is to solve the DC circuit fault isolation problem. The main R&D direction is the new converter topologies. Because the MTHVDC and HVDC grid may need the converters to realize the DC fault self-clearance and other functions. The hot issue of the current research is how to improve and even develop brand new converter topologies on the basis of the existing topological structures, so as to meet the application demands of these occasions. At present, the topologies which can be used in the multi-terminal system and have DC fault clearance function are not only the currently used full bridge sub-module topology, but also new type converter topological structures developed by many institutions and scholars, e.g. the one based on the full bridge or CDSM (clamping double submodules) [34]. The hybrid topology scheme proposed by Alstom can not only reduce the systematic costs efficiently, but also restrain the DC fault current so as to realize fault clearance, which is also one of the research hotpots now.

Second is to develop DC breakers to be applied in the DC system. Specific to the DC breakers, corresponding prototype design and development have been carried out in the worldwide. At present, ABB and Alstom have already finished the prototype tests. Companies in China are also carrying out relevant development and are expected to launch relevant products in the next 2 to 3 years [25].

Third is to solve the interconnection problems of DC grids at different voltage levels. The main solution is to research and develop appropriate DC transformers. Regarding to the DC transformers (DC/DC convertors), there are no relevant products which have been launched

yet in the world. Some research organizations in China are carrying out basic research and conceptual design now. It is expected to still need certain time to launch the products.

4.3 Long distance overhead line HVDC Flexible technology

HVDC Flexible technology in the cases that overhead lines are used as the transmission circuits has extensive application prospect as well. The use of overhead line transmission can not only increase the voltage level and system capacity, but also efficiently reduce the circuit investment and save the construction costs. As China has vast territory, the resource allocations of the electricity generation and utilization in various regions are serious unbalance. Therefore, the long distance overhead line power transmission plays an irreplaceable role in the process of the electric power development in China.

The use of overhead line transmission system needs the fault clearance ability which is required to solve the transient faults in the circuits. In addition to developing DC breakers with corresponding voltage levels, the problem can be solved by developing new type converter topologies which can clear the DC faults. This is the same as the requirements of the DC grids.

Meanwhile, normally the conventional HVDC system is used at the sending-end terminal and HVDC Flexible at the receiving-end terminal in the overhead line transmission system, or the conventional HVDC converter stations are converted into HVDC Flexible stations. It is an important development direction of using the HVDC Flexible in the overhead line transmission systems to solve the commutation failures caused by the system faults at the same time of saving the construction costs.

5 Conclusions and prospects

The increasing promotion of the requirements against the global climate change and the increasingly severe security situation of energy supply urgently require the construction of more intelligent, clean, efficient and reliable transmission grid, which has become the common goal of various countries in the world to develop the electric power industry. The HVDC Flexible technology attracts more and more researchers' attention due to its advantages of active and reactive power independent adjustment, black start capacity, easily constructing DC grid etc. Meanwhile, the constantly improving technical level of controllable switch devices, DC cables and other equipment has efficiently enhanced the transmission capacity of HVDC

Flexible and made HVDC Flexible become one of the main transmission modes used by the power grids. It is predicated that with the demands of integration of renewable energy sources and power grid transformation and upgrade in the future, the HVDC Flexible applications in China will gain increasingly wide development.

Due to the influence of regional and economic development, European countries generally face significant problems such as the power supply corridor tensions and energy structure adjustment. The construction of large scale DC grids is conductive to the optional configuration of energy sources and can solve the potential system safety hazards caused by wide-range wind power integration. The HVDC Flexible technology has a rapid development as the prior choice of wind power integration and DC grid construction. It is predicted that regional DC power grid with the foundation of HVDC Flexible, the backbone network of 500 kV or higher and the main transmission capacities of 10 GW and above will be built in Europe in the next 10 to 20 years.

In China and other countries with vast territories, the extra-high voltage AC/DC transmission technologies are still the efficient means to solve the long distance and high capacity electricity transmission problems. But for the grid integration and digestion and save problems of regionally new energy sources, the HVDC Flexible and DC grid technologies will be effective supplements. The next 10 years will be the rapid development and construction phase for the HVDC grid technology. As the capacity of the HVDC Flexible system is continuously increased, it is an irresistible trend that the HVDC Flexible system will gradually replace the conventional HVDC and HVAC transmission systems in some fields in the future.

Acknowledgement This work was supported by National Natural Science Foundation of China (No. 51261130471).

References

[1] State Grid Corporation of China (2009) Research report on the clean energy development advanced by SGCC, Beijing, China, 2009 (in Chinese)

[2] Tang GF, He ZY, Cao JZ et al (2012) A review of 2012 CIGRE on application and perspective in HVDC and power electronics. Autom Electr Power Syst 36(24):1–6 (in Chinese)

[3] Flourentzou N, Agelidis VG, Demetriades GD (2009) VSC-based HVDC power transmission systems: an overview. IEEE Trans Power Electr 24(3):592–602

[4] Tang GF (2010) Voltage source converter based HVDC transmission technology. China Electric Power Press, Beijing (in Chinese)

[5] ABB HVDC Light.. Accessed 31 Dec 2012

[6] Marquardt R, Lesnicar A (2004) New concept for high voltage—modular multilevel converter. In: Proceedings of the 2004 IEEE power electronics specialists' conference (PESC'06), Aachen, Germany, 20–25 June 2004, p 5

[7] Ding GJ, Ding M, Tang GF (2009) Analysis and comparison of VSC-HVDC topologies and corresponding modulation schemes. Automat Electr Power Syst 33(10):64–68 (in Chinese)

[8] Working Group B4.37 CIGRE (2005) DC transmission using voltage sourced converters. Technical Brochure No. 280, International Council on Large Electric Systems, Paris

[9] Ding GJ, Tang GF, Ding M (2009) Topology mechanism and modulation scheme of a new multilevel voltage source converter modular. P CSEE 29(36):1–8 (in Chinese)

[10] Ding GJ, Ding M, Tang GF et al (2009) A new hybrid PWM technology used in the VSC-HVDC transmission system. Power Syst Technol 33(7):7–13 (in Chinese)

[11] Henry S, Denis AM, Panciatici P (2010) Feasibility study of off-shore HVDC grids. In: Proceedings of the 2010 IEEE power and energy society general meeting, Minneapolis, 25–29 Jul 2010, p 5

[12] Wang SS, Zhou XX, Tang GF et al (2011) Modeling of modular multi-level voltage source converter. P CSEE 31(24):1–8 (in Chinese)

[13] Knaak HJ (2011) Modular multilevel converters and HVDC/FACTS: a success story. In: Proceedings of the 14th European conference on power electronics and applications (EPE'11), 30 Aug–1 Sept 2011, Ludvika, pp 2498–2503

[14] Jacobson B, Karlsson P, Asplund G, et al. (2010) VSC-HVDC transmission with cascaded two-level converters. B4-110, CIGRE, Paris

[15] Wang SS, Zhou XX, Tang GF et al (2011) Selection and calculation for sub-module capacitance in modular multi-level converter HVDC power transmission system. Power Syst Technol 35(1):26–32 (in Chinese)

[16] Wang SS, Zhou XX, Tang GF et al (2011) Influence of AC system strength on operating characteristics of MMC-HVDC system. Power Syst Technol 35(2):17–24 (in Chinese)

[17] ZhaoY HuXH, Tang GF et al (2011) Control strategy of modular multilevel converters based HVDC transmission. P CSEE 31(25):35–42 (in Chinese)

[18] Dong YL, Bao HL, Tian J et al (2011) Control and protection system for VSC-HVDC. Automat Electr Power Syst 35(19):89–92 (in Chinese)

[19] Wei YF, Wei ZN, Sun GQ et al (2012) New HVDC power transmission technology: MMC-HVDC. Electr Power Autom Equip 32(7):1–9

[20] Ying QL (2012) The prospect of development of DC submarine cables in China. Electr Wire Cable 3:1–7 (in Chinese)

[21] Liu D, Tang GF, Zheng JC et al (2012) Small signal modeling and analysis of open-loop response time constant of MMC. P CSEE 32(24):1–7 (in Chinese)

[22] Liu D, Tang GF, He ZY (2013) Hybrid teal-time simulation technology for MMC-HVDC. Electr Power Autom Equip 33(2):68–73 (in Chinese)

[23] Qiao WD, Mao YK (2011) Overview of Shanghai HVDC flexible transmission demonstration project. East China Electr Power 39(7):1137–1140 (in Chinese)

[24] Ge WC, Gu HQ, He ZY (2012) Overview of Dalian HVDC flexible transmission demonstration project. Northeast China Electr Power 2:1–5 (in Chinese)

[25] State Grid Corporation of China (2012) Zhejiang Zhoushan Islands multi-terminal HVDC flexible projects have been approved. Accessed 26 Nov 2012 (in Chinese)
[26] Rao H (2013) Research and application of the high-power electronic technology in China Southern Power Grid. Southern Power Syst Technol 7(1):1–5 (in Chinese)
[27] Guo CY, Zhao CY, Montanari A (2012) Investigation of hybrid bipolar HVDC system performances. P CSEE 32(10):98–104 (in Chinese)
[28] Li GK, Li GY, Liang HF et al (2011) Research on a novel hybrid HVDC system. Power Syst Technol 35(2):82–86 (in Chinese)
[29] Yuan XF, Cheng SJ, Wen JY (2006) Simulation study for a hybrid multi-terminal HVDC system based on VSC and CSC. Automat Electr Power Syst 30(20):32–36 (in Chinese)
[30] Tang GF, Luo X, Wei XG (2013) Multi-terminal HVDC and DC-grid technology. P CSEE 33(10):8–17 (in Chinese)
[31] Vrana TK, Torres-Olguin RE, Liu B, et al. (2010) The North Sea Super Grid—A technical perspective. In: Proceedings of the 9th IET international conference on AC and DC power transmission (ACDC'10), Glasgow, 19–21 Oct 2010, p 5
[32] Jovcic D, Van Hertem D, Linden K, et al. (2011) Feasibility of DC transmission networks. In: Proceedings of the 2nd IEEE PES international conference and exhibition on innovative smart grid technologies (ISGT Europe'11), Manchester, UK, 5–7 Dec 2011, p 8
[33] Feltes JW, Gemmell BD, Retzmann D (2011) From smart grid to super grid: Solutions with HVDC and FACTS for grid access of renewable energy sources. In: Proceedings of the IEEE Power and Energy Society general meeting, Detroit, 24–29 Jul 2011, p 6
[34] Marquardt R (2011) Modular multilevel converter topologies with DC short circuit current limitation. In: Proceedings of the IEEE 8th international conference on power electronics and ECCE Asia (ICPE & ECCE'11), Jeju 30 May–3 Jun 2011, pp 1425–1431

Guangfu TANG received the B.Eng. degree in electrical engineering from Xi'an Jiao Tong University, Shanxi, P.R. China, in 1990, and the M.Eng. degree and the Ph.D. degree in electrical engineering from Institute of Plasma Physics, The Chinese Academy of Sciences (ASIPP), in 1993 and 1996 respectively. During 1996–1998, he had a postdoctoral position with China Electric Power Research Institute (CEPRI), Beijing, P.R. China, and he was the vice director of China Energy Conservation Center. In 1998, he joined the CEPRI, where he led the Thyristor Controlled Series Compensator Group from 1998 to 1999, the Static Var Compensator from 2000 to 2001. Since 2002, he has been a professor level senior engineer of CEPRI. During the past 16 years, his research fields include the fault current limiter, the converter valve of high voltage or ultra high voltage in DC transmission system, and the voltage-source converter-high voltage dc (VSC-HVDC) transmission systems. Now, he is the vice chief engineer of CEPRI. He has published more than 110 papers, and won 70 patents in his research field. Dr. Tang is the recipient of the 2006 Second Award, and the 2008 First Award of the National Science and Technology Progress of P.R. China respectively. And he has obtained 6 Awards of the Provincial Scientific and Technological Progress Awards. He was the Convenor of 48 Working Group in the International Council on Large Electric Systems in 2007. Now, he sits on a committee of the IEEE/PES Narain Hingorani FACTS and Custom Power Award Committee.

Zhiyuan HE received the B.Eng. degree in electrical engineering from Sichuan University, P.R. China, in 2000, and the M.Eng. degree and the Ph.D. degree in electrical engineering from China Electric Power Research Institute (CEPRI), in 2003 and 2006 respectively. In 2006, he joined CEPRI, where he led the voltage-source converter based high voltage DC (VSC-HVDC) transmission systems Group, from 2008 to 2009, was the manager for CEPRI in the areas of HVDC Technology. Since January 2010, he has been the manager and chief engineer for CLP Power Engineering CO., LTD PURELL of CEPR. In the past six years, he has accomplished theoretical study on high power electronics technology for reliable operation of large interconnected power grids, relocatable DC de-ice system and VSC-HVDC transmission, including the first VSC-HVDC project commissioning in 2011 in China. He has published more than 50 papers, and won 36 patents in his research field. And he has obtained 2 awards of the Provincial Scientific and Technological Progress Awards. Dr. He was as a member of CIGRE B4 Working Group 48, and researched on "Components Testing of VSC System for HVDC applications" from 2006 to 2009. During 2009-2010, he got as a member of IEC SC22F Working Group 19, and researched on "High-Voltage Direct Current (HVDC) Power Transmission Using Voltage Sourced Converters (VSC).

Hui PANG received the B.Eng. degree and the M.Eng. degree in electrical engineering from Hefei University of Technology, P.R. China, in 2002 and 2005 respectively, and the Ph.D. degree in electrical engineering from China Electric Power Research Institute (CEPRI), in 2010. In 2010, he joined CEPRI, where he be a research of the voltage-source converter based high voltage DC (VSC-HVDC) transmission systems R&D department, from 2011 to 2012, was the project manager in the areas of VSC-HVDC electrical design. Since 2013, he has been the manager of system R&D of HVDC technology research department in Smart Grid Research Institute of SGCC. In the past 6 years, he has accomplished theoretical study on high power electronics technology, VSC-HVDC transmission and DC grid technology, including the first VSC-HVDC project commissioning in 2011 in China.

5

A new method of enhancing reliability for transmission expansion planning

Chengxin LI, Guo CHEN (✉), Junyong LIU

Abstract The reliability plays a significant role in power systems and it is an important objective or constraint in transmission expansion planning. Firstly, a DC optimization model was proposed to calculate the maximum arrival power at each load point. Compared to the network flow method, DC model is closer to the actual power flow and it is able to obtain more realistic reliability assessment results. Furthermore, a novel sensitivity index (SI) was also proposed to choose the most effective line so as to enhance the nodal and/or system reliability. The Monte Carlo simulation is used to simulate the system components state. This improved reliability evaluation method and SI can be used for transmission expansion planning or maintenance scheduling. Tests are performed using 6-bus system derived from the Garver's system and the IEEE 10-machine 39-bus system. The results show the effectiveness of the method.

Keywords Probabilistic reliability evaluation, Sensitivity analysis, Monte Carlo simulation, DC model

1 Introduction

Transmission system plays a significant role in electric power system. It is not only a linkage between the generation and distribution, but also provides a non-discriminative and reliable environment for demanders and suppliers [1]. With the growth of load demand, the generation expansion planning (GEP) and transmission expansion planning (TEP) become more and more important. The main objective of TEP is to develop the system as economically as possible [2] due to the load growth, and it is subject to a set of economic, technical and reliability constraints [1, 3]. In the regulated market, GEP and TEP are sub-tasks of a power system planning process performed by a regulated power utility. However, in the deregulated market environment, TEP is usually performed separately by transmission network service providers, while GEP becomes the task of generation companies or investors [4]. In the conventional monopolistic market, the power utilities have the social obligations to provide a reliable electricity supply. In a competitive market, the reliability of service is one of the important factors for market competitiveness [5, 6]. Therefore, a reliability level is an important constraint for TEP process. In addition, at the time of preparation for maintenance scheduling, a certain level of supplying reliability must be guaranteed after some power system components which are in outage and maintenance state [7, 8]. Thus, reliability evaluation and reinforcement are very meaningful works both in the power system planning or maintenance scheduling.

Generally, power system reliability includes two aspects: adequacy and security. Adequacy measures the generation and transmission capacities of the system under static conditions, without considering system disturbances [3]. In this paper, the reliability means adequacy. It is not only restricted by the capacities of the generators and transmission lines, but also subjected to the availabilities of them. The reliability analysis is carried out before stability and fault analyses in conventional TEP [9]. Thus, reliability evaluation should be incorporated in TEP. Otherwise there is no guarantee to have a trustworthy supply for demands [3].

C. LI, J. LIU, The School of Electrical Engineering and Information, Sichuan University, Chengdu 610065, China
C. LI
e-mail: Lcx36@scu.edu.cn
G. CHEN, The School of Electrical and Information Engineering, The University of Sydney, Camperdown, NSW 2006, Australia
(✉) e-mail: guo.chen@sydney.edu.au

Generally, TEP problem can be formulated as an optimization problem. Reliability is generally treated as a constraint [10] or a part of objective function to deal with [11]. The common reliability of the system is often assumed to be guaranteed via the demand constraints. That is, these constraints enforce that line capacities exceed the line loads based on an assumed demand profile [12]. In [11], the probabilistic load curtailment loss was handled as a part of objective function and a TEP approach considering the load and wind power generation uncertainties was proposed. The different reliability indices such as loss of load expectation (LOLE) [13], loss of load probability (LOLP) [14], hierarchical reliability assessment [15], loss of load cost (LOLC) [16] and expected energy not supplied (EENS) [17] have been used to investigate TEP. Among them, LOLE and EENS are two common reliability indices and they are easily obtained from the load duration curve (LDC). An extended effective load model, considering the capacities and uncertainties of generators and transmission lines, has been proposed [5, 6]. Then an extended nodal ELDC based on this extended effective load model can be obtained. Thus, the indices LOLE and EENS can be calculated.

Through reliability evaluation, if the system reliability indices do not satisfy the prescribed criteria, there is a reasonable question: where to add a transmission line is the most effective to improve the reliability indices. In [14], a method was developed for analyzing the reliability of composite power systems under the constraints of emissions. Some reliability indices are expressed by the function of the relevant factors such as the element forced outage rates (FOR) and the element capacities. Then their sensitivity with respect to various relevant factors are obtained by calculating the partial derivative of the reliability indices [14, 18–20]. A sensitivity index (SI) based constructive heuristic algorithm (CHA) has been applied for TEP [21, 22]. At each step of CHA, a component (circuit) of the feasible solutions must be added to the system. The choice of this component is determined by SI. The SI is based on the greatest active power flow of circuits, which is obtained from the solution to relaxing the integrality of DC investment model [21]. As an improvement, SI comes from the greatest apparent power flow of circuits, which is obtained from the solution to relaxing the integrality of AC investment model [22].

Following the previous work, this paper proposed a method to evaluate the composite power system reliability considering the FOR of the generators as well as the transmission lines. The Monte Carlo method is used to simulate the random behavior of the availability of the system components (generator or branch). A DC power flow model, instead of the network flow method [5, 6], is used to obtain more accurate maximum arrival power at load point. With the reliability evaluation process, an implicit sensitivity index based optimization method can be obtained, which can help to choose the most effective component so as to improve the system or bus reliability.

There are two main contributions of this paper. One is that the DC model optimization was used to improve the accuracy of maximum arrival power at each load point. The other one is that a novel SI based optimization method was presented to choose the most effective line so as to enhance the reliability.

The remainder of this paper is organized as follows. A DC model based reliability evaluation model of composite power system is presented in Section 2. Section 3 proposes the process of obtaining the ELDC using the Monte Carlo simulation and provides a sensitivity analysis index. Case studies are presented in Section 4. Finally, the paper is concluded in Section 5.

2 Reliability evaluation model of composite power system

The traditional ELDC considers only the FOR of the units, without the FOR of the transmission lines [5]. Firstly, a composite ELDC and the reliability indices LOLE and EENS [5, 6] are briefly introduced. Then, an optimization model of obtaining minimum outage power (i.e. maximum arrival power) at load point is presented, which is a DC constraint model.

2.1 Model of reliability evaluation in composite power system

The reliability indices, such as LOLE and EENS, can be assessed using a composite ELDC (CMELDC) based on the effective load model at the load point [5, 6]. When a load point curtails a certain amount of load due to outages of generator unit and/or transmission line, it is equivalent to the case because the same amount of probabilistic loads is added to the load point but no components outage [5, 6, 23]. Base on this concept, the effective inverted load duration curve can be obtained at load point k.

$$\Phi_i^k(x_e) = \Phi_{i-1}^k(x_e) \otimes f_{oi}^k(x_{oi}) = \int \Phi_{i-1}^k(x_e - x_{oi}) f_{oi}^k(x_{oi}) dx_{oi}$$

(1)

where $\otimes$ is the operator, indicating the convolution integral; k is the load point number; x_e is the random variable of the effective load; x_{oi} is the random variable of the probabilistic load caused by the forced outage of component i; $\Phi_i^k(x_e)$ is the effective inverted load duration curve caused by the forced outage of components 1 to i at load point k; $\Phi_{i-1}^k(x_e)$ is the effective inverted load duration curve caused by the forced outage of components 1 to $i-1$ at load point k; $f_{oi}^k(x_{oi})$ is the probability density function (PDF) of the outage capacities caused by the forced outage of component i at load point k.

From (1), it is important to note that $\Phi_i^k(x_e)$ is calculated by iterative manner. If the total probability density function of the outage capacities caused by the forced outage of components 1 to i at load point k is determined, $\Phi_i^k(x_e)$ can be obtained by (2).

$$\Phi_i^k(x_e) = \Phi_0^k(x_e) \otimes f_{osi}^k(x_{osi}) = \int \Phi_0^k(x_e - x_{osi})f_{osi}^k(x_{osi})dx_{osi} \tag{2}$$

where x_{osi} is the random variable of the synthesized fictitious probabilistic load caused by the forced outage of components 1 to i; $\Phi_0^k(x_e)$ is the original inverted load duration curve at load point k; $f_{osi}^k(x_{osi})$ is the total probability density function of the outage capacities caused by the forced outage of components 1 to i at load point k.

After obtaining the effective inverted load duration curve at load point, the reliability indices $LOLE_k$ and $EENS_k$ can be calculated at load point k.

$$LOLE_k = \Phi_{NE}^k(x) \,|\, x = AP_k \tag{3}$$

$$EENS_k = \int_{AP_k}^{AP_k + L_{Pk}} \Phi_{NE}^k(x)\,dx \tag{4}$$

where NE is the number of total elements (including generating units, transformer and transmission lines); AP_k is the maximum arrival power at load point k; L_{Pk} is the peak load at load point k.

The EENS of the entire system is equal to the summation of $EENS_k$ at all load points, as shown in (5). However, the approach of calculating LOLE is completely different from the EENS. The expected load curtailed (ELC) at the load point k must be calculated. Then, the ELC of the entire system can be obtained. Thus, the entire system LOLE index can be calculated.

$$EENS = \sum_{k=1}^{ND} EENS_k \tag{5}$$

$$ELC_k = EENS_k / LOLE_k \tag{6}$$

$$ELC = \sum_{k=1}^{ND} ELC_k \tag{7}$$

$$LOLE = EENS / ELC \tag{8}$$

where ND represents the number of load demand points.

2.2 Minimum outage power using DC model

To obtain the probability density function of the outage capacities caused by the forced outage of components, the load point outage power must be calculated first of all. There are several possible solutions when calculating the load point outage power for each component state. This problem can be formulated as an optimization model. The objective function for minimum outage power can be set up and an optimal solution can be obtained by the network flow method [5]. The network flow method takes into account Kirchhoff's current law (KCL) [24] at all nodes but neglecting Kirchhoff's voltage law (KVL) [25]. Thus, its accuracy is low. In this paper, transmission network is characterized by a simplified lossless DC load flow model. It not only satisfies KCL but also obeys Ohm's law. Thus, KVL is implicitly taken into account. Therefore, DC load flow model is more close to the actual situation and can improve the accuracy.

To minimize the outage power at load point, the mathematical model used to minimize the outage power at load point can be formulated as follows.

$$\sum_{i=1}^{ND} (L_{pi} - P_{Li})/L_{pi} \tag{9.1}$$

where L_{pi} is the peak load at bus i; P_{Li} is the decision variable meaning effective supplied power at bus i (therefore, they represent the arrival power at load point i).

Subject to

1) Power balance constraint at each node

$$P_{Gi} - P_{Li} = \sum_{j=1}^{n} P_{ij} \tag{9.2}$$

where P_{Gi} is the generation capacity at bus i; P_{ij} is the power flow between bus i and j; n is the number of transmission lines connected to bus i.

2) Constraint of line power

$$P_{ij} = b_{ij}(\theta_i - \theta_j) \tag{9.3}$$

where b_{ij} is the susceptance of the transmission lines between bus i and j; θ_i is the phase angle at bus i.

3) Limitation constraint of peak load

$$0 \le P_{Li} \le L_{Pi} \tag{9.4}$$

4) Limitation constraint of generation capacity

$$0 \le P_{Gi} \le P_{Gi}^{max} \tag{9.5}$$

where P_{Gi}^{max} is the maximum capacity of generation at bus i.

5) Limitation constraint of transmission capacity

$$-P_{ij}^{max} \le P_{ij} \le P_{ij}^{max} \tag{9.6}$$

where $P_{ij}^{\max}$ is the maximum transmission capacity between bus i and j.

3 Probabilistic reliability evaluation and sensitivity analysis

3.1 State probability calculation

There are two fundamental methods for probabilistic reliability evaluation: state enumeration and Monte Carlo simulation [2, 26]. Generally, if the outage probabilities of most of the components are very small (i.e. the system is reliable) or the number of components is very small, state enumeration method is usually more efficient. When the complex operating conditions are concerned or outage probabilities of most of the components cannot be ignored, it is almost impossible to enumerate all the system states. Thus, Monte Carlo simulation is an effective alternative to obtain all the approximate system states.

Monte Carlo simulation method treats the problem as a series of experiments. Generally, generator states are modeled using multiple state random variables. In this paper, for simplicity, the generating units are taken into account and their states are modeled using two-state (up and down) random variables as well as transmission circuit states. It is assumed that component outages are mutually independent events. Therefore, transmission line and generating unit outages are simulated by separate random numbers.

In a planning context, the probability is a measure of the likelihood that the power system will be in a given situation at a random time in the future, and it is also a function of the availability of every piece of equipment in the power system. This relationship can be represented as follows [27].

$$
\begin{aligned}
PRO &= \prod_{i \in \mathbf{U}} u(c_i) \prod_{j \in \mathbf{A}} a(c_j) \\
&= \underbrace{u(c_1)u(c_2)\ldots u(c_n)}_{\text{Unavailable}} \underbrace{a(c_1)a(c_2)\ldots a(c_m)}_{\text{Available}}
\end{aligned}
\tag{10}
$$

where $\mathbf{U}$ is the set of unavailable components; $\mathbf{A}$ is the set of available components.

Forced outage rate (FOR) of power system elements can be obtained through the historical data statistical method. Suppose that there are NE elements including such as generation units, transmission lines, transformers. The random number generation method can generate a series of random numbers $y_i(i = 1, 2,\ldots, NE)$ distributed uniformly under $\{0, 1\}$. s_i is the state of component i and FOR_i is its forced outage rate. s_i is expressed as (11).

$$
s_i = \begin{cases} 0 \ (\text{normal state}, y_i > FOR_i) \\ 1 \ (\text{abnormal state}, 0 \le y_i \le FOR_i) \end{cases}
\tag{11}
$$

The state of all components can be formulated by a vector $\mathbf{S}$ as (12).

$$
\mathbf{S} = [s_1, s_2, \ldots, s_i, \ldots s_{NE}]
\tag{12}
$$

The probability of component i appearing the state s_i can be expressed as follows.

$$
PRO(s_i) = s_i FOR_i + (1 - s_i)(1 - FOR_i)
\tag{13}
$$

According to (10), the probability of the state vector $\mathbf{S}$, $PRO(\mathbf{S})$, is calculated as (14).

$$
PRO(\mathbf{S}) = \prod_{i=1}^{NE} P(s_i)
\tag{14}
$$

It is assumed that the system states are sampled N times. If N is large enough, the state combinations of all the system components can be sampled. The component state vector $\mathbf{S}$ may be the same in the N times samples. After removing duplicate state $\mathbf{S}$, a state set $U(\mathbf{S})$ of all the system components and the corresponding probability values can be obtained.

3.2 Sensitivity index

The reliability indices are not only associated with the capacity of each component, but also related to the FOR. The sensitivity index (SI) based DC model [21] or AC model [22] does not consider the FOR of the component. The indices LOLE and EENS are strongly associated with the arrival power at load point. Therefore, based on the Monte Carlo simulation and the optimization model shown as (9), an implicit sensitivity index is presented.

Generally, at the initial stage of planning, human knowledge is needed to ensure rationality of the candidate line selection with practical engineering and management concerns [28]. The main focus of this paper is on the reliability evaluation and its sensitivity analysis. Therefore, it is assumed that the feasible candidate pool of transmission lines has been obtained properly. Assuming the candidate pool is expressed as follows.

$$
\mathbf{X} = \{x_1, x_2, \ldots, x_{NC}\}
\tag{15}
$$

where NC is the number of candidate transmission lines.

The state of all components can be expressed as a vector $\mathbf{S}$, and its corresponding probability can be calculated, shown as (12) and (14), respectively. Under the state $\mathbf{S}$, the arrival power vector $\mathbf{AP}_k$ ($k = 1,2,\ldots,ND$) at each load point can be obtained using the model (9) first of all. Next, assuming that one candidate line x_i ($i = 1,2,\ldots,NC$) is added to the power grid and then the corresponding arrival power vector $\mathbf{AP}_k^i$ at each load point can be obtained using the model (9) again. Thus, the incremental vector of the arrival power at each load point, caused by adding the line x_i, can be easily calculated as (16).

$$\Delta AP_k^i = AP_k^i - AP_k \qquad (16)$$

Due to the fact that the different system state has different probability, the above ΔAP_k^i should be taken into account with its system state probability. The incremental of the arrival power under the state probability, denoted as ΔP_k^i, can be expressed as (17).

$$\Delta P_k^i = PRO(\mathbf{S})\Delta AP_k^i \qquad (17)$$

Next, each of candidate lines is in turn added to the grid to get the corresponding incremental vector of the arrival power under the state probability. Then, the incremental arrival power matrix $\Delta \mathbf{P}$ under the current system state due to adding each of candidate lines to the gird is shown in (18).

$$\Delta \mathbf{P} = \begin{bmatrix} \Delta P_k^1 \\ \vdots \\ \Delta P_k^{NC} \end{bmatrix} = \begin{bmatrix} \Delta P_1^1 & \cdots & \Delta P_{ND}^1 \\ \vdots & \vdots & \vdots \\ \Delta P_1^{NC} & \cdots & \Delta P_{ND}^{NC} \end{bmatrix} \qquad (18)$$

For each state in the state set $U(\mathbf{S})$, the above calculating process of obtaining $\Delta \mathbf{P}$ is repeated. Every time, the obtained incremental arrival power matrix $\Delta \mathbf{P}$ is accumulated.

Finally, the sensitivity index for the entire system and for load point k can be calculated using (19) and (20), respectively.

$$SI_{sys} = \max\left\{ SI_{x_i} = \sum_{k=1}^{ND} \Delta P_k^i, (i = 1, 2, \ldots, NC) \right\} \qquad (19)$$

$$SI_{Bus}^k = \max\left\{ SI_{x_i}^k = \Delta P_k^i, (i = 1, 2, \cdots, NC) \right\} (k = 1, 2, \ldots, ND) \qquad (20)$$

3.3 The main procedure of reliability evaluation

As discussed above, reliability evaluation starts with Monte Carlo simulation to obtain the system components state set $U(\mathbf{S})$. Then, for a state $\mathbf{S}$, the arrival power at each load point can be obtained using the DC constraint optimization algorithm. At the same time, the incremental arrival power matrix $\Delta \mathbf{P}$ due to in turn adding one candidate line into grid can be gained too. Next, this calculating procedure is repeated until each state in state set $U(\mathbf{S})$ has been dealt with. Then, the probability density function of the outage power at each load point and the accumulated matrix $\Delta \mathbf{P}$ are obtained. Thus, the effective LDC at each load point can be achieved by calculating the convolution of the original inverted duration curve at load point and the pdf of the outage

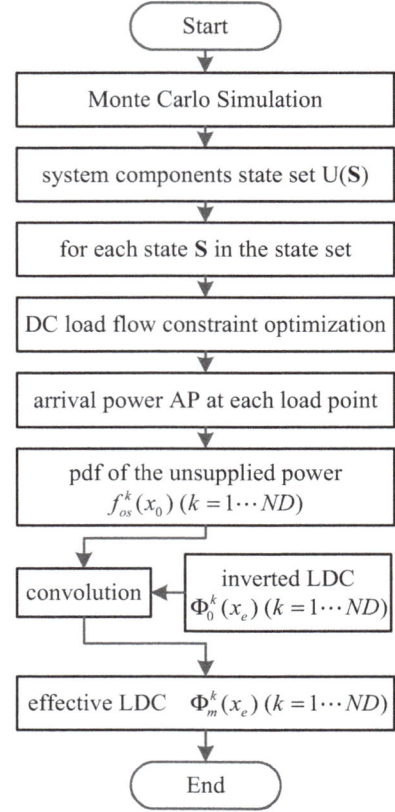

Fig. 1 Main procedure of obtaining the effective LDC

capacities. The main procedure of obtaining the effective LDC is shown in Fig. 1.

Next, the load bus reliability indices can be calculated according to (3) and (4), and the system reliability indices can be calculated according to (5)-(8). Moreover, the system and bus reliability indices can be augmented by adding the grid each time a line which is based on rules (19) and (20), respectively.

4 Case studies

The proposed method is tested on Garver's 6-bus system and IEEE 39-bus model system. The algorithm was implemented in Matlab 8.2, using a PC with Core i7-4770 CPU clocking at 3.4 GHz and 32 GB of RAM. The DC constraint optimization is solved using the YAMIP environment [29]. In the two cases, it is assumed that the inverted LDC of load point is a polyline, shown as in Fig. 2.

4.1 Garver's 6-bus system

This system has 6 buses and a demand of 760 MW [24]. The line data in [24] are remained unchanged. The load

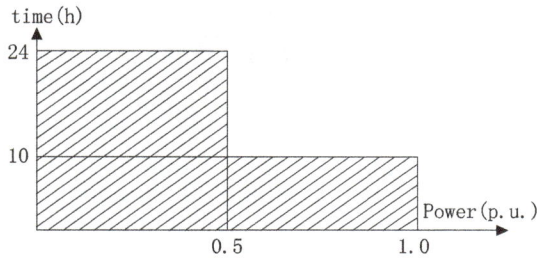

Fig. 2 Inverted load duration curve at load point

Table 1 Load and generation data for 6-bus system

Bus number	Load (MW)	Units (MW)
1	80	50
2	245	0
3	40	3×85
4	160	0
5	235	0
6	0	7×100

data and generator data are provided in Table 1. The generator is expressed using units due to the request for considering the FOR of the components. To exclude the impact on the reliability due to generator outage, the generator capacities are expanded.

The current system structure is shown in Fig. 3. For simplicity, FORs of all transmission lines and generating units are set to 0.5% and 1%, respectively.

Monte Carlo simulation method is used for state simulation. The reliability indices at the load points are evaluated using the proposed method and the results are shown in Table 2. Thus, the system reliability indices EENS and LOLE are calculated by (5) and (8), respectively. The system reliability indices are shown in Table 2 at last row.

Suppose that the candidate pool of transmission lines consists of the current transmission corridors, which have 8 right-of-way lines, shown in Table 3.

The incremental arrival power matrix ΔP by adding each candidate line is as follows.

$$\Delta P = \begin{bmatrix} 0 & -1.3212 & 0 & 0.6705 & 42.3723 \\ 0 & 2.3933 & 0 & -7.2632 & -7.3073 \\ 0 & 0.0351 & 0 & -2.3649 & 33.1669 \\ 0 & -0.9488 & 0 & -1.5299 & -25.3088 \\ 0 & 0.0457 & 0 & 13.6963 & 4.6497 \\ 0 & 3.1280 & 0 & 13.9975 & 1.7246 \\ 0 & -0.5101 & 0 & -3.8953 & 113.4931 \\ 0 & 1.6323 & 0 & 48.5670 & 38.5338 \end{bmatrix} \quad (21)$$

According to Table 2, the load point with the maximum EENS and LOLE is the bus 5. From (21), it can be found that the bus sensitivity index at bus 5 is $SI_{\text{Bus}}^5 =$

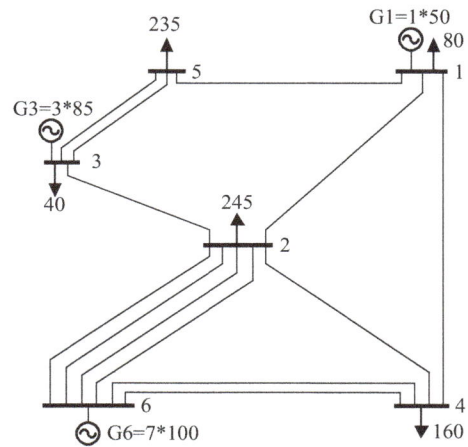

Fig. 3 The 6-bus system derived from the Garver's system

Table 2 Reliability indices of the base case

Load bus number	EENS (MWh/year)	LOLE (h/year)	ELC (MW/year)
1	0	0	0
2	120.7	2.5827	46.73
3	0	0	0
4	1777.1	38.72	45.90
5	4243.6	143.91	29.49
System	6141.4	50.29	122.12

$(SI)_{x_7}^5 = 113.4931$, and the corresponding candidate line is line 7 (line$_{3\text{-}5}$). The system sensitivity index is $SI_{\text{sys}} = SI_{x7} = 109.1877$ and the corresponding candidate line is also line 7.

The line 7 (line$_{3\text{-}5}$) is added to the grid and the reliability indices are evaluated again using the proposed method. The evaluation results are shown in Table 4.

From Table 4, it can be seen that the reliability indices of system and bus 5 have been greatly improved, after adding line$_{3\text{-}5}$ to the gird.

4.2 39-bus system

In this case, the proposed method is tested on New-England 10-machine 39-bus system, shown as in Fig. 4. The system data, shown in Table 5 and Table 6, are taken from the open source software Matpower 4.1 [30]. In Table 5, *NL* represents the number of Line; FB and TB are from and to buses of the line, respectively; and X stands for the reactance of the line.

After Monte Carlo simulation sampling times of 10^5, the load point ELDCs are obtained and the reliability indices are followed to be evaluated. For simplicity, the zero results are ignored and non-zero parts are shown in

Table 3 Candidate pool of transmission lines

Line number	From bus	To bus	Line number	From bus	To bus
1	1	2	5	2	4
2	1	4	6	2	6
3	1	5	7	3	5
4	2	3	8	4	6

Table 4 Reliability indices after adding line$_{3-5}$ to the gird

Load busnumber	EENS (MWh/year)	LOLE (hours/year]	ELC (MW/year)
1	0	0	0
2	139.3	3.28	42.51
3	0	0	0
4	1917.7	36.07	53.17
5	87.6	3.80	23.04
System	2147.6	18.06	118.73

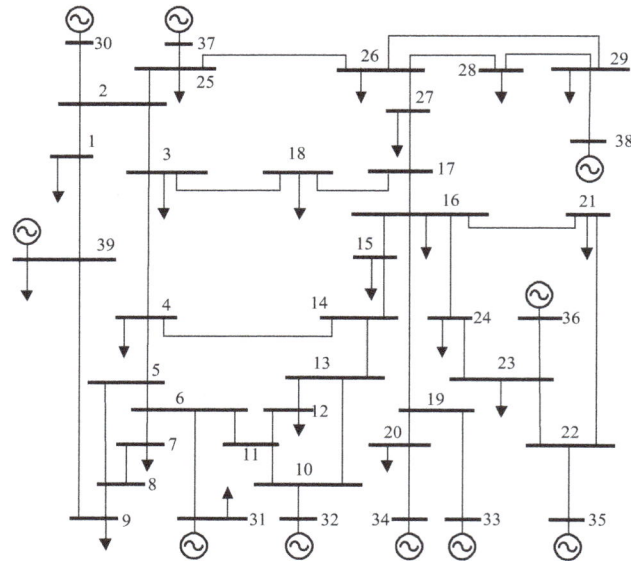

Fig. 4 The New-England 10-machine39bus system

Table 7. The corresponding sensitivity indices are calculated and shown in the last column of Table 7.

From Table 7, it can be seen that bus 39 has the maximum EENS and adding the line 1 (from bus 1 to bus 2) is the most effective for increasing the reliability of bus 39. However, adding the line 3 (from bus 2 to bus 3) is the most effective measures for increasing the system reliability. After adding the line 1, we can get the new reliability indices results, shown as in Table 8.

Table 5 Branch data for 39-bus system

NL	FB	TB	X (p.u.)	Pmax (MW)	NL	FB	TB	X (p.u.)	Pmax (MW)
1	1	2	0.0411	600	24	14	15	0.0217	600
2	1	39	0.0250	1000	25	15	16	0.0094	600
3	2	3	0.0151	500	26	16	17	0.0089	600
4	2	25	0.0086	500	27	16	19	0.0195	600
5	2	30	0.0181	900	28	16	21	0.0135	600
6	3	4	0.0213	500	29	16	24	0.0059	600
7	3	18	0.0133	500	30	17	18	0.0082	600
8	4	5	0.0128	600	31	17	27	0.0173	600
9	4	14	0.0129	500	32	19	20	0.0138	900
10	5	6	0.0026	1200	33	19	33	0.0142	900
11	5	8	0.0112	900	34	20	34	0.0180	900
12	6	7	0.0092	900	35	21	22	0.0140	900
13	6	11	0.0082	480	36	22	23	0.0096	600
14	6	31	0.0250	1800	37	22	35	0.0143	900
15	7	8	0.0046	900	38	23	24	0.0350	600
16	8	9	0.0363	900	39	23	36	0.0272	900
17	9	39	0.0250	900	40	25	26	0.0323	600
18	10	11	0.0043	600	41	25	37	0.0232	900
19	10	13	0.0043	600	42	26	27	0.0147	600
20	10	32	0.0200	900	43	26	28	0.0474	600
21	11	12	0.0435	500	44	26	29	0.0625	600
22	12	13	0.0435	500	45	28	29	0.0151	600
23	13	14	0.0101	600	46	29	38	0.0156	1200

Table 6 Load and generation data for 39-bus system

Bus number	Load (MW)	Generation (MW)	Bus number	Load (MW)	Generation (MW)
1	97.6	0	21	274	0
2	0	0	22	0	0
3	322	0	23	247.5	0
4	500	0	24	308.6	0
5	0	0	25	224	0
6	0	0	26	139	0
7	233.8	0	27	281	0
8	522	0	28	206	0
9	6.5	0	29	283.5	0
10	0	0	30	0	1040
11	0	0	31	9.2	646
12	8.53	0	32	0	725
13	0	0	33	0	652
14	0	0	34	0	508
15	320	0	35	0	687
16	329	0	36	0	580
17	0	0	37	0	564
18	158	0	38	0	865
19	0	0	39	1104	1100
20	680	0			

Table 7 Reliability indices and SI of the 39-bus system (non-zero parts)

Load bus number	EENS (MWh/year)	LOLE (h/year)	ELC (MW/year)	SI
4	809.8	6.60	122.59	0.1806(line$_3$)
8	73.5	0.74	98.84	0.0158(line$_3$)
18	17.8	0.16	111.92	0.0029(line$_7$)
20	27096.9	235.71	114.96	6.2797(line$_3$)
25	86.1	0.96	89.45	0.0170(line$_4$)
27	1.7	0.07	26.00	0.0050(line$_{31}$)
29	31.7	0.16	199.79	0.0051(line$_{46}$)
31	2.1	0.32	6.52	0.0030(line$_{14}$)
39	47209.7	171.16	275.82	10.7070(line$_1$)
System	75329.3	72.02	1045.89	15.6973(line$_3$)

Table 8 Reliability indices and SI of the 39-bus system, after adding line$_1$ (non-zero parts)

Load bus number	EENS (MWh/year)	LOLE (h/year)	ELC (MW/year)	SI
4	7302.6	93.73	77.91	0.1806(line$_3$)
8	6976.3	43.72	159.58	0.0158(line$_3$)
18	17.7	0.16	111.92	0.0029(line$_7$)
20	23806.7	193.88	122.79	6.2797(line$_3$)
25	99.5	1.12	89.17	0.0170(line$_4$)
27	6.9	0.26	26.00	0.0050(line$_{31}$)
29	128.9	0.70	183.41	0.0051(line$_{46}$)
39	12550.1	34.69	361.79	10.7070(line$_1$)
System	50888.7	44.93	1132.57	15.6973(line$_3$)

Comparing Table 8 with Table 7, it is interesting to note that adding line 1 greatly improves the reliability at the bus 39 and thus increases the system reliability, as well as decrease the reliability at some other buses (such as bus 4 and 8). The reason is that adding the line changes the system structure and then changes the power flow.

5 Conclusion

This paper addresses the reliability evaluation problem considering the forced outage rates of generators and transmission lines. Based on the works in [5, 6], the DC power flow model are used to obtain more accurate maximum arrival power for more accurate reliability indices. The system components states are simulated using Monte Carlo technique first of all. Next, the probability density functions of the outage capacities at load buses are obtained through calculating the maximum arrival power using DC model based optimization under all possible system states. Then, the effective load duration curves at every load bus are obtained and the reliability indices can be calculated. With the reliability evaluation process, the implicit sensitivity indices of the reliability can be obtained, which help to choose the most effective component to improve the reliability of the system or a load bus. The work is an important step in preparing a transmission expansion plan or maintenance scheduling employing probabilistic reliability evaluation methods to ensure the reliability of the electric power grid.

Acknowledgments This work is supported by China Scholarship Council, as well as Young Teacher Scientific Research Foundation of Sichuan University (No. 2012SCU11003).

References

[1] Lee CW, Ng SKK, Zhong J et al (2006) Transmission expansion planning from past to future. In: Proceedings of the 2005 IEEE PES power systems conference and exposition (PSCE'06), Atlanta, GA, 29 Oct–1 Nov 2006, pp 257–265
[2] Li WY, Choudhury P (2007) Probabilistic transmission planning. IEEE Power Energ Mag 5(5):46–53
[3] Hemmati R, Hooshmand RA, Khodabakhshian A (2013) Comprehensive review of generation and transmission expansion planning. IET Gener Transm Distrib 7(9):955–964
[4] Zhao JH, Foster J, Dong ZY et al (2011) Flexible transmission network planning considering distributed generation impacts. IEEE Trans Power Syst 26(3):1434–1443
[5] Choi J, Billinton R, Futuhi-Firuzabed M (2005) Development of a new nodal effective load model considering of transmission system element unavailabilities. IEE Proc Gener Transm Distrib 152(1):79–89
[6] Choi J, Tran T, El-Keib AA et al (2005) A method for transmission system expansion planning considering probabilistic reliability criteria. IEEE Trans Power Syst 20(3):1606–1615
[7] Ekpenyong UE, Zhang JF, Xia XH (2012) An improved robust model for generator maintenance scheduling. Electr Power Syst Res 92:29–36
[8] Zhang DB, Li WY, Xiong XF (2012) Bidding based generator maintenance scheduling with triple-objective optimization. Electr Power Syst Res 93:127–134
[9] Choi J, Mount TD, Thomas RJ (2007) Transmission expansion planning using contingency criteria. IEEE Trans Power Syst 22(4):2249–2261
[10] da Rochaa MC, Saraivab JT (2012) A multiyear dynamic transmission expansion planning model using a discrete based EPSO approach. Electr Power Syst Res 93:83–92
[11] Orfanos GA, Georgilakis PS, Hatziargyriou ND (2013) Transmission expansion planning of systems with increasing wind power integration. IEEE Trans Power Syst 28(2):1355–1362
[12] Shortle J, Rebennack S, Glover FW (2014) Transmission-capacity expansion for minimizing blackout probabilities. IEEE Trans Power Syst 29(1):43–52

[13] Shalash NA, Bin Ahmad AZ (2014) Agents for fuzzy indices of reliability power system with uncertainty using Monte Carlo algorithm. In: Proceedings of the IEEE 8th international power engineering and optimization conference (PEOCO'14), Langkawi, 24–25 Mar 2014, pp 258–264

[14] Benidris M, Mitra J (2014) Reliability and sensitivity analysis of composite power systems under emission constraints. IEEE Trans Power Syst 29(1):404–412

[15] Alizadeh B, Jadid S (2011) Reliability constrained coordination of generation and transmission expansion planning in power systems using mixed integer programming. IET Gener Transm Distrib 5(9):948–960

[16] da Silva AML, Rezende LS, Manso LAF et al (2010) Transmission expansion planning: A discussion on reliability and "N–1" security criteria. In: Proceedings of the IEEE 11th international conference on probabilistic methods applied to power systems (PMAPS'10), Singapore, 14–17 Jun 2010, pp 244–251

[17] Aghaei J, Amjady N, Baharvandi A et al (2014) Generation and transmission expansion planning: MILP-based probabilistic model. IEEE Trans Power Syst 29(4):1592–1601

[18] Sharaf TAM, Berg GJ (1982) Reliability optimization for transmission expansion planning. IEEE Trans Power App Syst 101(7):2243–2248

[19] Sharaf TAM, Berg GJ (1988) Reliability evaluation in power-system transmission planning: Practical considerations. IEEE Trans Reliab 37(3):274–279

[20] Zhao Y, Zhou NC, Zhou JQ et al (2006) Research on sensitivity analysis for composite generation and transmission system reliability evaluation. In: Proceedings of the international conference on power system technology (PowerCon'06), Chongqing, 22–26 Oct 2006, 5 pp

[21] Sánchez IG, Romero R, Mantovani JRS et al (2005) Transmission-expansion planning using the DC model and nonlinear-programming technique. IEE Proc Gener Transm Distrib 152(6):763–769

[22] Rider MJ, Garcia AV, Romero R (2007) Power system transmission network expansion planning using AC model. IET Gener Transm Distrib 1(5):731–742

[23] Sullivan RL (1977) Power system planning. McGraw-Hill, New York

[24] Romero R, Monticelli A, Garcia A et al (2002) Test systems and mathematical models for transmission network expansion planning. IEE Proc Gener Transm Distrib 149(1):27–36

[25] Jabr RA (2013) Optimization of AC transmission system planning. IEEE Trans Power Syst 28(3):2779–2787

[26] Li WY (2005) Risk assessment of power systems: Models, methods, and applications. IEEE Press, Piscataway

[27] EPRI (2003) Moving toward probabilistic reliability assessment methods: A framework for addressing uncertainty in power system planning and operation. EPRI Technical Results 1002639, Palo Alto

[28] Xu Z, Dong ZY, Wong KP (2006) A hybrid planning method for transmission networks in a deregulated environment. IEEE Trans Power Syst 21(2):925–932

[29] Löfberg J (2004) YALMIP: A toolbox for modeling and optimization in MATLAB. In: Proceedings of the 2004 IEEE international symposium on computer aided control systems design (CACSD'04), Taipei, 2–4 Sept 2004, pp 284–289

[30] Zimmerman RD, Murillo-Sánchez CE, Thomas RJ (2011) MATPOWER: Steady-state operations, planning and analysis tools for power systems research and education. IEEE Trans Power Syst 26(1):12–19

[31] Billinton R, Li WY (1994) Reliability assessment of electric power systems using Monte Carlo methods. Springer, New York, 352 pp

Chengxin LI was born in Chongqing, China, in 1976. He received his Master degree and PhD degree from Sichuan University in 2003 and 2012, respectively. He is now with the school of Electrical Engineering and Information in Sichuan University. He is currently conducting research as a visiting scholar in the University of Sydney. His research interests include power system planning, power system operation and control, reliability assessment, monitoring and control of power system low frequency oscillation.

Guo CHEN received the PhD degree from the University of Queensland in 2010. He is now a research fellow at the school of Electrical and Information Engineering in the University of Sydney. His research interests include power system planning, power system security, renewable energy and distributed generation, and planning and control in smart grids.

Junyong LIU received the PhD degree from Brunel University in 1998. He is a Professor and head of the School of Electrical Engineering and Information in Sichuan University. His main research interests include electricity market, power system planning, operation, stability and computer applications.

6

Coordinated optimization for controlling short circuit current and multi-infeed DC interaction

Dong YANG, Kang ZHAO, Yutian LIU (✉)

Abstract Due to increased penetration of renewable energies, DC links and other emerging technologies, power system operation and planning have to cope with various uncertainties and risks. In order to solve the problems of exceeding short circuit current and multi-infeed DC interaction, a coordinated optimization method is presented in this paper. Firstly, a branch selection strategy is proposed by analyzing the sensitivity relationship between current limiting measures and the impedance matrix. Secondly, the impact of network structure changes on the multi-infeed DC system is derived. Then the coordinated optimization model is established, which considers the cost and effect of current limiting measures, the tightness of network structure and the voltage support capability of AC system to multiple DCs. Finally, the non-dominated sorting genetic algorithm II combining with the branch selection strategy, is used to find the Pareto optimal schemes. Case studies on a planning power system demonstrated the feasibility and speediness of this method.

Keywords Operation and planning, Multiple DC infeed, Short circuit current, Sensitivity analysis, Multi-objective optimization

D. YANG, State Grid Shandong Electric Power Research Institute, Jinan 250002, China
K. ZHAO, Y. LIU, Key Laboratory of Power System Intelligent Dispatch and Control of Ministry of Education, Shandong University, Jinan 250061, China
(✉) e-mail: liuyt@sdu.edu.cn

1 Introduction

Modern power system operation and planning is undergoing dramatic changes. Due to increased penetration of renewable energies, DC systems and other emerging technologies, system operation and planning must now cope with various uncertainties and risks. Most of them belong to the multi-objective, non-linear, non-convex and mixed-integer programming problem. It calls for effective solutions to coordinate and optimize new and old technologies to improve overall system security and efficiency at large. The traditional optimization techniques are usually inefficient or even unable to handle these problems. Fortunately, the rapid development of modern optimization techniques provides the promising way to solve the difficulties and challenges in modern power system environments [1–5].

Exceeding short circuit current and multi-infeed DC interaction are two problems faced by the receiving-end power grid in China. Coordinated optimization for them is a typical multi-objective problem. The classical optimization algorithms suggest converting the multi-objective problem to a single-objective problem. When such an algorithm is applied to find multiple solutions, it has to be carried out many times, hopefully finding a set of optimal solutions. This process is usually inefficient and cost a lot of time. Recently, the emergence of evolutionary algorithms (EAs) can effectively solve this problem [6, 7]. EAs are well applicable to multi-objective problems because of their ability to find multiple solutions in one single simulation run. In the family of EAs, the non-dominated sorting genetic algorithm II (NSGA-II) outperforms others in terms of finding a diverse set of solutions and converging near the true Pareto optimal solutions [7].

Short circuit current can cause mechanical and thermal stresses proportional to the square of the current, and hence lead to the damage of the equipment in power system. This situation is worsened with the increase of interconnections in power system and high-capacity generators being injected to the grid. Short circuit current can be limited in many ways, which can be divided into two categories depending on the cost and effect. One is to open the switches of lines or buses, which is simple, economical and obviously effective. However, this method has a greater impact on the system stability. The other is to increase the installations of electrical equipment or upgrade them, such as installing the current limiting reactor or fault current limiter (FCL), even replacing with the high-impedance transformer. This method has little influence on the system stability, but the high cost is needed and the effect is limited.

Comparing various current limiting measures one by one is the traditional way to solve the problem of exceeding short circuit current, which is tedious and inefficient. When the requirement cannot be satisfied by using one single measure, several measures are integrated based on the engineering experience. The application of current limiting measures is mature, but the optimization of them using mathematical methods is unusual. In [8–11], the optimization for limiting short circuit current was mainly focused on the allocation of FCL installation location, quantity and impedance values. Reference [12] presented a method with genetic algorithm and particle swarm optimization to find the optimal location and the number of buses to be split. In [13], the optimization problem was modeled as a 0–1 mixed integer programming problem, and the model can be solved by a branch and bound algorithm to generate optimal current limiting strategies.

Multiple DC infeed is another important feature of the receiving-end power grid in China. From theoretical analysis and simulation, it is found that the biggest risk faced by the multi-infeed DC system is the voltage stability problem [14–16]. The short-circuit ratio was given to evaluate the voltage support capability of AC system to DC system in [17]. This index has not considered the interaction of multiple DCs, thus it is applicable to the system with just one DC. Reference [18] suggested an extension of the classical short-circuit ratio to multi-infeed DC systems, named the multi-infeed short-circuit ratio. This new index considers the AC system short circuit capacity, the multiple DC transmission capacity, and the electrical coupling relationship between DC inverter stations. Reference [19] showed that it is effective to use such new index for denoting the voltage stability level of the multi-infeed DC system. Reference [20] predicted the risk for voltage and power instability when the multi-infeed short-circuit ratio is low.

Controlling short circuit current and increasing power system stability is a contradiction [21]. The application of current limiting measures will change the power grid structure, and then affect the voltage support capability of AC system to multiple DCs [22]. Specifically, the current limiting measures, on the one hand, reduce short circuit current, on the other hand, stretch the electrical distance between DC inverter stations. As a result, the multi-infeed short-circuit ratio may be increased or decreased under different conditions. This fact indicates that there exists a coordinated optimization scheme, which can not only control short circuit current within a reasonable range, but also remain the multi-infeed short-circuit ratio at a high level. Such scheme should be obtained by establishing and solving the multi-objective optimization model. Unfortunately, most researches pay attention to the single-objective optimization for the cost of current limiting measures, and there are still no literatures that consider the influence of current limiting measures on the multi-infeed DC system.

A coordinated optimization method for controlling short circuit current and multi-infeed DC interaction is presented in this paper. Firstly, by analyzing the sensitivity relationship between current limiting measures and the impedance matrix, a branch selection strategy is proposed. Secondly, the impact of network structure changes on the multi-infeed DC system is derived. Then the coordinated optimization model, which considers the cost and effect of current limiting measures, the tightness of network structure and the voltage support capability of AC system to multiple DCs, is established. Finally, combining NSGA-II with the branch selection strategy, the Pareto optimal schemes are found. Taking the regional power grid in China for example, the feasibility and efficiency of the method is validated.

2 Current limiting measures sensitivity analysis

Opening the line and installing the current limiting reactor are two typical current limiting measures. Therefore, they are used in this paper to solve the optimization problem.

Three-phase short circuit is generally the most serious short circuit fault in power systems, and usually used to determine the rated breaking capacity of circuit breakers. Three-phase short circuit current is inversely proportional to the self-impedance. The following is the derivation of sensitivity relationship between current limiting measures and the self-impedance of overproof site. Assuming the original network forms an m-order impedance matrix $\mathbf{Z}_m$. If the branch z_{ij} is added into the network between node i and j, the impedance matrix changes into $\mathbf{Z}'_m$. According to the additional branch method, the element in matrix $\mathbf{Z}'_m$ can be derived as

$$Z'_{kl} = Z_{kl} - \frac{(Z_{ki} - Z_{kj})(Z_{li} - Z_{lj})}{Z_{ii} + Z_{jj} - 2Z_{ij} + z_{ij}} \tag{1}$$

where $k = 1,2,\ldots,m$; $l = 1,2,\ldots,m$; m is the number of network nodes; Z'_{kl} and Z_{kl} are the element of the impedance matrix $\mathbf{Z}'_m$ and $\mathbf{Z}_m$, respectively.

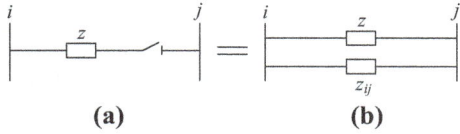

Fig. 1 Equivalent model of opening a line

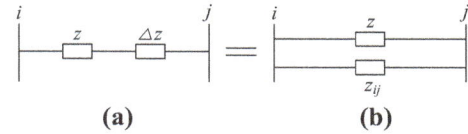

Fig. 2 Equivalent model of installing the current limiting reactor

2.1 Open the line

Opening a line is equivalent to adding a branch $z_{ij} = -z$, between node i and j, which is shown in Fig. 1.

If two buses connected by a bus coupler switch are regarded as independent nodes, splitting the buses is similar to opening a line. The sensitivity of opening the line with respect to the self-impedance is defined as

$$\begin{cases} \lambda_k = \dfrac{Z'_{kk} - Z_{kk}}{Z_{kk}} = \dfrac{\Delta Z^{(1)}_{kk}}{Z_{kk}} \\ \Delta Z^{(1)}_{kk} = -\dfrac{(Z_{ki} - Z_{kj})^2}{Z_{ii} + Z_{jj} - 2Z_{ij} - z} \end{cases} \quad (2)$$

where Z_{kk} and Z'_{kk} are the self-impedance of overproof site k before and after opening a line, respectively.

A greater λ_k indicates a better current limiting effect on overproof site k when opening the line. Considering the current limiting effect on all overproof sites, the weighted sensitivity is defined as

$$\begin{cases} \lambda = \displaystyle\sum_{k=1}^{N_e} \omega_k \lambda_k \\ \omega_k = \left(\dfrac{I_k}{I_k^{\max}} \right)^2 \end{cases} \quad (3)$$

where N_e is the number of overproof sites; ω_k is the weighted coefficient; I_k and $I_k^{\max}$ are the actual three-phase short circuit current and the maximum breaker interruptive current of overproof site k, respectively.

2.2 Install the current limiting reactor

Installing the current limiting reactor is equivalent to adding a branch $z_{ij} = -(z^2 + z\Delta z)\Delta z$, between node i and j, which is shown in Fig. 2.

The ideal fault current limiter does not affect the normal operation of power systems, and can quickly put into a large current limiting reactor when short circuit fault occurs. Therefore, installing the fault current limiter is similar to installing the current limiting reactor. The sensitivity of installing the current limiting reactor with respect to the self-impedance is defined as

$$\begin{cases} \gamma_k = \lim_{\Delta z \to 0} \dfrac{\Delta Z^{(2)}_{kk}}{\Delta z} = \left(\dfrac{Z_{ki} - Z_{kj}}{z} \right)^2 \\ \Delta Z^{(2)}_{kk} = -\dfrac{(Z_{ki} - Z_{kj})^2}{Z_{ii} + Z_{jj} - 2Z_{ij} - \frac{z^2 + z\Delta z}{\Delta z}} \end{cases} \quad (4)$$

A greater γ_k indicates a better current limiting effect on overproof site k when installing the current limiting reactor. Considering the current limiting effect on all overproof sites, the weighted sensitivity is defined as

$$\begin{cases} \gamma = \displaystyle\sum_{k=1}^{N_e} \omega_k \gamma_k \\ \omega_k = \left(\dfrac{I_k}{I_k^{\max}} \right)^2 \end{cases} \quad (5)$$

From (2) and (4), it can be concluded that for the same branch, $\Delta Z^{(1)}_{kk} = \Delta Z^{(2)}_{kk}$ when $\Delta z \to j \cdot +\infty$ and $\Delta Z^{(1)}_{kk} > \Delta Z^{(2)}_{kk}$ when $\Delta z > j \cdot 0$. It means that opening the line has a better current limiting effect than installing the current limiting reactor.

2.3 Branch selection strategy

As described above, for the same branch, the sensitivity of opening the line reflects the current limiting characteristic with $\Delta z \to j \cdot +\infty$; the sensitivity of installing the current limiting reactor reflects the current limiting characteristic with $\Delta z \to j \cdot 0$. Thus, the integrated sensitivity of one branch is defined as the average of these two sensitivities.

$$\begin{cases} \mu_l = \dfrac{\lambda_l^* + \gamma_l^*}{2} \\ \lambda_l^* = \dfrac{\lambda_l - \lambda^{\min}}{\lambda^{\max} - \lambda^{\min}} \\ \gamma_l^* = \dfrac{\gamma_l - \gamma^{\min}}{\gamma^{\max} - \gamma^{\min}} \end{cases} \quad (6)$$

where μ_l is the integrated sensitivity of the branch l; λ_l^* and λ_l are the normalized value and actual value of the sensitivity of opening the line for the branch l, respectively; γ_l^* and γ_l are the normalized value and actual value of the sensitivity of installing the current limiting reactor for the branch l, respectively.

Sort all the branches of the network in descending order of the integrated sensitivity, choose the first M branches, and then form the reduced-dimensional set of decision variables.

3 Multi-infeed short-circuit ratio analysis

3.1 Definition of the multi-infeed short-circuit ratio

The CIGRE DC working group proposed the definition of the multi-infeed short-circuit ratio [18]:

$$MISCR_i = \frac{S_{aci}}{P_{deqi}} = \frac{S_{aci}}{P_{di} + \sum\limits_{j=1,j\neq i}^{n} MIIF_{ji}P_{dj}}$$

$$= \frac{S_{aci}}{P_{di} + \sum\limits_{j=1,j\neq i}^{n} \frac{\Delta U_j}{\Delta U_i}P_{dj}} \tag{7}$$

where S_{aci} is the inverter bus short circuit capacity of the ith DC; P_{deqi} is the equivalent DC transmission capacity; n is the number of DCs; P_{di} is the rated transmission capacity of the ith DC; and $MIIF_{ji}$ is the multi-infeed interaction factor, which is defined as the ratio of the jth DC inverter bus voltage variation ΔU_j to the ith DC ΔU_i when a reactive power disturb is applied on the ith DC inverter bus.

Another practical definition of the multi-infeed short circuit ratio is based on the impedance matrix [19]. It can be expressed as

$$MISCR_i = \frac{U_{aci}^2/|Z_{eqii}|}{P_{di} + \sum\limits_{j=1,j\neq i}^{n} |Z_{eqij}/Z_{eqii}|P_{dj}} \tag{8}$$

where U_{aci} is the inverter bus voltage of the ith DC; Z_{eqij} is the ith row and j^{th} column element of the equivalent impedance matrix $\mathbf{Z}_{eq}$, which can be obtained by multi-port Thevenin equivalent method between DC inverter buses. And the value of Z_{eqij} is equal to the voltage of node i when the unit current being injected only to node j.

If the inverter bus rated voltage is set as the voltage base value, (8) is rewritten as

$$MISCR_i = \frac{1}{|Z_{eqii}|P_{di} + \sum\limits_{j=1,j\neq i}^{n} |Z_{eqij}|P_{dj}} = \frac{1}{\sum\limits_{j=1}^{n} |Z_{eqij}|P_{dj}} \tag{9}$$

3.2 Impact of network structure changes on the multi-infeed short-circuit ratio

Assuming the original network contains m nodes and n DCs, and the first n rows and n columns of the m-order impedance matrix $\mathbf{Z}_m$ are DC inverter buses. When a branch z_{kl} is added between node k and l, according to (1) and (9), the multi-infeed short-circuit ratio is changed into

$$MISCR_i' = \frac{1}{\sum\limits_{j=1}^{n} |Z_{ij}'|P_{dj}} = \frac{1}{\sum\limits_{j=1}^{n} \left|Z_{ij} - \frac{(Z_{ik}-Z_{il})(Z_{jk}-Z_{jl})}{Z_{kk}+Z_{ll}-2Z_{kl}+z_{kl}}\right|P_{dj}} \tag{10}$$

where $i = 1,2,\dots,n$; $j = 1,2,\dots,n$; n is the number of DCs; $k = 1,2,\dots,m$; $l = 1,2,\dots,m$; m is the number of network nodes; Z_{ij} and Z'_{ij} are the ith row and j^{th} column element of the impedance matrix before and after adding the branch, respectively.

The variation of the multi-infeed short-circuit ratio is:

$$\Delta MISCR_i = MISCR_i' - MISCR_i$$

$$= \frac{1}{\sum\limits_{j=1}^{n} \left|Z_{ij} - \frac{(Z_{ik}-Z_{il})(Z_{jk}-Z_{jl})}{Z_{kk}+Z_{ll}-2Z_{kl}+z_{kl}}\right|P_{dj}} - \frac{1}{\sum\limits_{j=1}^{n} |Z_{ij}|P_{dj}}$$

$$= \frac{\sum\limits_{j=1}^{n} \left(|Z_{ij}| - \left|Z_{ij} - \frac{(Z_{ik}-Z_{il})(Z_{jk}-Z_{jl})}{Z_{kk}+Z_{ll}-2Z_{kl}+z_{kl}}\right|\right)P_{dj}}{\sum\limits_{j=1}^{n} \left|Z_{ij} - \frac{(Z_{ik}-Z_{il})(Z_{jk}-Z_{jl})}{Z_{kk}+Z_{ll}-2Z_{kl}+z_{kl}}\right|P_{dj} \cdot \sum\limits_{j=1}^{n} |Z_{ij}|P_{dj}}$$

$$\tag{11}$$

The denominator of (11) is greater than 0. And if the elements of the impedance matrix are regarded as pure inductive reactance, the numerator of (11) can be rewritten as

$$\begin{cases} \Delta = \sum\limits_{j=1}^{n} \frac{(Z_{ik}-Z_{il})(Z_{jk}-Z_{jl})}{Z_{kk}+Z_{ll}-2Z_{kl}+z_{kl}}P_{dj} = \frac{\Delta'}{Z_{LL}} \\ \Delta' = \sum\limits_{j=1}^{n} (Z_{ik}-Z_{il})(Z_{jk}-Z_{jl})P_{dj} \end{cases} \tag{12}$$

For opening the line as is shown in Fig. 1, since that line reactance z is usually much larger than the self-impedance, the expression can be derived as

$$Z_{LL} = Z_{kk} + Z_{ll} - 2Z_{kl} - z < j \cdot 0 \tag{13}$$

For installing the current limiting reactor as is shown in Fig. 2, similarly, the expression can be derived as

$$Z_{LL} = Z_{kk} + Z_{ll} - 2Z_{kl} - \frac{z^2}{\Delta z} - z < j \cdot 0 \tag{14}$$

The Δ' expression shows that its value is determined by the network structure and component parameters. If Δ' is larger than 0, then $\Delta MISCR_i$ is larger than 0; if Δ' is smaller than 0, then $\Delta MISCR_i$ is smaller than 0. The current limiting measures, on the one hand, reduce short circuit current, on the other hand, stretch the electrical distance between DC inverter stations. As a result, the multi-infeed short-circuit ratio may be increased or decreased under different conditions. This fact indicates that there exists a coordinated optimization scheme, which can not only control short circuit current within a reasonable range, but also remain the multi-infeed short-circuit ratio at a high level. Such scheme should be obtained by establishing and solving the multi-objective optimization model.

4 Coordinated optimization method

4.1 Mathematical model

The decision variables consist of two parts. One is the control variable u_s which represents whether the measure is put into or not, and the other is the variable z_s which represents the specific parameters of current limiting equipment. The coordinated optimization not only considers the cost and effect of current limiting measures, but also tries to keep the tightness of network structure. Furthermore, the impact of current limiting measures on the multi-infeed short-circuit ratio should be considered.

The objective function f_1 is used to evaluate the economy of current limiting measures, which is expressed as the total cost

$$\min f_1 = \sum_{s=1}^{N_s} u_s(k_{as} + k_{bs}z_s) \tag{15}$$

where N_s is the number of put-into current limiting measures; $u_s = 1$ represents that the measure s is put into and $u_s = 0$ represents that the measure s is not put into; k_{as} and k_{bs} are the cost factors of the measure s [13]; z_s is the reactance value of the current limiting reactor or the short circuit voltage percentage increment of the high-impedance transformer.

The objective function f_2 is used to evaluate the tightness of network structure, which is expressed as the short circuit capacity margin:

$$\min f_2 = \sum_{k=1}^{N_b} \frac{scc_k^{\max} - scc_k}{scc_k} \tag{16}$$

where N_b is the total number of nodes; $scc_k^{\max}$ is the short circuit capacity upper limit of node k, and less than the maximum breaker interruptive capacity; scc_k is the short circuit capacity of node k after current limiting measures are put into.

The short circuit capacity reflects the anti-disturbance performance of each node and the network connection strength [21]. In this paper, current limiting measures are tried not to destroy the integrity and tightness of power grid. Thus, the minimum short circuit capacity margin is chosen as one target. At the same time, taking the current limiting effect into account, the short circuit current upper limit can be specified according to the engineering experience.

The objective function f_3 is used to evaluate the impact of current limiting measures on the multi-infeed short-circuit ratio, which reflects the voltage support capacity of AC system to multiple DCs. A greater value of f_3 indicates a stronger inherent strength of the AC system. It is expressed as the weighted multi-infeed short-circuit ratio:

$$\max f_3 = \sum_{i=1}^{n} \omega_i MISCR_i \tag{17}$$

where n is the number of DCs; $MISCR_i$ is the multi-infeed short-circuit ratio of the ith DC; ω_i is the weighted factor of the ith DC, which reflects the importance of the ith DC in multi-infeed DC systems. The greater influence of the ith DC on other DCs indicates the more importance of the ith DC. Thus, ω_i can be defined as:

$$\omega_i = \sum_{j=1, j\neq i}^{n} \left| \frac{Z_{\text{eq}ij}}{Z_{\text{eq}ii}} \right| \cdot \frac{P_{dj}}{P_{di}} \tag{18}$$

The constraints include that there are no isolated node in power grid, active power flow balance, reactive power flow balance, short circuit current within limit, branch power flow within limit, bus voltage within limit, the multi-infeed short-circuit ratio within limit, parameters of current limiting equipment within limit, etc. They are shown as (19):

$$\begin{cases} P_i = U_i \sum_{j=1}^{N_b} U_j(G_{ij}\cos\delta_{ij} + B_{ij}\sin\delta_{ij}) & i = 1, 2, \ldots, N_b \\ Q_i = U_i \sum_{j=1}^{N_b} U_j(G_{ij}\sin\delta_{ij} - B_{ij}\cos\delta_{ij}) & i = 1, 2, \ldots, N_b \\ I_k \leq I_k^{\max} & k = 1, 2, \ldots, N_b \\ S_l \leq S_l^{\max} & l = 1, 2, \ldots, N_l \\ U_k^{\min} \leq U_k \leq U_k^{\max} & k = 1, 2, \ldots, N_b \\ MISCR_i \geq MISCR_{\min} \\ z_s^{\min} \leq z_s \leq z_s^{\max} \end{cases} \tag{19}$$

where N_b is the total number of nodes; $I_k^{\max}$ is the short circuit current upper limit of node k; N_l is the total number of branches; $S_l^{\max}$ is the power upper limit of branch l; $U_k^{\max}$ and $U_k^{\min}$ are the voltage upper and lower limits of node k; $MISCR_{\min}$ is the lower limit of the multi-infeed short-circuit ratio; $z_s^{\max}$ and $z_s^{\min}$ are the upper and lower limits of parameters of current limiting equipment.

Additionally, the optimization schemes should meet the "N-1" constraints. In order to simplify the problem, the handing method in this paper is getting the preferred schemes firstly, and then checks them.

4.2 Multi-objective optimization algorithm

The core of multi-objective optimization is to coordinate the relationships between objective functions, and to find out the optimal solutions which make the value of each objective function as small as possible. The multi-objective optimization algorithm has three main performance indices: 1) the obtained solutions should be as close to the true Pareto optimal solutions as possible; 2) try to keep the distribution and diversity of the individuals; 3) avoid missing the Pareto optimal solutions in the solving process.

The classical optimization algorithms suggest converting the multi-objective problem to a single-objective problem by sorting or assigning weighted factors to multiple objectives. When such an algorithm is applied, only one single optimal solution can be obtained in each simulation run. It has to be carried out many times to find a set of optimal solutions. This process is usually inefficient and cost a lot of time. In addition, the solutions are badly influenced by weighted factors, which are determined according to the expert experience. Thus, these classical algorithms are not quit applicable to multi-objective problems.

Over the past decade, a number of multi-objective EAs have been suggested. EAs are well applicable to multi-objective problems because of their ability to find multiple solutions in one single simulation run. Since EAs work with a population of individuals, it can be extended to maintain a diverse set of solutions. With an emphasis for moving toward the true Pareto optimal region, an EA can be used to find the Pareto optimal solutions in one single simulation run. NSAG-II is one of such EAs and it outperforms others in terms of finding a diverse set of solutions and converging near the true Pareto optimal solutions. NSGA-II has three key technologies, which make it an excellent multi-objective optimization algorithm. They are the fast non-dominated sorting approach, individual crowding distance design and elitist strategy [7].

4.2.1 Individual encoding

There may be three states that exist on the same branch, including no current limiting measures applied, opening the line and installing the current limiting reactor. Perform integer encoding on the N individuals of the population. Each individual consists of M bits, which is shown in Fig. 3.

z_l is the value of the lth bit which can be any integer in $(0,[z_l^{min},z_l^{max}],z_l^{max}+1)$. $z_l = 0$ denotes no current limiting measure is applied; $z_l = z_l^{max}+1$ denotes opening the line for the branch l; if z_l is any integer in $[z_l^{min},z_l^{max}]$, it means installing a current limiting reactor with the value of z_l for the branch l.

4.2.2 Fitness function

Adding the constraints in (19) into the objective functions as a penalty, then the fitness functions can be expressed as

| z_1 | z_2 | $\cdots$ | z_l | $\cdots$ | z_M |

Fig. 3 Integer encoding structure

$$\begin{cases} f_1 = f_1 + W \\ f_2 = f_2 + W \\ f_3 = -f_3 + W \end{cases} \qquad (20)$$

If all the constraints are satisfied, the penalty W is assigned to 0; otherwise, W is assigned to a large value.

4.2.3 Fast non-dominated sorting approach

This approach sorts the individuals into different non-dominated fronts, and it guides the search process toward the Pareto optimal solutions. For each individual, we calculate two entities firstly: one is n_i, the number of individuals which dominate the individual i, and the other is S_i, a set of individuals that the individual i dominates. The sorting steps are as follows.

Step 1: Find out the individuals with $n_i=0$ in the initial population, put them into the set F_1 as the first non-dominated front, and set the non-dominated rank of all the individuals in this front as $i_{rank} = 1$.

Step 2: For each individual i in F_1, visit every individual l in S_i, and execute $n_l = n_l$-1. For any individual l, when n_l becomes zero, put it into the set F_2 as the second non-domination front. Set the non-dominated rank of all the individuals in this front as $i_{rank} = 2$.

Step 3: For the set F_2, repeat step 2. Put the individuals with $n_l = 0$ into the set F_3 as the third non-dominated front. Set the non-dominated rank of all the individuals in this front as $i_{rank} = 3$. Continue this process until all fronts are identified.

4.2.4 Individual crowding distance design

The design of individual crowding distance is proposed in NSGA-II to sort the individuals with the same i_{rank}. The crowding distance of the individual i is defined as the distance between its two adjacent individuals $i+1$ and $i-1$ in the target space. The calculating steps are as follows.

Step 1: Initialize the crowding distance of individuals in the same front, set $L[i] = 0$.

Step 2: Sort the individuals in the same front in ascending order of the value of the mth objective function.

Step 3: The boundary individuals are assigned an infinite value W:

$$\begin{cases} L[i_{min}] = L[i_{min}] + W \\ L[i_{max}] = L[i_{max}] + W \end{cases} \qquad (21)$$

where i_{max} and i_{min} are the individuals with the maximum and minimum value of the mth objective function, respectively.

Step 4: For all the intermediate individuals, the crowding distance can be derived from (21).

$$L[i] = L[i] + \frac{f_m^{i+1} - f_m^{i-1}}{f_m^{\max} - f_m^{\min}} \tag{22}$$

where f_m^{i+1} and f_m^{i-1} are the values of the mth objective function corresponding to the individual $i + 1$ and $i - 1$; $f_m^{\max}$ and $f_m^{\min}$ are the maximum and minimum value of the mth objective function, respectively.

Step 5: For other objective functions, repeat step 2–4. Then $L[i]$, the crowding distance of the individual i, can be obtained.

Prefer the individuals with the larger crowding distance, and then the solutions can be uniformly distributed in the target space. Thus, the diversity of the individuals can be preserved.

4.2.5 Elitism strategy

Elitism strategy can put the superior individuals of the parent generation into the child generation, which avoids the missing of the Pareto optimal solutions. The processing steps are as follows.

Step 1: Combine the parent generation P_t and the child generation Q_t to form one population $R_t = P_t \cup Q_t$, then apply the fast non-dominated sorting approach on R_t, and then calculate the individual crowding distance.

Step 2: Put the individuals of R_t into the new parent generation P_{t+1} in ascending order of the non-dominated rank. Stop it until the number of individual of P_{t+1} exceeds N (the number of the population), when all the individuals of F_j are added.

Step 3: Put the individuals of F_j into P_{t+1} in descending order of the individual crowding distance, until the number of P_{t+1} equals N.

4.2.6 Tournament selection

The selection approach should guide the optimization process towards the Pareto optimal solutions and keep the distribution and diversity of the individuals. The results of the tournament selection are the chosen individuals used to generate the child generation.

Tournament selection compares the individuals of parent generation in a random pairing mode. The individual i is thought to be superior to the individual j if $i_{\text{rank}} < j_{\text{rank}}$ or $i_{\text{rank}} = j_{\text{rank}}$ and $L[i] > L[j]$. That's to say, for two individuals with different non-dominated ranks, we prefer the individual with the lower rank; if both individuals belong to the same front, we prefer the individual that is located in the less crowded region.

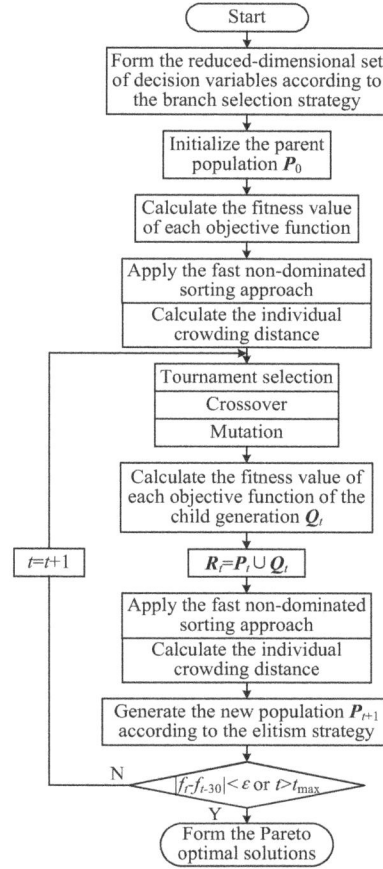

Fig. 4 Flow chart of coordinated optimization method

4.2.7 Crossover and mutation

Cooperation of crossover and mutation can give genetic algorithm good local and global search performances. Single point crossover and random mutation are performed on the population obtained by tournament selection, then the child population Q_t can be generated.

4.3 The process of method

In summary, the process of coordinated optimization method is shown in Fig. 4. Where f_t is the average of the fitness values of all the individuals in the first front; $t_{\max}$ is the maximum number of evolution generations.

5 Simulation of planning power system

Taking a planning power system for example, the feasibility and speediness of this method is validated. Considering a 5% margin, for the breakers with maximum interruptive current of 63 kA, the short circuit current upper limit is set to 59.85 kA. The impedance value of current limiting reactor is

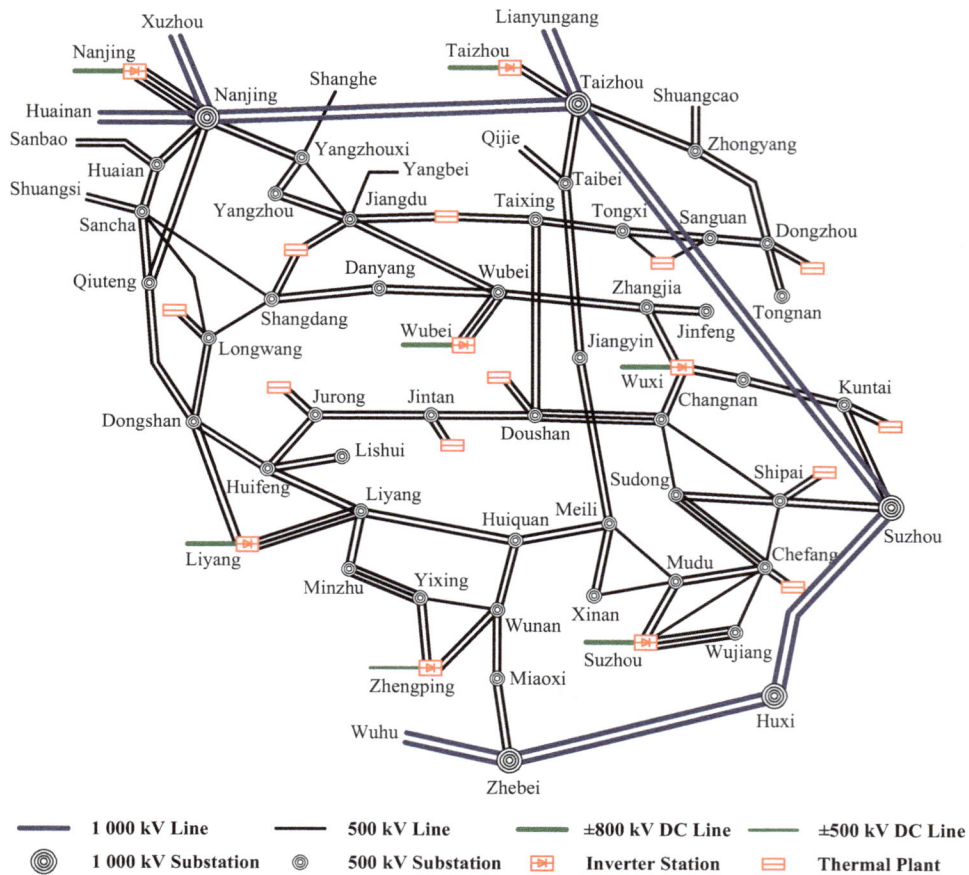

Fig. 5 Regional power grid structure in China

Table 1 MISCR of the multi-infeed DC system

Inverter Bus	Voltage/kV	Capacity/MW	MISCR
Suzhou	±800	Bipolar 7200	2.5618
Zhengping	±500	Bipolar 3000	2.9589
Liyang	±800	Bipolar 8000	2.4035
Nanjing	±800	Bipolar 8000	2.7150
Wubei	±800	Bipolar 8000	2.5708
Taizhou	±800	Bipolar 8000	3.2051
Wuxi	±800	Bipolar 8000	2.9194

set to the range of $0–10\,\Omega$. The minimum multi-infeed short-circuit ratio is set to 2.0 [18]. The cost factors of opening the line are as follows: $k_{as} = 60$, $k_{bs} = 0$, and the cost factors of installing the current limiting reactor are as follows: $k_{as} = 625$, $k_{bs} = 25$ [13].

The structure of regional power grid in China is shown in Fig. 5. There are 7 DC inverter stations (Suzhou, Zhengping, Liyang, Nanjing, Wubei, Taizhou and Wuxi) in this regional power grid, forming a typical multi-infeed DC system. The multi-infeed short-circuit ratio of this system are shown in Table 1.

Under this operating mode, the three-phase short circuit currents of 500 kV buses in Shipai, Changnan, Suzhou, Sudong, Doushan and Chefang substation are 77.43, 73.82, 70.64, 68.59, 67.28 and 66.46 kA, respectively. All of them are exceeding the short circuit current upper limit (59.85 kA). According to the branch selection strategy, all the lines in the network are arranged in the descending order of the integrated sensitivity. The results are shown in Table 2. The lines with $\mu_l > 0.01$ are put into the reduced-dimensional set, and the lines with $\mu_l < 0.01$ have no competitiveness.

The population size is set to 100, the maximum generation is set to 500, and the crossover rate is set to 0.9. In order to prevent the deviation of single optimization caused by random factors, 400 times calculations are carried out under different calculating parameters. The statistical results are shown in Table 3. The length of full-dimensional individual is 90, the length of reduced-dimensional individual is 44, and $P/\%$ is the average probability that the solutions of each calculation in accordance with the Pareto optimal solutions of 400 times calculations. As is shown in Table 3, the optimization using the reduced-dimensional

Table 2 Integrated sensitivity ranking of current limiting measures

No.	Line name	μ_l	No.	Line name	μ_l
1	Shipai-Changnan	1.000	23	Kuntai-Suzhou II	0.076
2	Shipai-Suzhou I	0.458	24	Wubei-Zhangjia I	0.073
3	Shipai-Suzhou II	0.458	25	Wubei-Zhangjia II	0.073
4	Changnan-Sudong	0.381	26	Meili-Huiquan I	0.046
5	Shipai-Sudong I	0.321	27	Meili-Huiquan II	0.046
6	Shipai-Sudong II	0.321	28	Changnan-Kuntai I	0.044
7	Doushan-Changnan I	0.219	29	Changnan-Kuntai II	0.044
8	Doushan-Changnan II	0.219	30	Meili-Xinan I	0.041
9	Doushan-Changnan III	0.219	31	Meili-Xinan II	0.041
10	Xinan-Mudu	0.208	32	Tongxi-Taixing I	0.032
11	Shiapai-Chefang	0.197	33	Tongxi-Taixing II	0.032
12	Taixing-Doushan I	0.195	34	Jurong-Jintan I	0.030
13	Taixing-Doushan II	0.195	35	Jurong-Jintan II	0.030
14	Chefang-Sudong I	0.142	36	Taibei-Jiangyin I	0.027
15	Chefang-Sudong II	0.142	37	Taibei-Jiangyin II	0.027
16	Chefang-Sudong III	0.142	38	Meili-Jiangyin I	0.019
17	Meili-Mudu	0.130	39	Meili-Jiangyin II	0.019
18	Jintan-Doushan I	0.101	40	Chefang-Wujiang	0.014
19	Jintan-Doushan II	0.101	41	Huifeng-Jurong I	0.012
20	Chefang-Mudu I	0.094	42	Huifeng-Jurong II	0.012
21	Chefang-Mudu II	0.094	43	Sanguan-Tongxi I	0.011
22	Kuntai-Suzhou I	0.076	44	Sanguan-Tongxi II	0.011

Table 3 Statistical results under different calculating parameters

No.	Individual dimension	Mutation rate	Calculating times	Average converged generation	$P/\%$
1	Reduced-dimensional	0.4	100	69	85
2	Full-dimensional	0.4	100	398	46
3	Reduced-dimensional	0.2	100	60	74
4	Full-dimensional	0.2	100	372	33

Table 4 Pareto optimal solutions

Scheme	Current limiting measure	
	Open the line	Install the current limiting reactor
1	Shipai-Changnan Changnan-Sudong Xinan-Mudu Meili-Mudu	None
2	Shipai-Changnan Doushan-Changnan I	Shipai-Suzhou I Line $+9\Omega$ Shipai-Suzhou II Line $+10\Omega$
3	Shipai-Changnan Xinan-Mudu Meili-Mudu Shipai-Suzhou II	None
4	Shipai-Changnan Doushan-Changnan I Shipai-Suzhou I	Shipai-Suzhou II Line $+2\Omega$
5	Shipai-Changnan Changnan-Sudong Shipai-Suzhou I	Shipai-Suzhou II Line $+2\Omega$

decision variables has less converged generation and better converged solutions. And using a larger mutation rate could get a closer solution compared with the Pareto optimal solutions, although it increases the converged generation.

The method combining the reduced-dimensional decision variables and a larger mutation rate is adopted to solve this optimization model. An optimization calculation is converged at the 70th generation. In order to keep the distribution and diversity of the individuals, sort the solutions in descending order of the crowding distance. The first 5 Pareto optimal solutions are shown in Table 4, and the corresponding fitness values are shown in Table 5. It

can be concluded that the total cost f_1 and the short circuit capacity margin f_2 are contradictory to each other. Taking scheme 1 and 2 for examples, the total cost of scheme 1 is

Table 5 Fitness values of Pareto optimal solutions

Scheme	Fitness value		
	f_1	f_2	f_3
1	240	17.232	−24.615
2	1845	10.643	−24.264
3	240	15.444	−24.582
4	855	11.674	−24.244
5	855	12.529	−24.308

the lowest and its short circuit capacity margin is largest, however, scheme 2 has the largest total cost and the smallest short circuit capacity margin. The contradiction between f_1 and f_2 is determined by the characteristics of current limiting measures. Opening the line is the one with best effect and lowest cost, but it will significantly reduce the tightness of network. Installing the current limiting reactor has a smoother current limiting effect and can keep the tightness of network, but the cost is high.

The weighted multi-infeed short-circuit ratio ($-f_3$) of scheme 1 and 3 are larger than the one (24.352) before the optimization is carried out. It is proved by the numerical simulations that such two schemes give the receiving-end AC system a stronger voltage support capacity to multiple DCs. When the same faults occur, the recovery rate of system voltage and DC power of scheme 1 is the fastest between all the schemes. The multi-infeed short-circuit ratio and short circuit current of all the schemes are shown

in Table 6 and Table 7. The power flow and stability calculation demonstrates that all the schemes can satisfy the "*N*-1" constraints.

If the single-objective optimization for the cost of current limiting measures is carried out, only scheme 1 and 3 in Table 4 can be obtained. The presented method in this paper can provide the Pareto optimal solutions. Therefore, the decision-makers can choose the one they prefer. For example, if the requirement is to invest the minimum cost of current limiting measures, scheme 1 and 3 can be chosen; if the requirement is to maintain the integrity of the electrical power grid, scheme 2 can be chosen; if the stronger voltage support capacity of AC system to multiple DCs is required, scheme 1 and 3 are the better choices; if the optimizing balance of all the objective functions is required, scheme 4 and 5 may be the preferred ones.

6 Conclusion

In order to solve the problems of exceeding short circuit current and improve the voltage stability in multi-infeed DC system, a coordinated optimization method is presented in this paper. The simulation results of the planning regional power grid in China demonstrated the feasibility and efficiency of the method. The branch selection strategy considering sensitivity ranking can effectively reduce the search range of decision variables, therefore avoid the optimization falling into the curse of dimensionality.

Table 6 MISCR of different schemes

Inverter bus	Scheme 1	Scheme 2	Scheme 3	Scheme 4	Scheme 5
Suzhou	2.3340	2.5505	2.3319	2.5497	2.5684
Zhengping	3.1283	2.9616	3.1271	2.9614	2.9637
Liyang	2.4651	2.4098	2.4627	2.4097	2.4078
Nanjing	2.7282	2.7113	2.7277	2.7113	2.7104
Wubei	2.5490	2.5459	2.5438	2.5459	2.5491
Taizhou	3.2919	3.2046	3.2929	3.2046	3.2047
Wuxi	2.8376	2.8521	2.8206	2.8523	2.8819

Table 7 Short circuit currents of different schemes

Bus name	Actual short circuit current (kA)				
	Scheme 1	Scheme 2	Scheme 3	Scheme 4	Scheme 5
Shipai	49.13	58.79	52.47	58.47	49.93
Changnan	48.77	59.37	59.74	59.37	48.62
Suzhou	58.88	59.64	58.96	59.32	58.85
Sudong	41.46	59.75	50.42	59.74	46.82
Doushan	51.77	57.66	59.28	57.65	51.64
Chefang	39.99	59.11	46.53	58.93	48.84

Combining this strategy with NSGA-II can improve the convergence characteristics of the optimizing process. Considering the total cost, short circuit capacity margin and weighted multi-infeed short-circuit ratio, the Pareto optimal schemes can be obtained, which provides the decision-makers with more comprehensive and enriched choices.

Acknowledgment This work was supported by State Grid Corporation of China, Major Projects on Planning and Operation Control of Large Scale Grid under Grant SGCC-MPLG020-2012.

References

[1] Zhang W, Liu YT (2008) Multi-objective reactive power and voltage control based on fuzzy optimization strategy and fuzzy adaptive particle swarm. Int J Electr Power Energ Syst 30(9): 525–532

[2] Granelli G, Montagna M, Zanellini F et al (2006) A genetic algorithm-based procedure to optimize system topology against parallel flows. IEEE Trans Power Syst 21(1):333–340

[3] Zhao CY, Guan YP (2013) Unified stochastic and robust unit commitment. IEEE Trans Power Syst 28(3):3353–3361

[4] Liu YT, Zhang P, Qiu XZ (2002) Optimal volt/var control in distribution systems. Int J Electr power Energ Syst 24(4): 271–276

[5] Hongesombut K, Mitani Y, Tsuji K (2003) Optimal location assignment and design of superconducting fault current limiters applied to loop power systems. IEEE Trans Appl Supercon 13(2–2):1828–1831

[6] Yannibelli V, Amandi A (2013) Hybridizing a multi-objective simulated annealing algorithm with a multi-objective evolutionary algorithm to solve a multi-objective project scheduling problem. Expert Syst Appl 40(7):2421–2434

[7] Deb K, Pratap A, Agarwal S et al (2002) A fast and elitist multiobjective genetic algorithm: NSGA-II. IEEE Trans Evolut Comput 6(2):182–197

[8] Shahriari SAA, Yazdian A, Haghifam MR (2009) Fault current limiter allocation and sizing in distribution system in presence of distributed generation. In: Proceedings of the 2009 IEEE power and energy society general meeting (PES'09), Calgary, 26–30 Jul 2009, 6 pp

[9] Kim SY, Bae IS, Kim JO (2010) An optimal location for superconducting fault current limiter considering distribution reliability. In: Proceedings of the 2010 IEEE power and energy society general meeting (PES'10), Minneapolis, 25–29 Jul 2010, 5 pp

[10] Nagata M, Tanaka K, Taniguchi H (2001) FCL location selection in large scale power system. IEEE Trans Appl Supercon 11(1–2):2489–2494

[11] Teng JH, Lu CN (2010) Optimum fault current limiter placement with search space reduction technique. IET Gener Transm Distrib 4(4):485–494

[12] Namchoat S, Hoonchareon N (2013) Optimal bus splitting for short-circuit current limitation in metropolitan area. In: Proceedings of the 10th international conference on electrical engineering/electronics, computer, telecommunications and information technology (ECTI-CON'13), Krabi, 15–17 May 2013, 5 pp

[13] Chen LL, Huang MX, Wu JY, et al (2010) An optimal strategy for short circuit current limiter deployment. In: Proceedings of the 2010 Asia-Pacific power and energy engineering conference (APPEEC'10), Chengdu, 28–31 Mar 2010, 4 pp

[14] Aik DLH, Andersson G (1997) Voltage stability analysis of multi-infeed HVDC systems. IEEE Trans Power Deliver 12(3):1309–1318

[15] Bui LX, Sood VK, Laurin S (1991) Dynamic interactions between HVDC systems connected to AC buses in close proximity. IEEE Trans Power Deliver 6(1):223–230

[16] Aik DLH, Andersson G (2013) Analysis of voltage and power interactions in multi-infeed HVDC systems. IEEE Trans Power Deliver 28(2):816–824

[17] IEEE Std 1204—1997. IEEE guide for planning DC links terminating at AC locations having low short-circuit capacities. 1997

[18] CIGRE Working Group B4.41 (2008) Systems with multiple DC infeed. CIGRE, Paris

[19] Lin WF, Tong Y, Bu GQ, et al (2010) Voltage stability analysis of multi-infeed AC/DC power system based on multi-infeed short circuit ratio. In: Proceedings of the 2010 international conference on power system technology (POWERCON'10), Hangzhou, 24–28 Oct 2010, 6 pp

[20] de Toledo PF, Bergdahl B, Asplund G (2005) Multiple infeed short circuit ratio-aspects related to multiple HVDC into one AC network. In: Proceedings of the 2005 IEEE/PES transmission and distribution conference and exhibition: Asia and Pacific, Dalian, 14–18 Aug 2005, 6 pp

[21] Tada Y, Okamoto H, Kurita A et al (1998) Analytical methods for determining a system configuration acceptable from viewpoints of both short circuit current and voltage stability. Electr Eng JPN 124(3):30–39

[22] Huang HY, Xu Z, Lin X (2012) Improving performance of multi-infeed HVDC systems using grid dynamic segmentation technique based on fault current limiters. IEEE Trans Power Syst 27(3):1664–1672

Dong YANG Received the B.E. and Ph.D. degrees from Shandong University in 2007 and 2013, respectively. He is now an engineer in State Grid Shandong Electric Power Research Institute. His research interests include power system operation and planning.

Kang ZHAO Received the B.E. degree from Huazhong University of Science and Technology in 2013. He is now a M.E. candidate in Shandong University. His research interests include power system operation and planning.

Yutian LIU received the B.E. and M.E. degrees from Shandong University in 1984 and 1990, respectively, and the Ph.D. degree from Xi'an Jiaotong University in 1994. He is now a professor with the School of Electrical Engineering, Shandong University. His research interests include renewable energy sources integrated into smart grid, power system security and stability, and power system restoration.

7

Derivation and conformity measurement of a popular explicit analytic Borowy 2C PV module model

Jianmin ZHANG (✉), Qianzhi ZHANG

Abstract A popular explicit analytic Borowy 2C PV module model is proposed for power generation prediction. The maximum power point and the open-circuit point which are calculated in this model cannot be equal to the data given by manufacturers under standard test condition (STC). The derivation of this model has never been mentioned in any literatures. The parameter forms of 2C model in this paper are more simplified, and the model is decomposed into a STC sub-model and an incremental sub-model. The STC model is derived successfully from an ideal single-diode circuit model. Relative error estimations are developed to do the conformity error measurements. The analysis results showed that though the biases at those critical points are very small, the conformity will depend on both of the two ratio values $I_{\mathrm{m}}/I_{\mathrm{sc}}$ and $V_{\mathrm{m}}/V_{\mathrm{oc}}$, which can be used to verify whether 2C model is applicable for the PV module produced by a particular manufacturer.

Keywords Photovoltaic module, Analytic model, Explicit model, Implicit model, Borowy model, Conformity check, Derivation of the model, Conformity error measurement

J. ZHANG, School of Automation, Hangzhou Dianzi University, Hangzhou 310017, China
(✉) e-mail: zhangjmhzcn@hdu.edu.cn
Q. ZHANG, School of Electrical, Computer, and Energy Engineering, Ira A. Fulton Schools of Engineering, Arizona State University, Tempe, AZ 85287-5706, USA
e-mail: zhang.qianzhi@asu.edu

1 Introduction

Analytic photovoltaic module modeling with the parameters provided by PV module manufacturer is critical for PV plant sizing [1, 2], simulation [3], testing [4], PV plant monitoring [5, 6], dispatching [7], and energy management [8]. A very popular PV module model introduced by Borowy & Salameh [2] named 2C model proposed in this paper, has been heavily cited by 229 papers from Google and 66 papers from IEEE-Xplore database, with no derivation process. [3–6] have used 2C model in their works. All known that the awareness of strong theoretic based or data well driven modeling process is necessary for a user, therefore, this paper aims to fill the blank and present not only a detailed derivation but also a conformity study of 2C model.

In Sect. 2, the 2C model is introduced, which can be decomposed into a sub-model under standard test condition (STC) and an incremental sub-model. Two new parameter sets $(\gamma_{I_{\mathrm{m}}}, \gamma_{V_{\mathrm{m}}})$ and $(\sigma_{I_m}, \sigma_{V_{\mathrm{m}}})$ are introduced to simplify the model and help the further derivation and conformity study. The contradictions of 2C model with manufacturer's datasheet have been explored in Sect. 3. In Sect. 4, 2C model under STC is derived from a single-diode circuit model. The conformity error measurements of 2C model with manufacturer's datasheet have been developed in Sect. 5. Finally conclusions are made in Sect. 6.

2 2C model and its decomposition

2.1 Basic formula

$$I = I_{\mathrm{SC}}\left[1 - C_1\left(\exp\left(\frac{V - \Delta V}{C_2 V_{\mathrm{OC}}}\right) - 1\right)\right] + \Delta I \qquad (1)$$

$$C_1 = \left(1 - \frac{I_m}{I_{SC}}\right) \exp\left(-\frac{V_m}{C_2 V_{OC}}\right) \tag{2}$$

$$C_2 = \left(\frac{V_m}{V_{OC}} - 1\right) / \ln\left(1 - \frac{I_m}{I_{SC}}\right) \tag{3}$$

$$\Delta I = \alpha \frac{R}{R_{ref}} \Delta T + \left(\frac{R}{R_{ref}} - 1\right) I_{SC} \tag{4}$$

$$\Delta V = -\beta \Delta T - R_s \Delta I \tag{5}$$

$$\Delta T = T_C - T_{ref} \tag{6}$$

where R and T_c are the solar irradiance and module temperature, respectively. C_1, C_2, ΔI, ΔV and ΔT are intermediate variables; I_{sc} is the short circuit current; V_{oc} is the open circuit voltage; I_m is the current at maximum power; V_m is the voltage at maximum power; α is the short circuit current temperature correlation (%/°C or A/°C); β is the open circuit voltage temperature correlation (%/°C or V/°C); R_{ref} and T_{ref} are the references of solar irradiance and ambient temperature and $R_{ref} = 1 \text{ kW/m}^2$, $T_{ref} = 25$ °C, respectively; R_s is internal series resistance.

2.2 New direct forms of C_1 and C_2

Assuming two parameter sets:

$$\begin{cases} \frac{I_m}{I_{SC}} = \gamma_{I_m}, & \frac{V_m}{V_{OC}} = \gamma_{V_m} \\ 1 - \frac{I_m}{I_{SC}} = \sigma_{I_m}, & 1 - \frac{V_m}{V_{OC}} = \sigma_{V_m} \end{cases} \tag{7}$$

From (7), the equation is obtained as follows:

$$C_1 = (\sigma_{I_m})^{\frac{1}{\sigma_{V_m}}} \tag{8}$$

$$C_2 = -\frac{\sigma_{V_m}}{\ln \sigma_{I_m}} \tag{9}$$

It can be proved that:

$$C_2 = -\frac{1}{\ln C_1} \tag{10}$$

$$0 < \sigma_{I_m} < 1, \quad 0 < \sigma_{V_m} < 1 \tag{11}$$

$$0 < C_1 < 1 \tag{12}$$

2.3 Decompose 2C model into two sub-models

2C model can be decomposed into two sub-models, one is the model under STC, the other one is the incremental model

1) Under STC, $R = R_{ref}$, $T_c = T_{ref}$, $\Delta I = 0$, $\Delta V = 0$, $\Delta T = 0$. (1) will be written as follows:

$$I(V) = I_{SC}\left[1 - C_1\left(\exp\left(\frac{V}{V_{OC} C_2}\right) - 1\right)\right] \tag{13}$$

$$V(I) = C_2 V_{OC} \ln\left[1 + \frac{1}{C_1}\left(1 - \frac{I}{I_{SC}}\right)\right] \tag{14}$$

2) With R and T_c changing in time, ΔI, ΔV and ΔT will also change according to (4)–(6).

$$\begin{cases} I^* = I - \Delta I \\ V^* = V - \Delta V \end{cases} \tag{15}$$

(1) will become as follows:

$$I^* = I_{SC}\left[1 - C_1\left(\exp\left(\frac{V^*}{C_2 V_{OC}}\right) - 1\right)\right] \tag{16}$$

Even if R and T_c are not the reference values, the characteristic curves of I and V are still the same under STC, since the pattern of (13) is same with (16). (15) is a coordinate shift operation to transform (V, I) into (V^*, I^*). (13) and (16) are both determined by I_{sc}, V_{oc} and C_1, C_2. Therefore, the parameter forms of C_1 and C_2 will determine 2C model not only for its STC model, but also for its incremental model.

3 Contradictions of 2C model under STC

3.1 2C model at $V = 0$ under STC

Assuming $I = I''_{SC}$ at $V = 0$, (16) can be written as:

$$I''_{SC} = I_{SC} \tag{17}$$

which means the short circuit point in model is exactly at the assumed short circuit point.

3.2 2C model at $V = V_m$ under STC

Assuming $I = I''_m$ at $V = V_m$, (16) can be written from (7, 8, 10) as:

$$I''_m = I_m + I_{SC} C_1 \tag{18}$$

From (18), the bias between $I = I''_m$ and I_m can be derived as:

$$\Delta I''_m = I''_m - I_m = I_{SC} C_1 > 0 \tag{19}$$

which means that the power point of the model (V_m, I''_m) always locates at the upper side of the maximum power point (V_m, I_m) provided by manufactory.

3.3 2C model at $I = I_m$ under STC

Assuming $V = V''_m$ at $I = I_m$, from (14), V''_m is given as:

$$V''_m = C_2 V_{OC} \ln\left[1 + \frac{1}{C_1}\left(1 - \frac{I_m}{I_{SC}}\right)\right] \tag{20}$$

Defining:

$$Z_2 = V''_m / V_m \tag{21}$$

From (8) and (9), (20) can be written as:

$$Z_2 = \ln\left(1 + \sigma_{I_m}^{1-\frac{1}{\sigma_{V_m}}}\right) / \ln\left(\sigma_{I_m}^{1-\frac{1}{\sigma_{V_m}}}\right) \tag{22}$$

Defining:

$$x_3 = \sigma_{I_m}^{1-\frac{1}{\sigma_{V_m}}} \tag{23}$$

Thus:

$$z_2 = \ln(1 + x_3)/\ln(x_3) \tag{24}$$

(23) can be derived as:

$$\ln(x_3) = \left(1 - \frac{1}{\sigma_{V_m}}\right)\ln(\sigma_{I_m}) \tag{25}$$

From (11), it can be derived that

$$\left(1 - \frac{1}{\sigma_{V_m}}\right) < 0, \quad \ln(\sigma_{I_m}) < 0 \tag{26}$$

$$\ln(x_3) > 0 \tag{27}$$

which means $X_3 > 1$, $(1 + X_3)/X_3 > 1$, so that $\ln[(1 + X_3)/X_3] > 0$, $\ln(1 + X_3) - \ln(X_3) > 0$. Finally, it can be derived that

$$\ln(1 + x_3) > \ln(x_3) \tag{28}$$

(28) can be expressed as:

$$\ln(1 + x_3)/\ln(x_3) > 1 \tag{29}$$

Substituting (29) into (24), it is shown that $Z_2 > 1$, so that $V_m'' > V_m$ at $I = I_m$, i.e., the power point (V_m'', I_m) of the model always locates at the right side of the maximum power point (V_m, I_m) provided by manufactory.

3.4 2C model at $V = V_{oc}$ under STC

Assuming $I = I_0''$ at $V = V_{oc}$, (13) can be written from (8) and (9) as:

$$I_0'' = I_{SC}\left[1 - \sigma_{I_m}^{\frac{1}{\sigma_{V_m}}}\left(\sigma_{I_m}^{\frac{-V_{OC}}{V_{oc}\sigma_{V_m}}} - 1\right)\right] = I_{SC}\sigma_{I_m}^{\frac{1}{\sigma_{V_m}}} = I_{SC}C_1 > 0 \tag{30}$$

which means the current at $V = V_{oc}$ is greater than 0, with a constant error $I_{sc}C_1$.

4 A derivation of C_1 and C_2 from an ideal single-diode PV circuit model

4.1 PV circuit model with an ideal single-diode

The circuit model based on semiconductor theory with a single diode is given by (1) [9].

$$I = I_{ph} - I_s\left[\exp\left(\frac{q(V + R_sI)}{nkT}\right) - 1\right] - \frac{V + R_sI}{R_{sh}} \tag{31}$$

where I_{ph} is the photonic current (principally depends on the solar irradiance); I_{s1} and I_{s2} are the reverse saturation currents of diode 1 and 2, respectively; q is the electron charge and $q = 1.60,217,646 \times 10^{-19}$ C; k is the Boltzmann constant and $k = 1.3,806,503 \times 10^{-23}$ J/K; T is the cell temperature in Kelvin; n is the diode ideality factors; R_s is the series resistance; R_{sh} is the shunt resistance. (31) is a complex equation without any explicit solutions for both voltage V and current I.

If neglecting R_s and R_{sh}, (1) will become:

$$I(V) = I_{ph} - I_s = I_{ph} - I_0\left[\exp\left(\frac{V}{V_T}\right) - 1\right] \tag{32}$$

where I_0 is the reverse saturation current; V_T is the energy equivalent [9].

At the short circuit point under STC, (32) becomes:

$$I(V)|_{V=0} = I_{ph} = I_{SC} \tag{33}$$

Thus:

$$I(V) = I_{SC} - I_0\left[\exp\left(\frac{V}{V_T}\right) - 1\right] \tag{34}$$

(14) can be expressed as:

$$V(I) = V_T \ln\left[1 + \left(\frac{I_{SC} - I}{I_0}\right)\right] \tag{35}$$

It is known that when $I = I_{SC} - I_0(e - 1)$, the observed voltage is equal to V_T, which can be expressed as follows:

$$V_T = V(I)|_{I=I_{SC}-I_0(e-1)} \tag{36}$$

4.2 A single-diode PV circuit model expressed by C_1 and C_2

Substituting (34) into the following form:

$$I(V) = I_{SC}\left\{1 - \frac{I_0}{I_{SC}}\left[\exp\left(\frac{V_{OC}}{V_T} \cdot \frac{V}{V_{OC}}\right) - 1\right]\right\} \tag{37}$$

Assuming there are two new parameters C_1* and C_2*, which are expressed as:

$$C_1^* = \frac{I_0}{I_{SC}}, \qquad C_2^* = \frac{V_T}{V_{OC}} \tag{38}$$

(37) can be derived as:

$$I(V) = I_{SC}\left\{1 - C_1^*\left[\exp\left(\frac{V}{C_2^*V_{OC}}\right) - 1\right]\right\} \tag{39}$$

4.3 Key equation to solve C_1* and C_2*

C_1* and C_2* can be derived by (39) at the open circuit point and MPP points.

1) At the open circuit point, assuming $V = V_{oc}$, $I = 0$, (39) can be derived as:

$$\exp\left(\frac{1}{C_2^*}\right) = 1 + \frac{1}{C_1^*} \tag{40}$$

For simplifying, defining

$$\begin{cases} C_3 = 1/C_1^* \\ C_4 = 1/C_2^* \end{cases} \tag{41}$$

(40) can be expressed as:

$$\exp(C_4) = 1 + C_3 \tag{42}$$

2) At the MPP point, assuming $V = V_m$, $I = I_m$, (38) can be derived as:

$$1 + (1 - \gamma_{I_m})C_3 = (1 + C_3)^{\gamma_{V_m}} \tag{43}$$

Defining

$$y = 1 + (1 - \gamma_{I_m})C_3 - (1 + C_3)^{\gamma_{V_m}} \tag{44}$$

which is the key equation to get the solution of C_3 when $y = 0$. The curve of (44) is shown in Fig. 1.

4.4 Solution of C_1^* and C_2^*

Normally, $C_3 \gg 1$ when $\gamma_{I_m} \in [0.85, 0.99]$; $\gamma_{V_m} \in [0.75, 0.85]$ [10]. Thus:

$$1 + C_3 \cong C_3 \tag{45}$$

(43) can be expressed by C_1^* according to (41) as:

$$C_1^*\left[C_1^* - \left(1 - \frac{I_m}{I_{SC}}\right)^{\frac{1}{\left(1 - \frac{V_m}{V_{OC}}\right)}}\right] = 0 \tag{46}$$

Since $C_3 \gg 1$, $C_1^* \neq 0$, the final solution is given by

$$C_1^* = \left(1 - \frac{I_m}{I_{SC}}\right)^{\frac{1}{\left(1 - \frac{V_m}{V_{OC}}\right)}} \tag{47}$$

which is same with C_1 in direct form given by (8), proving the indirect form given by (2). From (45), (42) can be written as:

$$\exp(C_4) = C_3 \tag{48}$$

Thus

$$C_2^* = (C_4^*)^{-1} = (\ln C_3^*)^{-1} = \left(\frac{V_m}{V_{OC}} - 1\right) / \ln\left(1 - \frac{I_m}{I_{SC}}\right) \tag{49}$$

which is also exactly the same with (3).

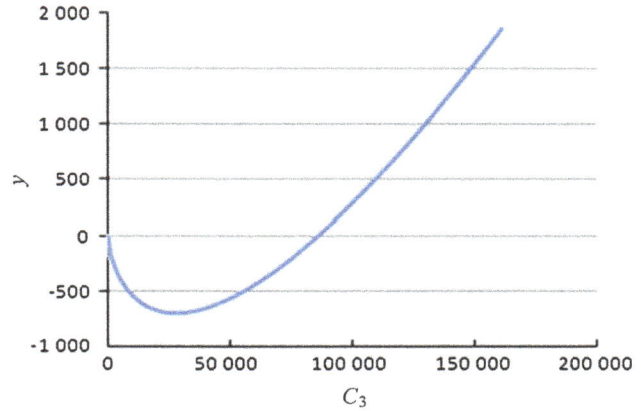

Fig. 1 Curve of (44)

4.5 I_0 and V_T in generic single-diode PV circuit model

According to (39) and (7), I_0 and V_T are given by

$$I_0 = C_1^* I_{sc} = I_{sc}(\sigma_{I_m})^{\frac{1}{\sigma_{V_m}}} \tag{50}$$

$$V_T = C_2^* V_{oc} = -V_{oc}(\sigma_{V_m}) / (\ln \sigma_{I_m}) \tag{51}$$

5 The conformity measurement of STC sub-model calculated data with manufacturer's datasheet

5.1 The error of the calculated maximum voltage in model and the provided maximum voltage

According to (20), the relative error is defined as:

$$\delta_{V_{oc}} = \frac{V_{oc}'' - V_{OC}}{V_{OC}} = \frac{V_{oc}''}{V_{OC}} - 1 = C_2 \ln\left(1 + \frac{1}{C_1}\right) - 1 \tag{52}$$

$$= -(\ln C_1)^{-1} \ln(1 + C_1^{-1}) - 1 \tag{53}$$

It is known that $\delta_{V_{oc}}$ is greater than 0 since C_1 is great than 0 from Fig. 2.

At the open-circuit point, V_{oc} calculated in 2C STC sub-model is greater than the the assumed value one, and the bias $\delta_{V_{oc}}$ greatly depends on the value of C_1. The bias will increase if C_1 increases.

Replacing C_1 by γ_{I_m} and γ_{V_m} into (53),

$$\delta_{V_{oc}} = -(1 - \gamma_{V_m}) \ln\left(1 + 1/(1 - \gamma_{I_m})^{1/(1-\gamma_{V_m})}\right) / \ln(1 - \gamma_{I_m}) - 1 \tag{54}$$

Fig. 3 shows that $\delta_{V_{oc}} \in [1.78e^{-15}, 6.67e^{-5}]$ when $\gamma_{I_m} \in [0.85, 0.99]$, $\gamma_{V_m} \in [0.75, 0.85]$. The error of the maximum voltage V_{oc}' which is calculated with the maximum

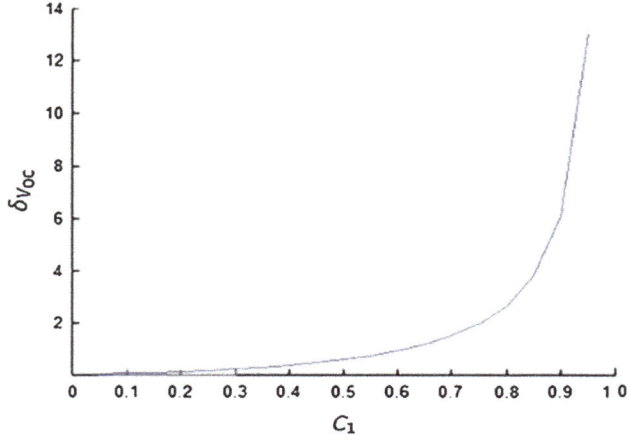

Fig. 2 Curve of (53)

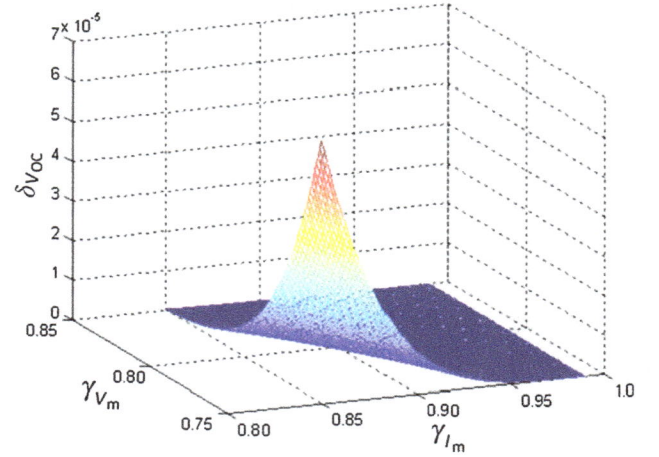

Fig. 3 Relative error $\delta_{V_{oc}}$ with γ_{I_m} and γ_{V_m} in (54)

voltage V_{oc} provided by manufactory is very small though $V'_{oc} > V_{oc}$.

5.2 The error of between I''_m and I_m at $V = V_m$ under STC

From (18) and (12), the relative error is given as:

$$\delta_{I_m} = \frac{\Delta I''_m}{I_m} = \frac{I''_m - I_m}{I_m} = \frac{I_{SC}}{I_m} C_1 > 0 \tag{55}$$

$$= \frac{1}{\gamma_{I_m}} \left(1 - \gamma_{I_m}\right)^{1/\left(1 - \gamma_{V_m}\right)} \tag{56}$$

From Fig. 4, it is shown that $\delta_{I_m} \in [4.69e^{-14}, 5.96e^{-4}]$ when $\gamma_{I_m} \in [0.85, 0.99]$, $\gamma_{V_m} \in [0.75, 0.85]$. The error of I''_m at $V = V_m$ and I_m at the MPP point is very small though $I''_m > I_m$

5.3 The error between V''_m and V_m at $I = I_m$ under STC

From (20) and (7), it can be obtained that:

$$\delta_{V_m} = -\frac{(1 - \gamma_{V_m})}{\gamma_{V_m} \ln(1 - \gamma_{I_m})} \ln\left[1 + (1 - \gamma_{I_m})^{\left(\frac{\gamma_{V_m}}{1 - \gamma_{V_m}}\right)}\right] \tag{57}$$

In Fig. 5, $\delta_{V_m} \in [1.78e^{-13}, 5.92e^{-4}]$ when $\gamma_{I_m} \in [0.85, 0.99]$, $\gamma_{V_m} \in [0.75, 0.85]$, which means the error between V''_m and V_m at the MPP point is very small though $V''_m > V_m$.

5.4 Prove of $V_m < V'_m < V''_m$ and $I_m < I'_m < I''_m$

dP/dV is decreasing monotonously and it is 0 at point (V''_m, I'_m). If dP/dV at point (V_m, I''_m) is greater than 0, and

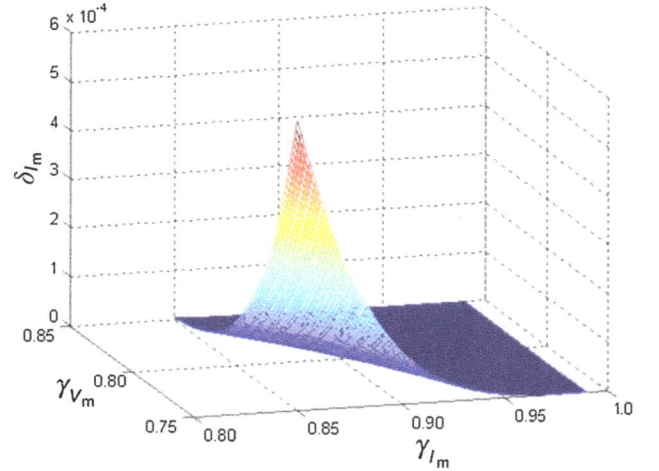

Fig. 4 Relative error δ_{I_m} with γ_{I_m} and γ_{V_m} in (56)

dP/dV at (V''_m, I_m) is smaller than 0, the above non-equality can be proved.

$$\frac{dP}{dV} = I_{SC}\left[\frac{I}{I_{SC}} - \frac{V}{C_2 V_{OC}}\left(1 + C_1 - \frac{I}{I_{SC}}\right)\right] \tag{58}$$

At point (V_m, I''_m),

$$\left.\frac{dP}{dV}\right|_{(V_m, I''_m)} = I_{SC}\left[\left(C_1 + \frac{I_m}{I_{SC}}\right) - \frac{V_m}{C_2 V_{OC}}\left(1 - \frac{I_m}{I_{SC}}\right)\right]$$

$$= I_{SC}\left\{1 + (1 - \gamma_{I_m})\left[(1 - \gamma_{I_m})^{\left(\frac{\gamma_{V_m}}{1 - \gamma_{V_m}}\right)} - 1\right.\right.$$

$$\left.\left. - \frac{\gamma_{V_m}}{1 - \gamma_{V_m}}\ln(1 - \gamma_{I_m})\right]\right\} \tag{59}$$

From Fig. 6, dP/dV is greater than 0 when $\gamma_{I_m} \in [0.85, 0.99]$, $\gamma_{V_m} \in [0.75, 0.85]$.

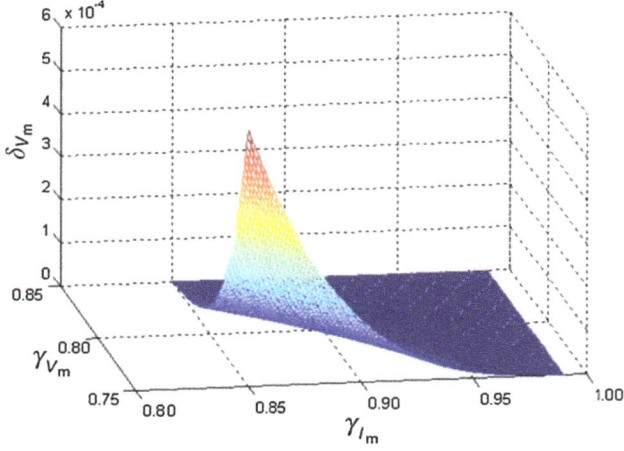

Fig. 5 Relative error δ_{V_m} with $\gamma_{I_m}\gamma_{V_m}$ and in (57)

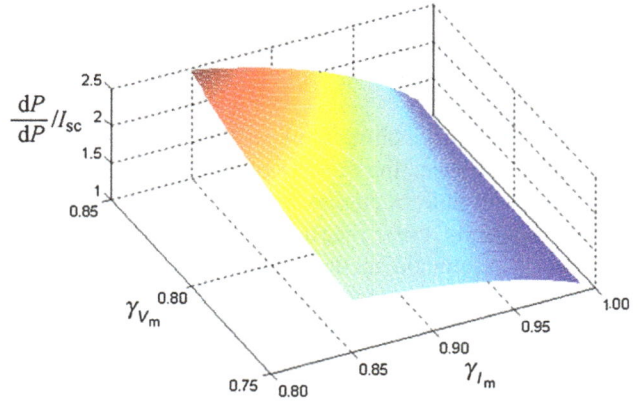

Fig. 6 Curve of dP/dV at point (V_m, I''_m) in (59)

From (17), it can be derived that

$$V''_m = C_2 V_{OC} \ln\left[1 + \frac{1}{C_1}\left(1 - \frac{I_m}{I_{SC}}\right)\right]$$
$$= C_2 V_{OC} \ln\left[1 + \frac{1}{C_1}(1 - \gamma_{I_m})\right] \quad (60)$$

Then (58) can be expressed as:

$$\frac{dP}{dV}\bigg|_{(V''_m, I_m)} = I_{SC}\left\{1 - (1 - \gamma_{I_m})\left\{1 + \left[1 + \left(1 + (1 - \gamma_{I_m})^{\frac{\gamma_{V_m}}{1-\gamma_{V_m}}}\right)\right]\right\}\right.$$
$$\left. \times \ln\left(1 + (1 - \gamma_{I_m})^{\frac{-\gamma_{V_m}}{1-\gamma_{V_m}}}\right)\right\} \quad (61)$$

The value of dP/dV at point (V''_m, I_m) is smaller than 0 when when $\gamma_{I_m} \in [0.85, 0.99]$, $\gamma_{V_m} \in [0.75, 0.85]$ in Fig. 7.

5.5 The error between the maximum power calculated in model and provided by manufactory

Defining the voltage and current at the MPP point are V'_m and I'_m $P_m = V_m I_m$, the error can be given as:

$$\delta_{P_m} = \frac{\Delta P}{P_m} = \frac{P'_m - P_m}{P_m} = (1 + C_1)\frac{I_{SC}}{I_m}\frac{V'_m}{V_m} - C_2\frac{I'_m}{I_m}\frac{V_{OC}}{V_m} - 1 \quad (62)$$

It is difficult to have an explicit expression of δ_{P_m}. It is proved above that

$$V_m < V''_m < V'''_m, \quad I_m < I''_m < I'''_m \quad (63)$$

From (56) and (57), it is derived that

$$\frac{I''_m}{I_m} = \delta_{I_m} + 1 < \max(\delta_{I_m}) + 1 = 1.000,596 \quad (64)$$

$$\frac{V''_m}{V_m} = \delta_{V_m} + 1 < \max(\delta_{V_m}) + 1 = 1.000,592 \quad (65)$$

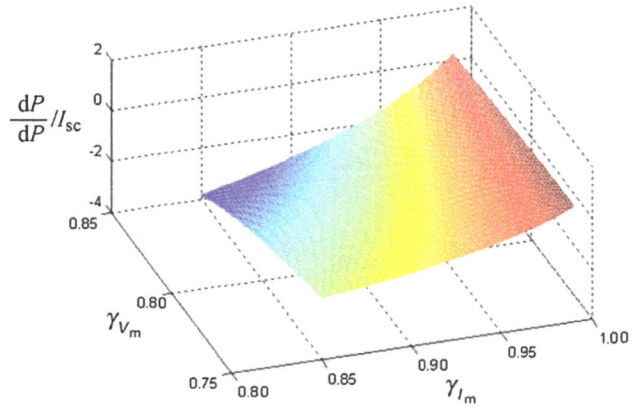

Fig. 7 Curve of dP/dV at point (V''_m, I_m) in (61)

Thus:

$$\delta_{P_m} = \frac{\Delta P}{P_m} = \frac{P''_m}{P_m} - 1 = \frac{I'_m V'_m}{I_m V_m} - 1 < \frac{I''_m V''_m}{I_m V_m} - 1$$
$$< (\max(\delta_{V_m}) + 1)(\max(\delta_{I_m}) + 1) - 1$$
$$= 1.000,596 * 1.000,592 - 1 = 0.001,19 = 0.119\% \quad (66)$$

which means the model relative error of the calculated maximum power is small and negligible.

6 Conclusion

1) 2C model is a single-diode circuit model, since it can be derived to a single-diode circuit model theoretically. (39) defines the original physical parameters of 2C, i.e., C_1 is the ratio of reverse saturation current I_0 to the short circuit current I_{sc}, C_2 is the ratio of energy equivalent V_T to the open circuit voltage V_{oc}.

2) The expressions of C_1 and C_2 in manufacturer's datasheet of 2C model are mostly correct under a reasonable approximation assumption for almost manufacturer's PV modules.

3) Though there are contradictions of 2C model with manufacturer's datasheet, the relative errors are very small, which can be negligible in engineering application.

4) The conformity error measurement method by 3D curves gives a systematic method for power community users to be aware of the conformity errors of their own PV modules by using 2C model in real application.

Therefore, the calculate data in 2C model is almost the same with the manufacturer's datasheet under STC. If applying 2C model in a real application, it is necessary to find other PV module models.

Acknowledgement This work was partially supported by Key Science, Technology Project of Zhejiang Province (LZ12E07001) and National Natural Science Foundation of China (51307038).

References

[1] Borowy BS, Salameh ZM (1994) Optimum photovoltaic array size for a hybrid wind/PV system. IEEE Trans Energ Convers 9(3):482–488

[2] Borowy BS, Salameh ZM (1996) Methodology for optimally sizing the combination of a battery bank and PV array in a Wind/PV hybrid system. IEEE Trans Energ Convers 11(2):367–375

[3] Mao MQ, Yu SJ, Su JH (2005) Versatile Matlab simulation model for photovoltaic array with MPPT function. J Syst Simul 17(5):1248–1251 (in Chinese)

[4] Mao MQ, Su JH, Chang LC et al (2009) Research and development of fast field tester for characteristics of solar array. In: Proceedings of the Canadian conference on electrical and computer engineering (CCECE'09), St John's, 3–6 May 2009, pp 1055–1060

[5] Zhang JM, Zhang QZ, Wang N et al (2011) Power generation model and its parameter calibration for grid-connected photovoltaic power plant energy data acquisition and supervisory system. Automat Electr Power Syst 35(13):22–26 (in Chinese)

[6] Zhang QZ, Zhang JM, Guo CX (2012) Photovoltaic plant metering monitoring model and its calibration and parameter assessment. In: Proceedings of the 2012 IEEE Power and Energy Society general meeting, San Diego, 22–26 July 2012, 7 pp

[7] Han XN, Ai X, Sun YY (2014) Research on large-scale dispatchable grid-connected PV systems. J Mod Power Syst Clean Energy 2(1):69–76 (in Chinese)

[8] Li YW, Nejabatkhah F (2014) Overview of control, integration and energy management of microgrids. J Mod Power Syst Clean Energy 2(3):212–222 (in Chinese)

[9] Masters GM (2004) Renewable and efficient electric power systems. Wiley, Hoboken

[10] Download solar panel datasheets.

Jianmin ZHANG received B.Sc. degree and M.Sc. degree from Huazhong University of Science and Technology (HUST), China, in 1984 and 1987, respectively. He received M.E. degree from Indian Institute of Technology (IIT, Roorkee) in electrical engineering in 1992. He joined Hangzhou Regional Center of Small Hydro Power (HRC) and National Institute of Rural Electrification, China, from 1987 to 1997. He is a full professor and academic leader of Electrical Engineering and Automation at Hangzhou Dianzi University. His research interests include electric power and energy system modeling, optimal operation and dispatching, intelligence engineering and automation and information system integration.

Qianzhi ZHANG received B.Sc. degree in electrical engineering from Shandong University of Technology in 2012. Presently he is a research assistant, pursuing master science of electrical engineering, in Ira A. Fulton Schools of Engineering, Arizona State University. His research interests include the renewable energy integration with power system and their planning, design, control and operation.

8

FLECH: A Danish market solution for DSO congestion management through DER flexibility services

Chunyu ZHANG, Yi DING (✉),
Niels Christian NORDENTOFT,
Pierre PINSON, Jacob ØSTERGAARD

Abstract Future electric power systems will face new operational challenges due to the high penetration of distributed energy resources (DERs). In Denmark, distribution system operator (DSO) expects a significant congestion increased in distribution grids. In order to manage these congestions and mobilize the DERs as economically efficient as possible in the future distribution grid, the brand new notion of flexibility clearing house (FLECH) is proposed in this paper. With the aggregator-based offers, the proposed FLECH market has the ability to promote small scale DERs (upto 5 MW) for actively participating in trading flexibility services, which are stipulated accommodating the various requirements of DSO. Accordingly, the trading setups and processes of the FLECH market are also illustrated in detail. A quantitative example is utilized to illustrate the formulation and classification of flexibility services provided by the DERs in the proposed FLECH market.

Keywords Electricity market, Distributed energy resources (DERs), Aggregator, Flexibility clearing house (FLECH), Flexibility services

1 Introduction

In the long term perspective, the renewable energy (wind, photovoltaic, etc.) and more efficient energy resources (combined heat and power, heat pump, electric vehicle, etc.) will critically improve the security of energy supply by drawing upon sustainable natural sources and reducing environmental impacts [1, 2]. The vast majority of previous and ongoing renewable energy resources and smart grid projects have focused on demonstrating the technical feasibility of these distributed energy resources (DERs) [3, 4] in the distribution grids. The high penetration of DERs is considerably observed worldwide. For instance, by 2020, the share of renewable energy in Denmark must be increased to at least 35 % of final energy consumption—50 % of electricity consumption supplied by wind power [5, 6].

In Denmark and other European Union (EU) countries, the rapid growth of these intermittent DERs will pose a significant challenge associated with congestion issues in distribution grids. Currently, depending on the existing electricity market structure, there are exclusively several approaches for congestion management specific to transmission system operator (TSO), which could be categorized into three types: (1) The optimal power flow (OPF) based method, which is based on a centralized optimization and is considered to be the most accurate and effective congestion management method. (2) The price area congestion control method, which eliminates congestions by generation rescheduling schemes according to congestion price-signals refer to the framework of OPF [7]. (3) The transaction based method, which incorporates the support of point-to-point tariffs for pool markets, based on pay as bid mechanism to provide price signals to promote the maximum use of the existing transmission network. However these congestion management approaches may not appropriate to the distribution grid with two principal reasons: (1) The existing market price system may not fully cover the benefit of DER owners. (2) Dispatching of distribution network is more complex due to a large amount of

C. ZHANG, Y. DING, P. PINSON, J. ØSTERGAARD, Center for Electric Power and Energy, Technical University of Denmark, 2800 Kgs. Lyngby, Denmark
(✉) e-mail: yding@elektro.dtu.dk
N. C. NORDENTOFT, Danish Energy Association, 1970 Frederiksberg, Denmark

small scale fluctuating DERs with high diversity and dispersal allocation.

Concerning that, the further development and innovation of electricity market will play an essential role on utilizing the DERs as economically efficient as possible.

Over the past decade, with the emerging new market player of virtual power plant (VPP), researches are mainly emphasized on enabling DERs to participate in the existing market, especially to provide different kinds of ancillary services to TSO. The EU project FENIX defines VPP as a flexible representation of a portfolio of DERs that can be used to make contracts in the wholesale market and to offer services to the system operator [8]. There are two types of VPP, the commercial VPP (CVPP) and the technical VPP (TVPP). The CVPP is a competitive market actor that manages the DER portfolios to make optimal decisions on participation in wholesale markets. The TVPP aggregates and models the response characteristics of a system containing DERs, controllable loads and networks within a single grid [9]. In other words, the CVPP optimizes its portfolio with reference to the wholesale markets, and passes DER schedules and operating parameters to the TVPP. The TVPP uses input from the CVPPs operating in its area to manage any local network constraints and determine the characteristics of the entire local network at the grid supply points [10]. Thus, the role of TVPP in distribution networks is the same as the TSO's role in transmission systems.

In recent years, aiming to stimulate the small scale DERs into the existing market structure, the real-time market demonstrated with several ongoing smart grid projects subdivides the time scale into 5 min intervals for providing balancing services to TSO, e.g. the EcoGrid EU market complies with a bidless clearing process, the Olympic Peninsula market and the PowerMatcher energy management system employ agents to submit bids to an auction for establishing an equilibrium price [2].

However, these frameworks may not be consistent with the future scenarios: The primary task of TSO is to avoid system-wide imbalance occurring, while the executive issue for DSO is to relieve the congestions in local network. In addition, the size limitations are often cited as another large barrier for small scale DERs (upto 5 MW) to access the wholesale electricity market, e.g. 10 MW in Nordpool market. Therefore, we have to pave a novel way for fully utilizing the advantages of small scale DERs—Focusing on the distribution grid, proper coordination and activation of consumers and DERs will provide more flexibility in ancillary services, which can enhance economical efficiency and reliability of distribution system.

In this paper, a FLECH market is proposed to give a shot for the feasibility of promoting small scale DERs to participate in flexibility services trading. With the aggregator-based service offers, the proposed FLECH market will

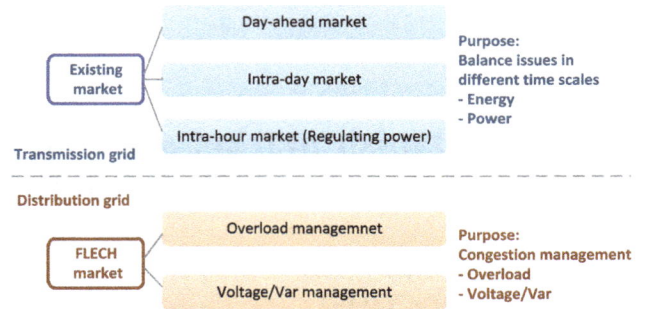

Fig. 1 Scope of the existing market and FLECH market

easily satisfy the congestion management of DSO and benefit DER owners, while further facilitate the integration of DERs into power system. Meanwhile, the flexibility services provided by DERs will expand the properties of existing ancillary services, conducive to the security and stability of distribution network.

2 FLECH market

2.1 Scope

The existing wholesale electricity trade in Nordic power system is NordPool, which consists of day-ahead market, intra-day market, and intra-hour (regulating power) market corresponding to different balancing issues in various time scales, as shown in Fig. 1. The first settlement between energy supply and demand in a given hour (in next 24 h) of operation happens in the day-ahead market, then the price where the expected production meets the predicted consumption is obtained. This price commonly referred to system price or spot price. Obviously, as the hour of operation approaches, this targeted balance might need adjustment as the expectations regarding power plant schedule changes or demand fluctuations. Therefore, a new settlement between production and consumption is obtained, first on the intra-day energy market and then on the intra-hour regulating power market. The regulating power market is responsible for physical electricity trading in 15–60 min prior to the hour of operation.

With the high penetration of DERs and massive growth of smart grid projects, congestion becomes one of the most challenging operation issues in Danish distribution grid. The proposed FLECH market is a parallel running market with the existing markets specializing in the distribution grid, in order to assist DSO to mitigate the congestions and revitalize the DER economy, which is depicted in Fig. 1 as well. Generally, the main congestion management could be classified as follows:

1) Feeder overload management: The high power flow comes over feeder capacity-limit, which can be caused by regular growth in electricity consumption, mobilization of reserve capacity in the grid, activation of regulating power for the TSO, or very low prices of electricity.

2) Feeder voltage/Var management: The oscillations of voltage and issues of reactive power can be exceeded the band of feeder deviation limits, which normally caused by the variations of local generation or demand.

2.2 Architecture

The parallel architecture of the proposed FLECH market with existing market is shown in Fig. 2. The demand-side DERs and consumers can be re-categorized as flexibility provider and ordinary electricity consumer. The aggregators will harvest the individual flexibility to formulate various types of flexibility services and trade in FLECH market to satisfy the appropriate requirements of DSO congestion management, illustrated in the left dashed box of Fig. 2. As indicated in the right portion, the wholesale electricity produced by conventional supply and delivered through TSO is still settled in the existing market with balance responsible party, further consumed by the ordinary electricity demand via retailers as usual. Therefore, the FLECH market and the existing market could coexist in time and space focusing on different issues of congestion and balance, respectively. It could be also observed that the aggregator and flexibility clearing house (FLECH) are the brand new participants.

1) The aggregator, which is a new commercial player, has three basic functions:

   - Assemble and mobilize the flexibility of DERs, pack and schedule flexibilities from individual DER, and provide the service offers to the highest possible bidder with contract.
   - Have thorough knowledge of the electricity markets, put the right price on the flexibility services, and represent DERs to trade in FLECH market.
   - Paid by the DSO for delivering flexibility services. From this payment the aggregator will pay his affiliated DERs according to their contractual agreements.

2) The FLECH: which is an independent non-profit driven entity, also responsible for:

   - Make the standardized contracts with DSO and aggregators by stipulating service category.

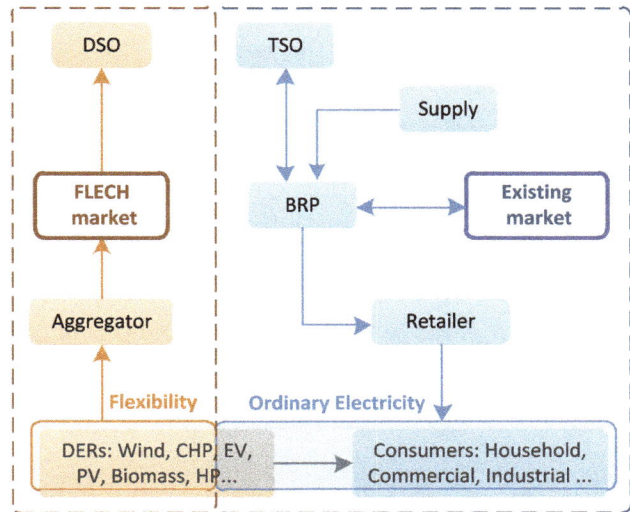

Fig. 2 Parallel architecture of the FLECH market and the existing market

Table 1 The liquidity features of FLECH market

Factors	Liquidity
Energy/power flow	DERs → Aggregator → DSO
Capital flow	DSO → FLECH → Aggregator → DERs
Control signal	DSO → Aggregator → DERs
Information signal	DSO ↔ FLECH ↔ Aggregator ↔ DERs
Physical network	DERs → Distribution gird

   - Ensure the FLECH market integrity by mitigating counterparty default risk, and monitors the contracts are being carried out more targeted and efficiently.
   - Provide clearing of all contracts traded on the exchange, which is an ex post financial settlement.

It could be further observed that the framework of FLECH market is concise and efficient, in which the supplier is DERs while the consumer is DSO, totally inverts the roles with the prevailing market. Correspondingly, the new liquidity features of FLECH market can be summarized in Table 1.

3 Market trading

3.1 Trading setups

The core missions for FLECH are contracts regulation and ex post financial settlement, there are two possible trading setups and identified as follows:

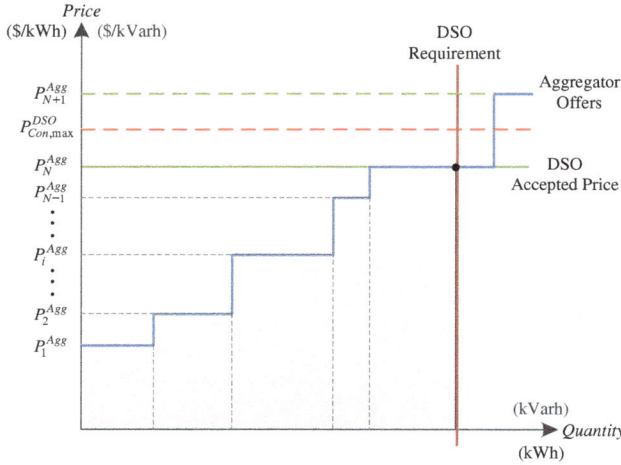

Fig. 3 Single-side aggregator auctions (SAA)

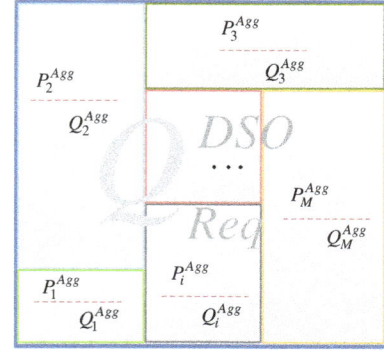

Fig. 4 Super market (SM)

1) Single-side aggregator auctions (SAA): With the vast volume in trading flexibility services, an auction-based setup will arise. The DSO proposes the request quantity of each sort of flexibility service, and the aggregators will submit the offers to satisfy the corresponding services of DSO. Finally, the DSO chooses the available offers appropriately from the aggregators, and the standardized contracts are automatically formed according to the market-rules of FLECH.

More precisely, this trading setup could be referred to SAA. Offers from aggregators are ranked in increasing order and accepted beginning with the least expensive and continuing until the DSO is satisfied, as what illustrated in Fig. 3.

The marginal price for each aggregator offer block can be calculated by

$$P_i^{Agg} = \left(C_i^R + C_i^A + C_i^O + C_i^P + C_i^{Au}\right)\Big/Q_i^{Agg}, \quad i \in M \tag{1}$$

where M is the total number of the aggregators to participating in SAA; for ith aggregator, P_i^{Agg} is the marginal price of a certain type of flexibility service ($\$/kWh$)/($\$/kVarh$); C_i^R is the service reservation cost ($\$$); C_i^A is the total amount of activation cost ($\$$); C_i^O is the operation cost on assembling, scheduling, mobilizing, and transacting with affiliated DERs ($\$$); C_i^P is the possible penalty cost if the service failure or against the rules of contract ($\$$); C_i^{Au} is the uncertainty cost according to communication, policies, enforcement, etc. ($\$$); and Q_i^{Agg} is the maximum production of active or reactive energy (kWh)/(kVarh).

Herein, P_N^{Agg} is the last accepted block and its marginal price, P_{N+1}^{Agg} is the first rejected offer, the uniform price for clearing is then set equal to P_N^{Agg}. Note that, the selection of P_N^{Agg} should firstly meet the DSO willing price $P_{Con,max}^{DSO}$ (the red dotted line), i.e. $P_N^{Agg} \leq P_{Con,max}^{DSO}$. This willing price is applicable to all categories of flexibility services, could be computed by

$$P_{Con,max}^{DSO} = \left(C_{Con}^I + C_{Con}^L + C_{Con}^O + C_{Con}^{Du}\right)\Big/Q_{Con}^{DSO} \tag{2}$$

where, for dealing with the congestions in the whole range of distribution grid, $P_{Con,max}^{DSO}$ is the highest price for DSO willing to pay ($\$/kWh$)/($\$/kVarh$); C_{Con}^I is the investment cost by updating new feeders, substations or Var devices ($\$$); C_{Con}^L is the load curtailment cost ($\$$); C_{Con}^O is the operation cost for re-dispatching and grid losses ($\$$); C_{Con}^{Du} is the uncertainty cost according to communication, policies, transactions, etc. ($\$$); and Q_{Con}^{DSO} is the maximum requirement of active power or reactive power in this distribution grid (kWh)/(kVarh).

2) Super market (SM): In opposite to the SAA auctions, the aggregators have the initiative in SM setup. Considering the historical data, the aggregators will be able to estimate where and how much the DSO might be interested in purchasing the desired flexibility services. Then, the aggregators could propose and price various services, just like in the "super market", the DSO is the consumer of these flexibility services willing to choose their favourite commodities.

Furthermore, this SM trading setup could be formulated by a portfolio optimization problem [11], as indicated in Fig. 4. Q_{Req}^{Agg} is the total required active or reactive energy of DSO (kWh)/(kVarh). The expected value of DSO willing price (i.e. clearing price) can be expressed as the weighted average of the individual expected price of each aggregator,

$$P_{Exc}^{DSO} = \sum_{i=1}^{K} P_i^{Agg} \omega_i \tag{3}$$

$$\omega_i = Q_i^{Agg}\Big/Q_{Req}^{DSO} \tag{4}$$

where, for a type of flexibility service to deal with the congestions in the whole range of distribution grid, P_{Exc}^{DSO} is the DSO excepted price ($\$/kWh$)/($\$/kVarh$); K is the total number of the aggregators to participating in SM; for ith aggregator, P_i^{Agg} is the expected price (\$/kWh)/(\$/kVarh); ω_i is the share in this portfolio (%); Q_i^{Agg} is the production of active or reactive energy (kWh)/(kVarh).

Subsequently, the optimal value of ω_i can be calculated by an optimization problem with the objective of minimizing the DSO portfolio investment risk, denoted as σ_p and described as

$$\text{Min } \sigma_p = \left[\sum_{i=1}^{K} \sum_{j=1}^{K} \omega_i \omega_j \sigma_i \sigma_j \rho_{ij} \right]^{1/2} \quad (5)$$

$$\text{s.t. } \sum_{i=1}^{K} \omega_i = 1 \quad (6)$$

$$0 \le \omega_i \le \omega_{i,\max} \quad (7)$$

$$\omega_{i,\max} = Q_{i,\max}^{Agg} / Q_{Req}^{DSO} \quad (8)$$

$$\rho_{ij} = 1, \quad \text{if } \omega_i = \omega_j \quad (9)$$

where (6)–(9) are the constraints of proportional allocation of individual aggregator in this portfolio. For ith aggregator, $Q_{i,\max}^{Agg}$ is the maximum output of this type of flexibility service, $\omega_{i,\max}$ is the upper limit of the proportion in this portfolio. σ_i and σ_j are the standard deviations corresponding to the holding period returns of annual costs of

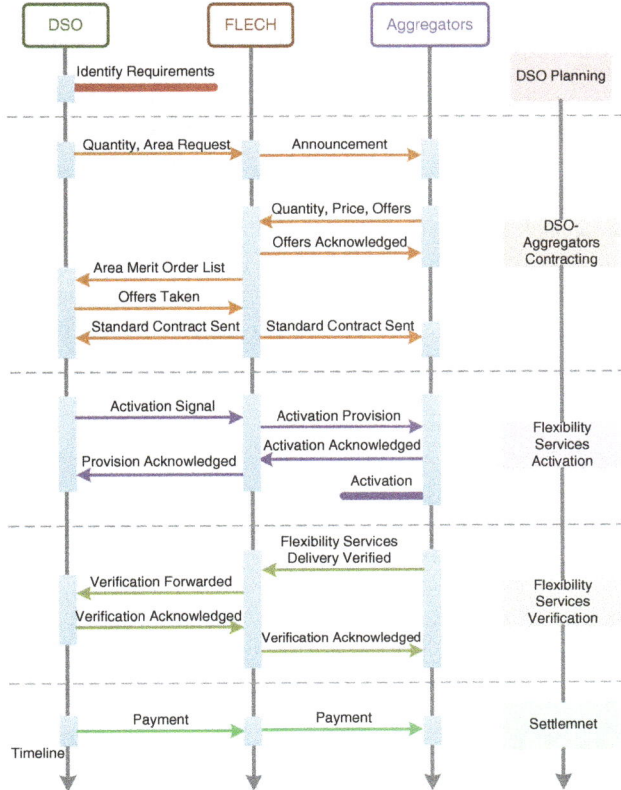

Fig. 5 SAA-based trading processes in FLECH [12]

the ith and jth aggregators, respectively, and ρ_{ij} is the correlation among them.

These two trading setups do not necessarily replace each other, but will be a mutually beneficial co-existence according to their own merits.

3.2 Trading processes

1) SAA-based trading process:
   The SAA-based trading processes in FLECH are shown in Fig. 5 and depicted below:

   - DSO planning: during the DSO year-ahead planning and scenario analysis, the requirement of flexibility services can be identified, including service category, area, location, quantity, and activation numbers.
   - DSO-aggregators contracting: the DSO posts the desired flexibility services at FLECH with a deadline for aggregators to submit offers. Then FLECH announces this information on the website. Accordingly, aggregators will pre-schedule the affiliated DERs and submit flexibility service offers to FLECH with explicit quantity, price and maximum activation numbers. DSO gets the area merit order list and assesses the feasibility of various offers based on OPF. By trading off the grid reinforcement or flexibility service purchase by (2), if DSO could see the substantial benefits in mobilizing flexibility services, the desirable offers will be taken and standard contracts will be made.
   - Flexibility services activation: when the contractual service period is coming, the DSO will activate flexibility services as specified in the contract, if necessary. Aggregators will schedule, optimize and coordinate DERs refer to contracted flexibility services.
   - Flexibility services verification: till the contractual service period is over, a bilateral verification between DSO and aggregators will be carried out to authorize the exactly delivered flexibility services.
   - Settlement: according to the authorized flexibility services, settlement between the DSO and involved aggregators will be accomplished, simultaneously, the mutual contractual obligations have been fulfilled.

2) SM-based trading process:
   The processes of SM-based trading in FLECH are similar with SAA-based trading procedures in flexibility services activation, verification, and settlement, but differ over the first two processes, as illustrated in Fig. 6.

- Aggregator forecasting: on basis of the historical data, the aggregators will forecast the DSO favourite flexibility services in prior day/month and identify the productions by pre-scheduling the affiliated DERs.
- DSO-aggregators contracting: the aggregators submit the flexibility services bids to FLECH with service category, area, location, quantity, price and maximum activation numbers. FLECH will announce the bids information on the website after acknowledgement. Then the DSO will make flexibility services portfolio optimization adapted to his willing price refer to (3)–(9). Whereby, the preferred bids are taken, the DSO will stipulate the types of flexibility services and sign standard contracts with aggregators in FLECH.

4 Flexibility services

As mentioned above, aiming to explore FLECH market solution to relieve the congestions in distribution grid, contractual flexibility services offered by aggregators are

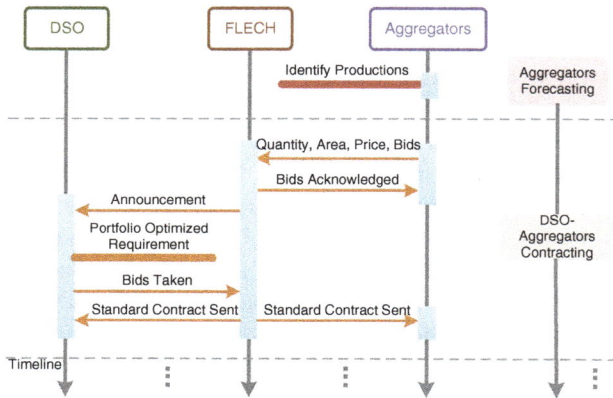

Fig. 6 First two sections of SM-based trading processes in FLECH

the backbone. Therefore, the stipulated flexibility services due to different congestion management categories are elaborated in this section.

4.1 Overload management

For satisfying DSO requests of feeder overload management, five types of flexibility services feasibly provided by aggregators are defined as FS_{OP}, FS_{OU}, FS_{OR}, FS_{OC} and FS_{OM}. The expected service effectiveness for individual flexibility service is shown in solid orange line in Fig. 7, additionally, the desired quantity of each service for eliminating peak load, i.e. Q_{FSOP}, Q_{FSOU}, Q_{FSOR}, Q_{FSOC} and Q_{FSOM}, is also shown in the shaded area, respectively.

- FS_{OP} is suitable for handling the predictable peak loads for periodically daily capacity issues, if the distributions grid experiences the highest load and the hourly load patterns could be forecasted at each feeder, then the DSO will desire the load reduction service from aggregators hourly. This service will ensure the whole overload situation under 70 % capacity limit, previously activated before the load touching this limit and terminated later than the overload gone, show in Fig. 7a.
- FS_{OU} is an event-based flexibility service, which looks similar to FS_{OP} but will be activated sharply when overload starts, shown in Fig. 7b. This service will be less frequently activated every day during contractual period.
- FS_{OR} has the ability of exploiting the new reserve-supply within the feeder capacity limit of 70 %–100 %. Moreover, in view of un-locking this expansion of available capacity, it is necessary to reduce loads when facing such a situation. However, this type of flexibility service will be rarely activated as it will only be served when a neighbouring feeder get faulted plus the load

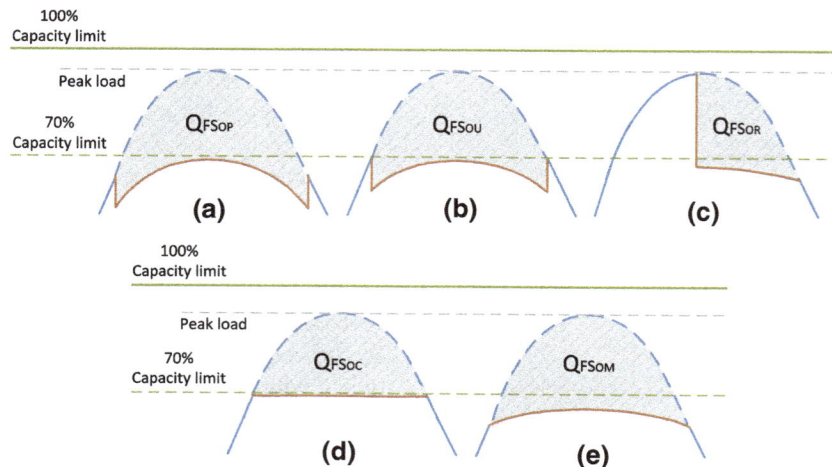

Fig. 7 Flexibility services offered by aggregators for overload management. **a** FS_{OP}. **b** FS_{OU}. **c** FS_{OR}. **d** FS_{OC}. **e** FS_{OM}

Table 2 Quantitative sample of flexibility services

	Service	FS_{OP}	FS_{OU}	FS_{OR}	FS_{OC}	FS_{OM}	FS_{VS}	FS_{VV}
1	Contract valid	Dec. 1–31	Dec. 1–31	Dec. 1–31	Dec. 1–31	Dec. 1–31	Jan. 1–31	Jan. 1–31
2	Area-location	240–24791	270–27791	420–46791	310–34891	324–324791	261–265891	108–189125
3	Activation-time	–	15 min	5 min	10 min	–	5 min	5 min
4	Activation-duration	4 h	3 h	3 h	3 h	4 h	2 h	2 h
5	Activation-quantity	200 kWh	–	–	–	200 kWh	250 kVarh	500 kVarh
6	Activation-number	20	15	1	10	20	30	18
7	Price	3 $/kWh	10 $/kWh	6 $/kWh	4 $/kWh	3 $/kWh	2 $/kVarh	1 $/kVarh
8	Failure-penalty	4 $/kWh	15 $/kWh	8 $/kWh	5 $/kWh	4 $/kWh	4 $/kVarh	2 $/kVarh

exceeding the 70 % capacity limit during the exactly hours of a year. Once this situation occurs, DSO can open the spare capacity of this feeder as a reserve to keep an adequate supply. In case the surge beyond the 100 % capacity limit, FS_{OR} service should be activated timely to hold the peak load back under 70 % capacity limit, as shown in Fig. 7c.

- FS_{OC} pledges a feeder capacity limit specified by the DSO will not be violated, see the flat solid orange line nearly 70 % in Fig. 7d.
- FS_{OM} means that the aggregators have the obligation to guarantee their local portfolio will not exceed a certain quantity which identified by DSO, as shown in Fig. 7e.

4.2 Voltage management

For serving the DSO with voltage stability support, there are two flexibility services can be offered to ensure that the respective feeders stay within a proper voltage band (e.g. ±10 %).

- FS_{VS} will be specified by DSO in different voltage levels with the best knowledge of grid state, and the contracted aggregators have to ensure these voltages will not beyond the limits.
- FS_{VV} mobilizes the aggregators to cooperate with the reactive power control of DSO, primarily for the voltage of transformers maintained in the particular limits.

4.3 Quantitative example

For each flexibility service, a set of contractual prerequisites should be explicitly stipulated to achieve an efficient and economic operation, including service price, area, location, duration, activations, failure and penalty

statement, etc. Certain feeders in a 10 kV distribution grid are taken for a sample to further illustrate the stipulations of these services, shown in Table 2.

5 Conclusion

A FLECH market is proposed to deal with the rising requirements of DSO congestion management. The parallel running structure with existing market makes it possible to perform their duties, namely, eliminate the congestions in distribution grid and circumvent imbalances in transmission grid, respectively. Furthermore, the SAA and SM trading setups and processes of FLECH market are analyzed exhaustively. In addition, the typically defined flexibility services, i.e. FS_{OP}, FS_{OU}, FS_{OR}, FS_{OC}, FS_{OM}, FS_{VS} and FS_{VV}, are categorized and described comprehensively with a quantitative sample. The proposed FLECH market shows its superiority to benefit DER owners and facilitate DSO dispatch and operation, which will contribute to improve the stability and reliability of distribution grid even transmission grid.

Acknowledgments The authors gratefully acknowledge the financial supports of Danish national project iPower and the great contributions of Lars Henrik Hansen, Poul Brath and Peder Dybdal Cajar from DONG Energy.

References

[1] Omu A, Choudhary R, Boies A (2013) Distributed energy resource system optimisation using mixed integer linear programming. Energ Policy 61:249–266

[2] Ding Y, Pineda S, Nyeng P et al (2013) Real-time market concept architecture for EcoGrid EU—A prototype for European smart grids. IEEE Trans Smart Grid 4(4):2006–2016

[3] Hansen H, Hansen LH, Johannsson H et al (2013) Coordination of system needs and provision of services. In: Proceedings of the 22nd international conference and exhibition on electricity distribution(CIRED'13), Stokholm, Sweden, 10–13 Jun 2013, 4 pp

[4] Romanovsky G, Xydis G, Mutale J (2011) Participation of smaller size renewable generation in the electricity market trade in UK: Analyses and approaches. In: Proceedings of the 2nd IEEE PES international conference and exhibition on innovative smart grid technologies (ISGT Europe'11), Manchester, UK, 5–7 Dec 2011, 5 pp

[5] O'Connell N, Wu QW, Østergaard J et al (2012) Day-ahead tariffs for the alleviation of distribution grid congestion from electric vehicles. Electr Power Syst Res 92:106–114

[6] Danish climate and energy policy. Danish Energy Agency, 2012

[7] Glatvitsch H, Alvarado F (1998) Management of multiple congested conditions in unbundled operation of a power system. IEEE Trans Power Syst 13(3):1013–1019

[8] Kieny C, Berseneff B, Hadjsaid N et al (2009) On the concept and the interest of virtual power plant: Some results from the European project Fenix. In: Proceedings of the IEEE Power and Energy Society general meeting(PES'09), Calgary, Canada, 26–30 Jul 2009, 6 pp

[9] Mashhour E, Moghaddas-Tafreshi SM (2011) Bidding strategy of virtual power plant for participating in energy and spinning reserve markets–Part I: problem formulation. IEEE Trans Power Syst 26(2):949–956

[10] Peik-Herfeh M, Seifi H, Sheikh-El-Eslami MK (2013) Decision making of a virtual power plant under uncertainties for bidding in a day-ahead market using point estimate method. Int J Electr Power Energ Syst 44(1):88–98

[11] Fang Y, Lai KK, Wang S (2008) Fuzzy portfolio optimization: theory and methods. Springer, Berlin

[12] Nordentoft NC, Ding Y, Zhang CY et al (2013) Development of a DSO market. Final Report of iPower project WP3.8, Copenhagen, Denmark

Chunyu ZHANG received the B.Eng. and M. Eng. degrees from North China Electric Power University, Beijing, China, in 2004 and 2006, respectively, both in electrical engineering. Since then he joined National Power Planning Center of China as a senior engineer till 2012. He is currently pursuing the Ph.D. degree at the Center for Electric Power and Energy, Technical University of Denmark (DTU), Denmark. His research interests include power systems planning and power market innovation.

Yi DING received the B.Eng. degree from Shanghai Jiaotong University, China, and the Ph.D. degree from Nanyang Technological University, Singapore, both in electrical engineering. He is an Associate Professor in the Department of Electrical Engineering, Technical University of Denmark (DTU), Denmark. His research interests include power systems reliability/performance analysis, and smart grid performance analysis.

Niels CHRISTIAN NORDENTOFT received the M.Sc. degree from Technical University of Denmark (DTU), Denmark, in 2007, in electrical engineering. He is currently working as a consultant at Danish Energy Association. His research interests include Distribution grids, smart grid and power market innovation.

Pierre PINSON received the M.Sc. degree in applied mathematics from INSA Toulouse, France, and the Ph.D. degree in Energy from Ecole des Mines de Paris. He is a Professor in modeling of Electricity Markets at the Technical University of Denmark, Department of Electrical Engineering. His research interests include among others forecasting, uncertainty estimation, optimization under uncertainty, decision sciences, and renewable energies.

Jacob ØSTERGAARD received the M.Sc. degree in electrical engineering from the Technical University of Denmark (DTU), Lyngby, Denmark, in 1995. He was with Research Institute of Danish Electric Utilities for 10 years. Since 2005, he has been Professor and Head of Centre for Electric Technology, DTU. His research interests cover smart grids with focus on system integration of renewable energy and distributed energy resources, control architecture for future power system, and flexible demand.

9

A self-learning TLBO based dynamic economic/environmental dispatch considering multiple plug-in electric vehicle loads

Zhile YANG, Kang LI (✉), Qun NIU,
Yusheng XUE, Aoife FOLEY

Abstract Economic and environmental load dispatch aims to determine the amount of electricity generated from power plants to meet load demand while minimizing fossil fuel costs and air pollution emissions subject to operational and licensing requirements. These two scheduling problems are commonly formulated with non-smooth cost functions respectively considering various effects and constraints, such as the valve point effect, power balance and ramp rate limits. The expected increase in plug-in electric vehicles is likely to see a significant impact on the power system due to high charging power consumption and significant uncertainty in charging times. In this paper, multiple electric vehicle charging profiles are comparatively integrated into a 24-hour load demand in an economic and environment dispatch model. Self-learning teaching-learning based optimization (TLBO) is employed to solve the non-convex non-linear dispatch problems. Numerical results on well-known benchmark functions, as well as test systems with different scales of generation units show the significance of the new scheduling method.

Keywords Economic dispatch, Environmental dispatch, Plug-in electric vehicle, Self-learning, Teaching learning

based optimization, Peak charging, Off-peak charging, Stochastic charging

1 Introduction

One of the key operational activities in the power system is to schedule power production according to the predicted load demands. Dynamic economic and environmental dispatches (DEED) are both crucial objectives in this scheduling task because a small percentage improvement may potentially bring significant cost savings and operational improvements [1, 2]. The goal of economic dispatch is to minimize the cost by determining the power production of thermal power plant units, while managing system constraints, balancing power production and load demand, and meeting plant operational requirements, e.g. ramp rates. Similarly, the objective of environmental dispatch is to minimize emissions under the same system constraints. Both dispatch problems are difficult to solve due to the non-smooth non-convex DEED formulations in cost functions and constraints, especially for large power systems [3].

Fast development of renewable power sources and changes in load demand bring more planning and operational uncertainties to the grid [4]. Plug-in electric vehicles (PEVs) with high penetrations are potentially important participants in the power system due to the additional large load requirements [5, 6]. The stochastic charging of PEVs may also significantly affect the distribution grid as well as increase the generation costs and pollutants emissions if not managed efficiently. It is therefore of importance to measure the impact of different PEV charging scenarios on the power system and intelligently schedule and dispatch power generation using an optimized DEED system approach.

Z. YANG, K. LI, School of Electrical, Electronics and Computer Science, Queen's University Belfast, Belfast BT9 5AH, UK
(✉) e-mail: k.li@qub.ac.uk
Z. YANG
e-mail: zyang07@qub.ac.uk
Q. NIU, Shanghai Key Laboratory of Power Station Automation Technology, School of Mechatronic Engineering and Automation, Shanghai University, Shanghai 200072, China
Y. XUE, NARI Group Corporation (State Grid Electric Power Research Institute), Nanjing 211106, China
A. FOLEY, School of Mechanical and Aerospace Engineering, Queen's University, Belfast BT9 5AH, UK

In this paper, four different PEV charging scenarios are modeled using charging time probability distribution, based on PEV charging data from Electric Power Research Institute (EPRI), a full peak charging scenario and a full off-peak charging scenario and a stochastic charging scenario. These four charging time probability distributions are measured with a certain number of charging PEVs and integrated in the power demand of a 5-unit system and a 15-unit system respectively. Both the economic and environmental impacts are evaluated by solving the dynamic dispatch problems.

The objective functions and constraints of DEED problems are significantly non-smooth and non-convex due to factors like the valve point effects and exponential nature of emission output curve. Meta-heuristic approaches have consequently been employed to solve these problems. Numerous meta-heuristic algorithms have been used to solve the economic dispatch problem such as particle swarm optimization (PSO) [7], differential evolution (DE) [8], harmony search (HS) [9], biogeography-based optimization (BBO) [10], and imperialist competitive algorithm (ICA) [11], etc.

Based on our previous work [12], a new variant of teaching learning based optimization (TLBO), namely self-learning teaching-learning based optimization (SL-TLBO) is proposed and adopted to solve DEED problems. Some state-of-the-art variants of TLBO have been numerically compared to 10 well-known benchmarks and DEED problem for 5-unit and 15-unit power systems respectively. The dispatch results of four charging scenarios are also comparatively studied. The results show the significance of the new SL-TLBO scheduling method when applied to both benchmarks test and two scales of DEED test systems in terms of both the convergence speed and accuracy. For the four charging profiles, the comparative studies show that the off-peak charging scenario is the most economical and an environmental friendly choice.

2 Problem formulations

The DEED problem is a multi-objective problem combining the economic dispatch objective denoted as F_1 and environmental dispatch objective denoted as F_2. This two-objective problem can also be transformed into a single objective problem as

$$\min F = \omega_{ee} F_1 + (1 - \omega_{ee}) F_2 \tag{1}$$

where ω_{ee} is a weighting factor, being a constant between 0 and 1. The overall cost F denotes the single objective to be minimized. In terms of the formulations of F_1 and F_2, both non-linear models for economic and environmental dispatch are considered.

2.1 Dynamic economic load dispatch model

The dynamic economic load dispatch problem is to minimize the total economic cost of the fossil fuel in a whole day time. The decision variable is the dispatched power $P_{i,t}$ in each time interval t. The problem is formulated as

$$F_1 = \sum_{t=1}^{T} \sum_{i=1}^{N_u} F_i(P_{i,t}) = \sum_{t=1}^{T} \sum_{i=1}^{N_u} [(a_i + b_i P_{i,t} + c_i P_{i,t}^2)] + |e_i \sin(f_i(P_{imin} - P_{i,t}))| \tag{2}$$

where a_i, b_i and c_i are the fuel cost coefficients of the ith generator; e_i and f_i the fuel cost coefficients for evaluating ripples in the cost curve caused by the valve-point effect; and F_1 accumulates the cost of N_u generators in totally T intervals. In addition to the cost functions, there are several system constraints associating with the objective functions as follows.

1) Power output limits:

$$P_{imin} \leq P_{i,t} \leq P_{imax} (i = 1, 2, \ldots, N_u) \tag{3}$$

where the power output should be within the capacity of each specific power generator P_{imax} and P_{imin}.

2) Power balance limits:

$$\sum_{i=1}^{N_u} P_{i,t} = P_{D,t} + P_{L,t+} + L_{ev,t} \quad t = 1, 2, \ldots, T \tag{4}$$

The total power generated in each time interval should meet the power load demand in the corresponding time period. In this paper, the power demand constraint considers the original load demand $P_{D,t}$ in the time interval t, associated with the transmission losses $P_{L,t}$ and the PEV charging load $L_{ev,t}$. This PEV charging load is a new load type and will be further addressed in Section 3. The transmission losses are also considered and approximated with the widely used B-coefficients method [13] denoted as

$$P_{L,t} = \sum_{i=1}^{N_u} \sum_{j=1}^{N_u} P_{i,t} B_{i,j} P_{j,t} + \sum_{i=1}^{N_u} B_{0,i} P_{i,t} + B_{00} \tag{5}$$
$$t = 1, 2, \ldots, T$$

where $B_{i,j}$, $B_{0,i}$ and B_{00} are loss coefficients. The handling approach is implemented by the method proposed in [14].

3) Ramp rate limits:

$$\begin{cases} P_{i,t} - P_{i,t-1} \leq UR_i \\ P_{i,t-1} - P_{i,t} \geq DR_i \end{cases} \tag{6}$$

Thermal generators are subject to the power ramp rate limitation that the power outputs cannot dramatically change between two adjacent intervals. The D_{Ri} and U_{Ri} are the ramp-up and ramp-down rate limits of the ith generator respectively. The dispatched power of a generator in the tth time interval $P_{i,t}$ should be limited by the previously dispatched power $P_{i,t-1}$ at time interval $t-1$ within ramp-up and ramp-down rate limits D_{Ri} and U_{Ri}.

2.2 Dynamic environmental load dispatch model

The environmental load dispatch problem minimizes the emissions of environment pollutants including sulphur dioxide (SO_2) and nitrogen oxides (NO_x) [15]. A quadratic polynomial formulation is associated with an exponential term to model the emissions as

$$F_2 = \sum_{t=1}^{T} \sum_{i=1}^{N_u} F_i(P_{i,t}) = \sum_{t=1}^{T} \sum_{i=1}^{N_u} [(\alpha_i + \beta_i P_{i,t} + \gamma_i P_{i,t}^2)] + \eta_i \exp(\delta_i P_{i,t})$$

(7)

where α_i, β_i, γ_i, η_i, and δ_i are the emission coefficients of the ith generation unit. The power output limit, balance limit as well as the ramp rate limit is also taken into consideration in forming the constraints.

3 Plug-in electric vehicle load profiles

Plug-in electric vehicles are an unusual power demands and unlike traditional household and industry loads, the simultaneous charging of the 20 kW household chargers and 120 kW superchargers will possibly form huge ripples or even spikes on the daily power demand curve. Such effects could be avoided by the coordination and scheduling of charging. With the development of smart grid technology the introduction of smart PEV chargers to coordinate and control PEV charging looks highly alike. In this section, four different charging scenarios, including an EPRI predicted profile based on the assuming behaviors of drivers, an off-peak charging profile and a peak charging profile and a stochastic charging profile are modeled to compare and evaluate the impact on both economic and environmental aspects for the power system operation.

3.1 EPRI profile

The Electric Power Research Institute (EPRI) is a US funded non-profit organization founded in 1973. It is one of the leading organisations in the world, which produces publications and reports on the electric power industry. In an environmental assessment of PEVs [16], EPRI proposes

an aggregate distribution of charging profiles to assess GHG emissions, where a probability distribution of charging profile is proposed as shown in Table 1.

In this charging scenario, over 60% power is delivered in 7 hours in the evening during 22:00 to 4:00. The other time slots see low charging rates and cover the rest of energy delivery.

3.2 Off-peak profile

In [17], two charging scenarios, peak and off-peak charging by assuming flat load demand for PEV charging during peak and off-peak time in the whole Ireland are proposed as a case study. The probability distribution in each hour for the off-peak case is illustrated in Table 2.

This profile assumes three charging levels with 18.5% of power is delivered every hour during 23:00 to 02:00, 9% of power is delivered in each time interval from 02:00 to 04:00 and the rest of charging is completed in 06:00. It is apparently an ideal case that besides the 8 hours charging, other time slots are forbidden for PEVs to get charged.

3.3 Peak profile

Similar to off-peak charging, peak charging assumes a flat load curve with three levels of charging power to describe a certain number of EVs getting charged during the peak load time on the wholesale electricity market. The probability distribution of each hour for the peak case is listed in Table 3.

This is another extreme case where all the PEV charging power requests are provided during peak time of electricity consumption during daytime from 13:00 to 20:00. The charging period and power in this paper are based on [16], where the peak time for charging is slightly shifted due to the profile of the load fleet.

Table 1 EPRI charging probability distribution

Time	Probability/%					
01:00–06:00	10	10	9.5	7	5	3
07:00–12:00	1	0.3	0.3	1.3	2.1	2.1
13:00–18:00	2.1	2.1	2.1	1	0.5	0.5
19:00–24:00	1.6	3.6	5.4	9.5	10	10

Table 2 Off-peak charging probability distribution

Time	Probability/%					
01:00–06:00	18.5	18.5	9	9	4	4
07:00–12:00	0	0	0	0	0	0
13:00–18:00	0	0	0	0	0	0
19:00–24:00	0	0	0	0	18.5	18.5

Table 3 Peak charging probability distribution

Time	Probability/%					
01:00–06:00	0	0	0	0	0	0
07:00–12:00	0	0	0	0	0	0
13:00–18:00	18.5	18.5	18.5	18.5	9	9
19:00–24:00	4	4	0	0	0	0

3.4 Stochastic profile

Considering the uncertainties of drivers' charging behavior, a stochastic charging profile is proposed in this paper. The stochastic load profile simulates some urgent group charging or distributed fast charging at random time throughout the whole day. The random probability follows the normal distribution with the mean value as 5%. The probability distribution of each hour for the stochastic case is presented in Table 4.

The probability of stochastic charging profile in each hour ranges from 1.1% to 9.7%. It changes randomly regardless of the peak or off-peak load time.

The four different PEV load charging distributions are illustrated in Fig. 1 respectively. These four profiles will impose extra load $L_{ev,t}$ in the power demand constraints (3). The new dispatch problems are tightly constrained and show strong non-convex, calling for more powerful computational tools to solve.

Table 4 Stochastic charging probability distribution

Time	Probability/%					
01:00–06:00	5.7	4.9	4.8	2.4	2.6	9.7
07:00–12:00	8.7	4.8	1.1	3.2	2.1	5.7
13:00–18:00	3.8	2.2	2.1	6.1	3.2	2.2
19:00–24:00	2.8	2.2	5.5	2.5	3.5	8.2

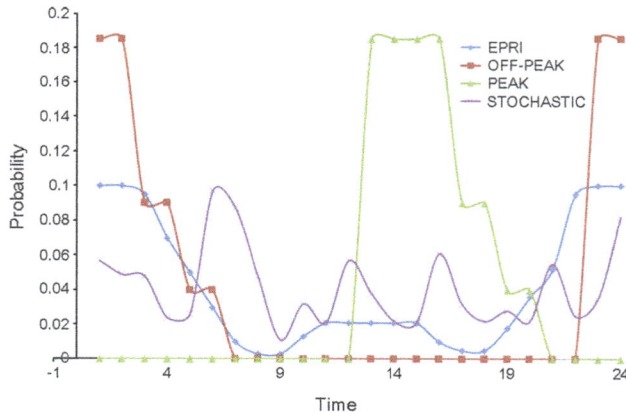

Fig. 1 Four different PEV load distributions

4 Self-learning teaching-learning based optimization

Teaching-learning based optimization (TLBO) is a new meta-heuristic algorithm proposed in 2011 and has been utilized in solving some engineering problems [18], [19], [20], [21]. Two phases are designed for each evolution iteration in the original TLBO, named teaching phase and learning phase respectively. Both the convergence speed and exploitation ability of TLBO have been tested on well-known benchmarks and its effectiveness have been confirmed [18, 19]. Though TLBO is powerful in solving many optimization problems, it can be further improved for specific problems. In this paper, a new self-learning phase is incorporated into original TLBO, aiming to continuously improve the exploitation ability.

4.1 Teaching phase

The teaching phase mimics a class teaching process that a teacher shares his/her knowledge to the students. A teacher will be first selected from the whole population by sorting the fitness function value. The deviation between the teacher and the mean of students will be calculated as:

$$DM_i = rand_1(T_i - T_F Mean_i) \tag{8}$$

where DM_i is the value difference in the ith iteration; $Mean_i$ the mean value; T_i the selected teacher; and T_F a teaching factor. According to the original paper of TLBO, the T_F can either be 1 or 2 denoted as

$$T_F = round(1 + rand_2(0, 1)) \tag{9}$$

Each learner in the class will gain knowledge from the value difference and update themselves as

$$X_{ij}^{new} = X_{ij}^{old} + DM_i \tag{10}$$

where X_{ij}^{new} and X_{ij}^{old} are the jth old and new learners of ith iteration. The new learners will compete with his/her predecessor and replace them if a better fitness value is achieved.

4.2 Learning phase

Followed the teaching phase, a learning phase provides a chance for each student to learn from a classmate. In this phase, each solution would randomly select another solution to compare the fitness, and update the knowledge storage according to the interaction. The phase is denoted as

$$X_{ij}^{new} = \begin{cases} X_{ij}^{old} + rand_3(X_{ik} - X_{ij}) & f(X_{ik}) < f(X_{ij}) \\ X_{ij}^{old} + rand_3(X_{ij} - X_{ik}) & f(X_{ij}) < f(X_{ik}) \end{cases} \tag{11}$$

where the jth learner X_{ij} and k^{th} learner X_{ik} are randomly selected from the population in the ith iteration. Through a competition, the initial learner X_{ij} will refresh his/her knowledge based on the deviation of the two learners.

Similar with teaching phase, the refreshed student will have to compete with his/her predecessor and the better one will remain in the class.

4.3 Self-learning phase

The original TLBO performs well on weakly constrained or completely unconstrained problems. For strong constrained ill-conditioned optimization problems, the global optimal solutions are sensitive to slight changes. It is likely that the global optimum would be missed out due to the low exploitation ability. A self-learning phase is designed to further exploit the space near the particle position for the promising global optimum. Each student will get a chance to learn from his/her self-surrounding spaces. The self-learning phase is illustrated as

$$X_{ij}^{new} = X_{ij}^{old}(1 + (rand_4 - 0.5)\omega) \qquad (12)$$

where a self-learning weight ω is designed to determine the self-learning range of each particle. The parameter tuning of this factor will be discuss in Section 5.

5 Numerical experiments

In order to evaluate the performance of the new SL-TLBO algorithm, 10 well-known benchmark functions with 30 dimensions from [22] were tested. The SL-TLBO was compared with some TLBO variants including original TLBO, elite TLBO [23], and a modified TLBO [24]. In addition, some other commonly used algorithms which include weighted PSO [25], PSO-CF [26] and classical DE/rand/1/bin [27] are selected for the comparative study.

5.1 Benchmark functions

The 10 benchmark functions are defined in [22] and listed with the dimensions and boundaries respectively as follows.

Sphere function (f_1): dimension = 30, [−100, 100];
Schwefel's problem 1.2 (f_2): dimension = 30, [−100, 100];
Rosenbrock function (f_3): dimension = 30, [−30, 30];
Ackley's function (f_4): dimension = 30, [−32, 32];
Griewank function (f_5): dimension = 30, [−600, 600];
Rastrigin function (f_6): dimension = 30, [−5.12, 5.12];
Step function (f_7): dimension = 30, [−100, 100];
Schwefel's problem 2.21 (f_8): dimension = 30, [−100, 100];
Schwefel's problem 2.26 (f_9): dimension = 30, [−500, 500];
Quartic function (f_{10}): dimension = 30, [−1.28, 1.28].

5.2 Determination of control parameters in SL-TLBO

Though there are no algorithm specific parameters to be tuned in original TLBO, the new SL-TLBO method has introduced a self-learning weighting factor ω to adjust the learning range. It is therefore important to find proper settings of ω. Three benchmarks have been tested with the ω ranging from 0.1 to 10. The maximum generation is set as 100 and the particle number is 30. To eliminate the experimental incidents, 30 different run were employed. The initialization values were randomly generated within the boundary and were taken as the same input for all the algorithms. The searching results are showed in Table 5 with mean values and standard deviations for each parameter setting respectively.

Table 5 Benchmark tests result with different weighting factor

	f_1	f_4	f_9
$\omega = 0.1$	8.4878e−27 ± 7.9217e−26	2.3488e−11 ± 3.7617e−12	−5.0731e03 ± 2.6459e03
$\omega = 0.3$	2.5354e−29 ± 3.0583e−28	6.5648e−13 ± 2.7191e−12	−5.0010e03 ± 2.7628e03
$\omega = 0.2$	1.1154e−27 ± 1.0138e−26	1.9191e−13 ± 6.9575e−13	−4.9442e03 ± 2.5526e03
$\omega = 0.4$	2.2029e−30 ± 4.5483e−29	3.5113e−14 ± 1.5700e−13	−5.0013e03 ± 2.2619e03
$\omega = 0.5$	5.0894e−32 ± 6.1376e−31	9.6515e−15 ± 3.3226e−14	−4.9387e03 ± 2.3824e03
$\omega = 0.6$	1.1328e−33 ± 8.3986e−33	4.6777e−15 ± 4.8539e−15	−5.1075e03 ± 2.6581e03
$\omega = 0.7$	9.4206e−35 ± 1.5798e−33	4.4409e−15 ± 0.0000	−5.3365e03 ± 2.4129e03
$\omega = 0.8$	4.3055e−37 ± 3.3442e−36	4.3225e−15 ± 3.4930e−15	−5.0814e03 ± 2.7728e03
$\omega = 0.9$	1.6973e−38 ± 1.7813e−37	4.4409e−15 ± 0.0000	−5.0571e03 ± 2.7343e03
$\omega = 1$	1.8046e−40 ± 3.4272e−39	4.4409e−15 ± 0.0000	−5.3451e03 ± 3.8784e03
$\omega = 2$	4.5126e−78 ± 3.4475e−79	8.8816e−16 ± 0.0000	−5.4113e03 ± 2.3002e03
$\omega = 3$	8.2091e−89 ± 1.6473e−87	8.8816e−16 ± 0.0000	−5.0689e03 ± 3.1955e03
$\omega = 5$	1.0296e−68 ± 2.9208e−67	8.8816e−16 ± 0.0000	−4.9982e03 ± 2.5672e03
$\omega = 10$	2.1032e−48 ± 6.0272e−47	8.8816e−16 ± 0.0000	−5.2526e03 ± 2.5658e03

It could be observed that the results get better with the increase of ω in both on f_1 and f_4. The best results are achieved when ω is 3 in the f_1 test and from 1 to 10 in the f_4 test. On the other hand, the final results change stochastically with the increase of ω in the f_9 test where the best result was achieved when ω is 2. These results show that the performance of the parameter settings for the new SL-TLBO algorithm is problem specific. We choose $\omega = 3$ for the further comparative study on algorithms performance in sub-section 5.3.

5.3 Simulation results and discussions

In this sub-section, the new SL-TLBO algorithm were tested and comparatively studied with some counterparts on the ten benchmark functions tests mentioned in 5.1. To fairly compare the algorithm performance, the number of the function evaluation (FES) is introduced as the iteration criteria. Each generation of PSO and DE method accounts for one FES. The original TLBO has two phases and accounts for double times of function evaluation, whereas the SL-TLBO has one more self-learning phase in each generation and triple times are used to calculate the FES.

In terms of the parameters settings for the algorithms, the population number is 30 while the FES is set as 12,000. The weighted PSO uses $c_1 = 1, c_2 = 3, w_{max} = 0.9, w_{min} = 0.4$; for PSO-CF, $c_1 = c_2 = 2.05, K = 0.729$; for classical DE, $F = 0.7, CR = 0.5$; for elite TLBO, the elite number is set as 5; the self-learning weighting factor ω is 3; no specific parameters are required to be set for original TLBO and mTLBO. Similarly, 30 different runs were carried out for each benchmark function by each algorithm. The mean values and standard deviations for each test are illustrated in Table 6.

From the results, it is shown that the new SL-TLBO performs best in the limited FES among 7 of totally 10 tests. In the tests on f_1, f_4, f_7, and f_8, the SL-TLBO significantly outperforms other counterparts. Comparable performance for the new algorithm is displayed on f_2, f_5, f_6. In the cases of f_3, f_9, and f_{10}, SL-TLBO is outperformed by the original TLBO, mTLBO, and eTLBO, respectively.

However, in these three cases, the performances of the four TLBO variants are relatively the same, which indicates that the general problem solving capability of SL-TLBO appears to be satisfactory. Therefore, this new method is utilized in solving the DEED problem with PEV loads.

6 Simulation results on DEED problem and discussions

6.1 Case 1: 5-unit economic dispatch without PEV

In order to show the significance of the new SL-TLBO algorithm, the original 5-unit considering valve-point effect and transmission loss is tested. The FES is set as 60000, and the population number for the SL-TLBO method is 50. The 5-unit economic dispatch benchmark data is taken from [28]. According to the experimental tests, ω is set as 0.05, which is much smaller than the previous setting in benchmark functions tests. This is because the DEED problem is highly constrained due to which a big learning rate will easily cause the violation of constraints. On the other hand, a small learning factor would increase the searching ability in nearby solution space, which is more adaptable for DEED problem to search better solutions in a limited available space.

It can be seen from the Table 7 that the new SL-TLBO can obtain the best results comparing with some previous methods. This method is then utilized in PEV integration analysis.

6.2 Case 2: 5-unit economic dispatch with PEV

In this case, a 5-unit system combined with the four PEV load profiles considering valve-point effect and transmission losses is investigated to dynamically dispatch the generation production from the economic perspective.

Considering the local population and load situation of the benchmark system, an extra load of 30000 different types of PEVs is integrated, where 45% of these PEVs are low hybrid vehicles equipped with 15 kWh batteries.

Table 6 Benchmark tests results for different algorithms

	wPSO	PSO-CF	DE	TLBO	eTLBO	mTLBO	SL-TLBO
f_1	$4.869e02 \pm 1.069e03$	$1.904e03 \pm 3.895e03$	$3.562e03 \pm 1.130e04$	$2.090e{-}45 \pm 2.617e{-}44$	$2.774e{-}76 \pm 2.905e{-}75$	$8.200e{-}91 \pm 1.521e{-}89$	$2.747e{-}115 \pm 4.087e{-}114$
f_2	$5.155e00 \pm 5.236e01$	$4.390e{-}02 \pm 5.259e{-}01$	$1.601e{-}01 \pm 1.690e00$	$0.000e00 \pm 0.000e00$	$5.342e{-}31 \pm 1.268e{-}29$	$0.000e00 \pm 0.000e00$	$0.000e00 \pm 0.000e00$
f_3	$1.369e03 \pm 1.128e04$	$1.714e00 \pm 1.074e01$	$3.829e00 \pm 9.934e00$	$1.547e00 \pm 7.442e00$	$3.077e00 \pm 5.274e00$	$3.513e00 \pm 3.118e00$	$1.746e00 \pm 4.955e00$
f_4	$2.114e01 \pm 2.957e{-}01$	$1.946e01 \pm 1.398e01$	$1.446e01 \pm 4.263e01$	$2.828e{-}08 \pm 8.340e{-}07$	$2.072e{-}15 \pm 9.173e{-}15$	$9.063e00 \pm 5.678e01$	$8.882e{-}16 \pm 0.000e00$
f_5	$2.493e01 \pm 3.151e01$	$1.390e01 \pm 9.142e{-}01$	$1.491e01 \pm 4.792e00$	$0.000e00 \pm 0.000e00$	$0.000e00 \pm 0.000e00$	$0.000e00 \pm 0.000e00$	$0.000e00 \pm 0.000e00$
f_6	$2.297e02 \pm 2.516e02$	$1.092e02 \pm 8.724e01$	$4.953e01 \pm 2.736e02$	$4.970e{-}08 \pm 1.466e{-}06$	$0.000e00 \pm 0.000e00$	$0.000e00 \pm 0.000e00$	$0.000e00 \pm 0.000e00$
f_7	$2.076e01 \pm 4.834e01$	$7.076e00 \pm 1.700e01$	$1.229e01 \pm 2.125e01$	$5.084e00 \pm 3.918e00$	$5.546e00 \pm 4.097e00$	$5.878e00 \pm 3.798e00$	$4.744e00 \pm 3.225e00$
f_8	$2.751e00 \pm 2.513e00$	$1.089e00 \pm 1.370e00$	$2.170e00 \pm 3.321e00$	$5.162e{-}24 \pm 2.490e{-}23$	$2.568e{-}39 \pm 3.930e{-}38$	$1.022e{-}46 \pm 9.416e{-}46$	$1.881e{-}59 \pm 3.248e{-}58$
f_9	$-4.321e03 \pm 2.680e02$	$-5.223e03 \pm 2.449e03$	$-5.898e03 \pm 1.752e03$	$-6.083e03 \pm 2.319e03$	$-5.667e03 \pm 2.339e03$	$-6.382e03 \pm 1.733e03$	$-6.302e03 \pm 2.588e03$
f_{10}	$2.489e01 \pm 3.211e01$	$1.571e01 \pm 7.902e00$	$1.821e01 \pm 6.346e00$	$9.465e00 \pm 2.256e00$	$9.323e00 \pm 2.601e00$	$9.429e00 \pm 1.488e00$	$9.549e00 \pm 2.747e00$

Besides, 25% of PEVs are medium hybrid vehicles using 25 kWh batteries, and 30% PEVs are pure battery vehicles of which the total power are provided by 40 kWh batteries as in Table 8. It is also assumed that 50% SOC are the energy necessity for PEVs in a 24-hour period [32]. The total PEV load for one day is calculated as $L_{ev} = 30000 \times (15 \times 45\% + 25 \times 25\% + 40 \times 30\%) \times 0.5 = 375$ MW.

Fig. 2 illustrates the four different load demand profiles with corresponding situations of PEV charging probability in a 24-hour time period. Differences lie on the noon time between 12:00-17:00 and midnight between 22:00-06:00. The peak load has increased significantly in the peak charging situation shown as the purple line in Fig. 2 lasts till the next load valley. On the contrary, the valley load is increased by both off-peak charging and EPRI charging (see green and red line in Fig. 2). The stochastic charging profile generates a small new sub-peak during 6:00 to 9:00 in the morning showed in the blue line again in Fig. 2.

The four circumstances are tested by seven algorithms respectively. The particle number is set as 50 and the FES is 20000. The algorithms parameter settings are almost the same with the tests in 5.3, while the only difference lies on the setting of the self-learning weighting factor ω in SL-TLBO as mentioned in case 1.

Table 9 shows the dispatched results of four PEV charging profiles solved by seven algorithms. The SL-TLBO outperforms all its counterparts and achieves the best results on all the four charging profiles. Moreover, comparing the four charging patterns, the off-peak charging costs 46508.86 \$/day which is the lowest among all the situations. In contrast, the peak charging costs reach to 47367.17 \$/day and becomes the highest cost. The EPRI charging profile ranks the second lowest in terms of the cost and outperforms the stochastic charging behavior which ranks the third place with the costs of 46770.71 \$/day and 47158.86 \$/day respectively. The off-peak charging profile costs 858.31 \$/day lower than the peak charging profile, which implies that under the same charging

Table 7 Comparison of total fuel cost over 30 runs (Case 1: 5-unit without PEV)

Method	Fuel cost (\$/day)		
	Min	Ave	Max
SA [28]	47356.00	NA	NA
PS [29]	46530.00	NA	NA
EP [30]	46777.00	NA	NA
PSO [31]	44253.24	45657.06	46402.52
SL-TLBO	44199.98	45655.74	46113.64

Table 8 Multiple types of PEV

EV type	Battery capacity/kWh	Proportion/%
Low hybrid	15	45
High hybrid	25	25
Pure battery	40	30

Table 9 Economic dispatching result for case 1 (\$/day)

	L_{EPRI}	L_{Offp}	L_{Peak}	L_{Sto}
wPSO	49004.13	48587.97	50875.78	49333.11
PSO-CF	51482.18	51231.77	51682.02	51292.57
DE	51457.32	51238.97	51310.22	51283.18
TLBO	49649.47	48884.45	48775.31	49292.38
eTLBO	49049.49	49306.12	49270.68	49549.59
mTLBO	48974.99	47656.89	48459.7	48970.59
SL-TLBO	46770.71	46508.86	47367.17	47158.86

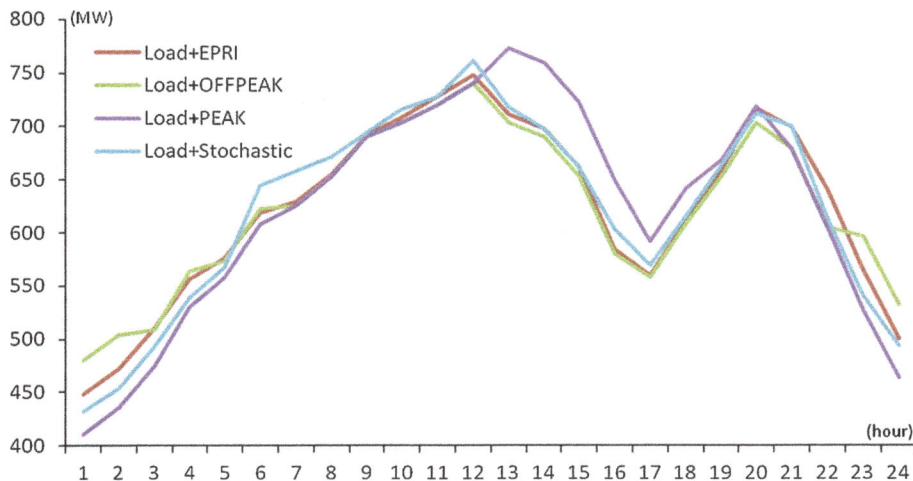

Fig. 2 Power load demands with four PEV profiles

demand, charging preferentially at off-peak time could save 1.85% in terms of the economic cost.

6.3 Case 3: 5-unit environmental dispatch with PEV

In this case, the environmental emission is considered rather than the economic cost. Similarly, the 5-unit system is employed and the data is taken from [33]. The objective function is presented as (7) and the constraints are the same with economic dispatch in the previous subsection. The PEV number and the power necessity are the same with the aforementioned economic case. All the four charging profiles are also comparatively tested by seven algorithms including wPSO, PSO-CF, classical DE, original TLBO, elite TLBO, modified TLBO, and SL-TLBO. The parameter initializations for these algorithms are the same with case 2.

Table 10 shows the environmental dispatch results for the four different charging profiles solved by seven algorithms. The lowest environmental emission results are still produced by the proposed SL-TLBO method on all the four PEV charging scenarios. Among these charging profiles, the off-peak charging profile again emits the least air pollutants with the amount of 18659.24 lb/day achieved by SL-TLBO, while the peak charging profile produces most emissions with 19227.18 lb/day. Therefore, a reduction of 3.0% emission production with 567.94 lb/day from the generation side would be achieved by shifting the PEV charging time. It is the same situation as in case 2, the EPRI charging profile and the stochastic charging profile rank the 2nd and the 3rd with the emissions of 18820.78 lb/day and 18963.69 lb/day respectively.

6.4 Case 4: 5-unit economic and environmental dispatch with PEV

Since the economic aspect and the environmental aspect are both vital for power system dispatch, a combination of these two objectives is also investigated. The ω_{ee} in the objective function (1) is defined as 0.5 in this case to comprehensively consider the economic and environmental profit and trade-off the both situations. The 5-unit test system is again utilized of which the load demand and four PEV charging situations remain the same with previous cases. Dealing with the objective function with significant high non-linear characteristics with both sinusoidal and exponential terms, the case would be more complex for algorithms to solve. In order to evaluate the algorithms performances and compare the four charging situations, the seven algorithms with same parameter initializations are again implemented.

Table 11 shows all the dispatching results considering both economic and environmental aspects. Not

Table 10 Environmental dispatching result for case 2 (lb/day)

	L_{EPRI}	L_{Offp}	L_{Peak}	L_{Sto}
wPSO	19189.83	18998.30	19443.99	19112.25
PSO-CF	20232.49	19794.18	20440.12	20183.56
DE	20030.88	19658.45	20493.63	20109.08
TLBO	19002.82	18887.37	19483.80	19195.84
eTLBO	19170.59	18930.36	19390.75	19154.89
mTLBO	19112.78	18879.95	19379.09	19369.38
SL-TLBO	18820.78	18659.24	19227.18	18963.69

Table 11 Economic/environmental dispatching result for case 3 (0.5 · $/day + 0.5 · lb/day)

	L_{EPRI}	L_{Offp}	L_{Peak}	L_{Sto}
wPSO	35785.15	34705.22	35514.07	36022.33
PSO-CF	36611.46	36601.88	36814.66	36632.76
DE	36534.69	36421.38	36657.76	36581.62
TLBO	35037.04	34959.16	35112.58	35269.76
eTLBO	35064.55	35167.21	35048.14	35654.87
mTLBO	35300.92	35355.49	35162.68	34857.65
SL-TLBO	33998.31	33924.62	34731.92	34245.83

incidentally, SL-TLBO again achieved the best fitness values. The fitness value of off-peak charging profile is 33924.62 while it is 34731.92 for the peak charging scenario. There is a 2.4% deviation between these two values. If the performances of all the algorithms are compared in all the cases, SL-TLBO always gives the best performance.

6.5 Case 5: 15-unit economic dispatch with PEV

The 5-unit system is only a small scale system with 120 variables for a dynamic dispatch in 24-hour period of time. However, dispatching tasks for larger system widely exist. In this case, a 15-unit system is implemented of which the total load demand is 60981 MW [34]. The number of variables is tripled and reaches to 360 in one-day time. To simplify the situation, the valve point-effect is neglected in the objective function as

$$F = \sum_{t=1}^{T} \sum_{i=1}^{N_u} F_i(P_{i,t}) = \sum_{t=1}^{T} \sum_{i=1}^{N_u} [(a_i + b_i P_{i,t} + c_i P_{i,t}^2)] \quad (13)$$

where the transmission loss and ramp rate constraints are also considered. The system data is taken from [31]. To evaluate the economic impact of the four PEV charging profiles in this test system, the total PEV number is proportionally increased to 90000. While the PEV types remain the same as in Table 8, resulting in that the total PEV charging power necessity is tripled to 1,125 MW in a single day. This extra PEV charging load is distributed

Table 12 Economic dispatching result for Case 4 ($/day)

	L_{EPRI}	L_{Offp}	L_{Peak}	L_{Sto}
wPSO	783004.14	783650.51	783863.93	784610.33
PSO-CF	784391.24	784532.96	785851.62	785491.74
DE	784354.55	784313.52	785512.30	785273.31
TLBO	781644.49	783002.47	784004.33	783962.29
eTLBO	782323.93	782320.70	783383.72	783280.51
mTLBO	781562.91	781179.19	782922.74	782138.87
SL-TLBO	781001.23	780862.82	781961.91	781459.24

within the four charging scenarios and integrated with the original load demand. The seven algorithms are again employed to solve the large scale problem for comparative study where the specific parameter settings for these algorithms are the same with those in case 2.

The results provided in Table 12 illustrate that the SL-TLBO method again achieved the optimal solutions, outperforming with all the other six algorithms in all the four load scenarios. The lowest cost appears in the off-peak charging scenario with the cost of 780862.82 $/day, while the peak charging scenario costs 1099.09 $/day more than the off-peak one, reaching to 781961.91 $/day.

7 Conclusion

Dynamic economic dispatch has long been an intractable problem for power system operators and the complexity is ever increasing with new participants such as PEV entering in the equation. In this paper, the non-convex dynamic economic and environmental dispatch has been comparatively investigated with the integrations of various PEV charging scenarios. A new self-learning teaching learning based optimization method is proposed to solve the economic and environmental dispatch problems. A small scale 5-unit system and a large scale 15-unit system are tested in 24-hour time period. Four different PEV charging scenarios including EPRI predicted charging, off-peak charging, peak charging and stochastic charging profiles with different number of PEV have been integrated in the load demands of both systems. The numerical results show that the new SL-TLBO algorithm is a viable alternative approach for solving both small and large scale dynamic dispatch problems and outperforms other popular heuristic methods and state-of-the-art TLBO variants in the tests on well-known benchmarks and DEED problem with proper parameter tuning. In terms of the four PEV charging scenarios, the off-peak charging scenario, as expected, has the advantage in reducing the economic cost and environmental pollutant emissions.

In future studies, renewable energy sources such as photovoltaic panels and wind power as well as the vehicle

to grid (V2G) will be introduced in the system dispatch to comprehensively analyze the interaction between the PEV and power systems.

Acknowledgments Zhile YANG would like to thank UK Engineering and Physical Sciences Research Council (EPSRC) for sponsoring his research. The authors would also like to thank UK EPSRC under grant EP/L001063/1 and China NSFC under grants 51361130153 and 61273040.

Reference

[1] Wong KP, Fung CC (1993) Simulated annealing based economic dispatch algorithm. IEE Proc Gener Transm Distrib 140(6):509–515

[2] Victoire TAA, Jeyakumar AE (2005) Reserve constrained dynamic dispatch of units with valve-point effects. IEEE Trans Power Syst 20(3):1273–1282

[3] Abido MA (2003) Environmental/economic power dispatch using multiobjective evolutionary algorithms. IEEE Trans Power Syst 18(4):1529–1537

[4] Zhang Y, Yao F, Lu HHC et al (2013) Sequential quadratic programming particle swarm optimization for wind power system operations considering emissions. J Mod Power Syst Clean Energy 1(3):231–240

[5] Saber AY, Venayagamoorthy GK (2012) Resource scheduling under uncertainty in a smart grid with renewables and plug-in vehicles. IEEE Syst J 6(1):103–109

[6] Yang Z, Li K, Foley A et al (2014) Optimal scheduling methods to integrate plug-in electric vehicles with the power system: a review. In: Proceedings of the 19th world congress of the International Federation of Automatic Control (IFAC' 14), Cape Town, South Africa, 24–29 August 2014, pp 8594–8603

[7] Mohammadi-Ivatloo B, Rabiee A, Soroudi A et al (2012) Iteration PSO with time varying acceleration coefficients for solving non-convex economic dispatch problems. Int J Electr Power Energy Syst 42(1):508–516

[8] Yuan XH, Wang L, Zhang YC et al (2009) A hybrid differential evolution method for dynamic economic dispatch with valve-point effects. Expert Syst Appl 36(2–2):4042–4048

[9] Niu Q, Zhang HY, Wang XH et al (2014) A hybrid harmony search with arithmetic crossover operation for economic dispatch. Int J Electr Power Energy Syst 62:237–257

[10] Bhattacharya A, Chattopadhyay PK (2010) Biogeography-based optimization for different economic load dispatch problems. IEEE Trans Power Syst 25(2):1064–1077

[11] Mohammadi-Ivatloo B, Rabiee A, Soroudi A et al (2012) Imperialist competitive algorithm for solving non-convex dynamic economic power dispatch. Energy 44(1):228–240

[12] Yang Z, Li K, Niu Q et al (2014) Non-convex dynamic economic/environmental dispatch with plug-in electric vehicle loads. In: Proceedings of the IEEE symposium series on computational intelligence (SSCI' 14, accepted)

[13] Wood AJ, Wollenberg BF (2012) Power generation, operation, and control. Wiley, New York

[14] Niu Q, Zhang HY, Li K et al (2014) An efficient harmony search with new pitch adjustment for dynamic economic dispatch. Energy 65:25–43

[15] Gong DW, Zhang Y, Qi CL (2010) Environmental/economic power dispatch using a hybrid multi-objective optimization algorithm. Int J Electr Power Energy Syst 32(6):607–614

[16] EPRI executive summary: environmental assessment of plug-in hybrid electric vehicles, Volume 2: United States air quality analysis based on AEO-2006 assumptions for 2030 (2007). Electric Power Research Institute, Palo Alto

[17] Foley A, Tyther B, Calnan P et al (2013) Impacts of electric vehicle charging under electricity market operations. Appl Energy 101:93–102

[18] Rao RV, Savsani VJ, Vakharia DP (2011) Teaching–learning-based optimization: a novel method for constrained mechanical design optimization problems. Comput Aided Des 43(3):303–315

[19] Rao RV, Savsani VJ, Vakharia DP (2012) Teaching–learning-based optimization: an optimization method for continuous non-linear large scale problems. Inf Sci 183(1):1–15

[20] Niu Q, Zhang HY, Li K (2014) An improved TLBO with elite strategy for parameters identification of PEM fuel cell and solar cell models. Int J Hydrog Energy 39(8):3837–3854

[21] Sultana S, Roy PK (2014) Optimal capacitor placement in radial distribution systems using teaching learning based optimization. Int J Electr Power Energy Syst 54:387–398

[22] Yao X, Liu Y, Lin GM (2013) Evolutionary programming made faster. IEEE Trans Evol Comput 3(2):82–102

[23] Rao RV, Patel V (2013) Comparative performance of an elitist teaching–learning-based optimization algorithm for solving unconstrained optimization problems. Int J Ind Eng Comput 4(1):29–50

[24] Satapathy SC, Naik A (2014) Modified teaching–learning-based optimization algorithm for global numerical optimization—a comparative study. Swarm Evol Comput 16:28–37

[25] Shi Y, Eberhart RC (1998) Parameter selection in particle swarm optimization, evolutionary programming VII. Springer, Berlin, pp 591–600

[26] Clerc M, Kennedy J (2002) The particle swarm—explosion, stability, and convergence in a multidimensional complex space. IEEE Trans Evol Comput 6(1):58–73

[27] Das S, Suganthan PN (2011) Differential evolution: a survey of the state-of-the-art. IEEE Trans Evol Comput 15(1):4–31

[28] Panigrahi CK, Chattopadhyay PK, Chakrabarti RN et al (2006) Simulated annealing technique for dynamic economic dispatch. Electr Power Compon Syst 34(5):577–586

[29] Alsumait JS, Qasem M, Sykulski JK et al (2010) An improved pattern search based algorithm to solve the dynamic economic dispatch problem with valve-point effect. Energy Convers Manag 51(10):2062–2067

[30] Basu M (2007) Dynamic economic emission dispatch using evolutionary programming and fuzzy satisfying method. Int J Emerg Electr Power Syst 8(4):1–15

[31] Hemamalini S, Simon SP (2011) Dynamic economic dispatch using artificial bee colony algorithm for units with valve-point effect. Eur Trans Electr Power 21(1):70–81

[32] Saber AY, Venayagamoorthy GK (2011) Plug-in vehicles and renewable energy sources for cost and emission reductions. IEEE Trans Ind Electron 58(4):1229–1238

[33] Wang LF, Singh C (2007) Environmental/economic power dispatch using a fuzzified multi-objective particle swarm optimization algorithm. Electr Power Syst Res 77(12):1654–1664

[34] Gaing ZL, Ou TC (2009) Dynamic economic dispatch solution using fast evolutionary programming with swarm direction. In: Proceedings of the 4th IEEE conference on industrial electronics and applications (ICIEA'09), Xi'an, China, 25–27 May 2009, pp 1538–1544

Zhile YANG is pursuing his Ph.D. degree at the School of Electrical, Electronics and Computer Science, Queen's University Belfast. He received the BEng in Electrical Engineering and the MSc degree in Control Theory and Control Engineering from Shanghai University (SHU) in 2010 and 2013 respectively. His research interests include non-linear modeling and meta-heuristic optimization on power system and electric vehicle.

Kang LI received the B.Sc. degree from Xiangtan University, Hunan, China, in 1989, the M.Sc. degree from the Harbin Institute of Technology, Harbin, China, in 1992, and the Ph.D. degree from Shanghai Jiaotong University, Shanghai, China, in 1995. He is currently a Professor of intelligent systems and control with the School of Electronics, Electrical Engineering, and Computer Science, Queen's University Belfast, Belfast, U.K. His research interests include nonlinear system modeling, identification, and control, bio-inspired computational intelligence, fault diagnosis and detection, with recent applications on power systems and renewable energy, polymer extrusion processes, bioinformatics with applications on food safety, healthcare, and biomedical engineering.

Qun NIU is received the B.Sc. degree in automation and the Ph.D. degree in control theory and control engineering from the East China University of Science and Technology, Shanghai, China, in 2002 and 2007 respectively.

From 2009 to 2011, she was a Postdoctoral Research Fellow with Queen's University Belfast, Belfast, U.K. She is currently an Associate Professor with the Shanghai Key Laboratory of Power Station Automation Technology, School of Mechatronics and Automation, Shanghai University, Shanghai. Her main research interests include meta-heuristic optimization and intelligent computing with applications to power system scheduling and economic dispatch, and manufacturing scheduling.

Yusheng XUE received his Ph.D. degree in Electrical Engineering from the University of Liege (Belgium) in 1987. He became a Member of Chinese Academy of Engineering in 1995. He is now the Honorary President of State Grid Electric Power Research Institute (SGEPRI), State Grid Corporation of China. His research interests include nonlinear stability, control and power system automation.

Aoife FOLEY is a Lecturer in the School of Mechanical and Aerospace Engineering, Queen's University Belfast. She has a BE (Hons) (1996), a PhD (2011) from University College Cork and an MSc (Hons) (1999) from Trinity College Dublin. She was awarded an Environmental Protection Agency STRIVE Fellowship (2009) to model the impact of Electric Vehicles on GHG emissions. She has more than 12 year's industrial experience with ESB International, Siemens and SWS. She is a Fellow of Engineers Ireland (2011), a Member of the IEEE PES (2009) and Secretary of the IEEE VTS UK and Republic of Ireland Chapter (2011). She has authored and co-authored almost 50 peer reviewed technical publications. Her research interests include wind power integration using controllable loads and energy storage, electricity markets and power systems operations.

10

A review on applications of heuristic optimization algorithms for optimal power flow in modern power systems

Ming NIU, Can WAN, Zhao XU (✉)

Abstract Optimal power flow (OPF) is one of the key tools for optimal operation and planning of modern power systems. Due to the high complexity with continuous and discrete control variables, modern heuristic optimization algorithms (HOAs) have been widely employed for the solution of OPF. This paper provides an overview of the latest applications of advanced HOAs in OPF problems. The most frequently applied HOAs for solving the OPF problem in recent years are covered and briefly introduced, including genetic algorithm (GA), differential evolution (DE), particle swarm optimization (PSO), and evolutionary programming (EP), etc.

Keywords Heuristic optimization algorithm, Optimal power flow, Multi-objective optimization, Constraint optimization

1 Introduction

For the stable and economic operation of power systems, optimal power flow (OPF) aims to find the optimal settings of a given power system network that optimize a certain objective function while satisfying its power flow equations, system security, and equipment operating limits [1].

The most commonly used objective is the minimization of the overall fuel cost function. In addition, there are several other objectives are to minimize the active power loss, bus voltage deviation, emission of generating units, number of control actions, and load shedding [2]. In practical power system operation, the OPF problem adjusts the continuous control variables (e.g. real power outputs and voltages) and discrete control variables (e.g. transformer tap setting, phase shifters, and reactive injections) to reach the optimal objective function while satisfying a set of physical and operational constraints. Therefore, it is a highly constrained, mixed-integer, non-linear and non-convex optimization problem.

Conventional techniques such as linear programming (LP), quadratic programming (QP), and non-linear programming (NLP), were developed to solve the OPF problems with some theoretical assumptions, such as convexity, differentiability, and continuity, which may not be suitable for the actual OPF conditions. In addition, the convergence to the global or local optimal solution is highly dependent on the selected initial guess [3]. Moreover, continuous LP, QP, and NLP formulations cannot accurately model discrete control variables, such as transformer tap ratios or switched capacitor banks. Mixed integer linear programming (MILP) techniques were introduced to solve this problem [4]. However, the nonlinearity of the power system cannot be fully represented by MILP formulations, and therefore cause inherent inaccuracy. To overcome these drawbacks, HOAs such as genetic algorithm (GA), evolutionary programming (EP), simulated annealing (SA), tabu search (TS), particle swarm optimization (PSO), differential evolution (DE), etc., have been employed for the solution of the OPF problem. The results reported in previous literatures were promising and encouraging for further research in this area.

This paper provides an extensive coverage of the major research work that make use of HOAs to the OPF problem.

M. NIU, C. WAN, Z. XU, The Hong Kong Polytechnic University, Hong Kong, China
(✉) e-mail: zhao.xu@polyu.edu.hk
M. NIU
e-mail: zimo.niu@connect.polyu.hk
C. WAN
e-mail: can.wan@connect.polyu.hk

The remainder of this paper is structured as follows. A brief introduction of the most frequently used HOAs (GA, EP, PSO, DE, etc) for the OPF problem is given in Section 2. The generalized formulation of the OPF problem is expressed in Section 3. Section 4 provides a detailed investigation to OPF related research work applying HOAs. Section 5 gives a discussion on the status and trend of HOA application on OPF problem. Finally, this paper is summarized in Section 6.

2 Modern heuristic optimization algorithms

The modern HOAs represent a group of intelligent algorithms that either make analog of the natural evolution process based on Darwinian principles or mimic a certain natural phenomenon in searching for an optimal solution. They have been successfully applied to a wide range of power system optimization problems where non-differentiable regions exist and the global solution are extremely difficult to be gauged. The most popularly used HOAs in solving OPF problem are compactly introduced as the follows.

1) Genetic algorithm: GA is one of the most popular and famous approaches in evolutionary computation. Founded on the mechanism of natural genetics and Darwinian principles of evolution and natural selection, this novel algorithm showed strong capabilities and advantages for solving a wide range of problems as introduced in [5]. GA can be considered as a population-based approach, the search process of which is conducted by means of transforming a set of points (individuals) to another set of points in the search space. In the original GA, each individual is represented via a fixed-length binary string. This method maps the points in the search space into the instances of artificial chromosome. Desired precision can be simply approximated through tuning the length of binary string. The strong preference to the binary representation of GA probably derives from Schema Theorem [6] which tries to investigate the mathematical foundation of GA.

2) Particle swarm optimization: PSO, which was introduced in [7, 8] in 1995, is one of the most important swarm intelligence paradigms. The PSO uses a simple mechanism that mimics swarm behavior in birds flocking and fish schooling to guide the particles to search for globally optimal solution. As PSO is easy to implement, it has rapidly progressed in recent years and with many successful applications in solving real-world optimization problems.

3) Differential evolution: The DE approach is firstly proposed in a technical report in 1995 [9]. It is a population-based method and is generally considered a parallel stochastic direct search optimizer that is simple yet powerful. DE is a stochastic population-based optimization algorithm with real parameters and real-valued functions. The core idea behind DE is a scheme for generating trial parameter vectors. DE generates new parameter vectors by weighing the difference vector between two population members and then adding that to a third member. If the resulting vector yields a lower objective function value than a previously determined population member, the newly generated vector replaces the vector to which it was compared. In comparisons to most other HOAs, the DE algorithm is much simpler and more straightforward to implement. The main body of the algorithm takes four or five lines of code in any programming language. Despite its simplicity, the gross performance of DE in terms of accuracy, convergence rate and robustness makes it attractive for applications to various real-world optimization problems [10–12], where finding an approximate solution in a reasonable amount of computational time is of considerable importance. The spatial complexity of DE is lower than that of some highly competitive real parameter optimizers. This feature helps in extending DE to handle expensive and large-scale optimization problems.

4) Evolutionary programming: EP was first introduced in the research of artificial intelligence [13]. In order to achieve intelligent behavior, there came an idea of defining the environment as a sequence of symbols (in a finite alphabet) and evolving an algorithm to predict the next symbol to appear based on the former observed sequence of symbols. Finite state machine (FSM) is chosen to be the form of individuals, as it provides a meaningful representation for the required behaviors in the environment. While the original form of EP was applied in discrete problems due to the FSM representation, EP was extended into the real-valued continuous optimization problem [14]. Both the mutation mode and the number of mutations per offspring FSM are with respect to a probability distribution, which means some individual may mutates more than once in one generation.

In addition to the abovementioned HOAs, ant colony optimization (ACO) [15], artificial neural network (ANN) [16], simulated annealing (SA) [17], tabu search (TS) [18], quantum-inspired evolutionary algorithm (QEA) [19, 20], and artificial bee colony (ABC) [21] are also popular HOA applications with regard to the OPF problem. Thus, corresponding literature reviews are also included in Section 4.

3 Problem formulation

This section provides a generalized OPF problem description. It should be noted that OPF formulations differ greatly depending on the different selection of variables, objectives and constraints. Due to this aspect of OPF problem, the formulation selected often has impacts for both algorithm design and solution accuracy.

From the perspective of mathematics, the OPF optimize a constructed objective subject to different sets of equality and inequality constraints. Without loss of generality, the problem can be formulated as follows:

$$\text{Min} f(x, u) \tag{1}$$

Subject to

$$g(x, u) = 0 \tag{2}$$

$$h(x, u) \leq 0 \tag{3}$$

where $f(\ )$ is the objective function to be minimized; x denotes the dependent or state variables including active power output of the slack bus P_{G_1}, voltage of the load buses V_L (magnitudes and phase angles), generator reactive power outputs Q_G, and transmission line flow S_l, expressed as

$$x^T = [P_{G_1}, V_{L_1} \cdots V_{L_{NL}}, Q_{G_1} \cdots Q_{G_{NG}}, S_{l_1} \cdots S_{l_{nl}}] \tag{4}$$

where NL, NG and nl denote the number of load buses, generators and transmission lines, respectively; u represents the independent variables, also known as control variables consisting of voltages at generation buses V_G (magnitudes), active power generation P_G at PV buses, transformer tap settings T, and reactive power injections Q_C by the shunt volt-amperes reactive (VAR) compensations, represented as

$$u^T = [P_{G_2} \cdots P_{G_{NG}}, V_{G_1} \cdots V_{G_{NG}}, Q_{C_1} \cdots Q_{C_{NC}}, T_1 \cdots T_{NT}] \tag{5}$$

where NC and NT are the number of the VAR compensators and regulating transformers, respectively. The control variables can be divided into two categories, discrete u_d and continuous u_c, given by

$$u_d^T = [T_1 \cdots T_N, Q_{C_1} \cdots Q_{C_N}] \tag{6}$$

$$u_c^T = [P_{G_2} \cdots P_{G_N}, V_{G_2} \cdots V_{G_N}] \tag{7}$$

In practice, the OPF problem has two types of constraints, i.e., the equality and inequality constraints. The equality constraints, defined by (2), are a set of non-linear power flow equations, represented as

$$P_{G_i} - P_{D_i} - P_{B_i}(V) = 0 \tag{8}$$

$$Q_{G_i} - Q_{D_i} - Q_{B_i}(V) = 0 \tag{9}$$

where P_{G_i} and Q_{G_i} are active and reactive power generated at bus i, respectively; P_{D_i} and Q_{D_i} represent the load demand at the same bus; and P_{B_i} and Q_{B_i} denote the total sending and receiving power at each bus.

The inequality constraints, represented in (3), are a set of continuous and discrete constraints that define the system operational and security limits, including:

1) Generation constraints:
   To ensure stable operation, generation bus voltages, active power and reactive power outputs are restricted by their lower and upper limits as

$$V_{G_i}^{\min} \leq V_{G_i} \leq V_{G_i}^{\max}, \quad i = 1, \cdots, NG \tag{10}$$

$$P_{G_i}^{\min} \leq P_{G_i} \leq P_{G_i}^{\max}, \quad i = 1, \cdots, NG \tag{11}$$

$$Q_{G_i}^{\min} \leq Q_{G_i} \leq Q_{G_i}^{\max}, \quad i = 1, \cdots, NG \tag{12}$$

2) Transformer constraints:
   The discrete transformer tap settings are restricted by their limits as

$$T_i^{\min} \leq T_i \leq T_i^{\max}, \quad i = 1, \cdots, T_N \tag{13}$$

3) Shunt VAR constraints:
   The discrete reactive power injections due to capacitor banks are restricted by their lower and upper limits defined as

$$Q_{C_i}^{\min} \leq Q_{C_i} \leq Q_{C_i}^{\max}, \quad i = 1, \cdots, C_N \tag{14}$$

4) Security constraints:
   The constraints of voltages at load buses as well as the transmission lines loading should be ensured as

$$V_{L_i}^{\min} \leq V_{L_i} \leq V_{L_i}^{\max}, \quad i = 1, \cdots, NL \tag{15}$$

$$S_{l_i} \leq S_{l_i}^{\max}, \quad i = 1, \cdots, nl \tag{16}$$

4 Modern heuristic algorithms for the OPF

This section provides a detailed survey on the OPF related research works with the applications of GA, PSO, DE, EP, and some other commonly used HOAs.

4.1 Genetic algorithm based approach

GA was successfully used for optimal reactive power planning in [22] to search for a global optimal solution. It has been verified on practical 51-bus and 224-bus systems to indicate its feasibility and capability. An improved genetic algorithm (IGA) with the dynamical hierarchy of the coding system was developed to solve the OPF problem [23]. The IGA demonstrate the ability to code a large number of control variables in a practical system. It was tested on the IEEE 30-bus system with both normal and contingent operation states. OPF problem for a multi-node auction market was studied by means of GA in [24] to maximize the total participants' benefit at all nodes in the power system. In [25], a self-adaptive real coded genetic algorithm (SARGA) was developed to solve OPF problem, where the self-adaptation in real coded genetic algorithm was reached through simulated binary crossover operator. A novel evolutionary algorithm was developed combing a new decoupled quadratic load flow (DQLF) solution with enhanced genetic algorithm (EGA) to solve the multi-objective OPF problem [26]. A strength pareto evolutionary algorithm (SPEA) based approach was employed to obtain the Pareto-optimal set. The proposed multi-objective evolutionary algorithm demonstrates superiority in comparisons to PSO–Fuzzy approach. An adaptive genetic algorithm (AGA) was developed to solve OPF problems and voltage control [27], where the probabilities of crossover and mutation were adjusted in terms of the fitness values of the solutions and the normalized fitness distances between the solutions in the evolution process. In [28], a refined genetic algorithm (RGA) was developed for solving OPF problem. This GA can code a large number of control variables and has less sensitivity to starting points. GA was also used to deal with power system security enhancement based OPF in [29] considering the actions to possible overloads in the network due to contingencies. An enhanced genetic algorithm (EGA) with advanced and problem-specific operators was introduced for solving OPF with both continuous and discrete control variables [30]. An efficient real-coded mixed-integer genetic algorithm (MIGA) was presented in [31] to solve non-convex OPF problems with security constraints. According to the numerical studies on 26-bus and the IEEE 57-bus systems, the MIGA performs better than the EP. A novel hybrid method integrating a GA with a nonlinear interior point method (IPM) was proposed for OPF problem [32]. In the hybrid approach, GA was responsible for solving the discrete optimization with the continuous variables, and the IPM is responsible for solving the continuous optimization with the discrete variables. Numerical simulations were implemented on IEEE 30-bus, IEEE 118-bus and realistic Chongqing 161-bus test systems. Reference [33] discussed the effects of various combination of control variables on the convergence of simple genetic algorithm. Statistical parameter based study was conducted to visualize the effects of the selection of control variables on OPF convergence in terms of the computation time and the accuracy improvement. The experiment results proved that the set of control variables with the voltage of slack bus, the active/reactive power outputs of generators, and the reactive power outputs of controllable buses can be the most effective in obtaining the global solution under normal and contingent conditions. In [34], it was claimed that the main disadvantages of GAs was the high CPU execution time and the qualities of the solution deteriorate with practical large-scale OPF problems. An efficient parallel GA was developed for the solution of large-scale OPF problem with

the consideration of practical generators constraints. The length of the original chromosome was reduced on basis of the decomposition level and adapted with the topology of the new partition. Partial decomposed active power demand was added as a new variable and searched within the active power generation variables of the new decomposed chromosome. The strategy of the OPF problem was decomposed into two sub-problems, of which the first sub-problem was related to active power planning to minimize the fuel cost function and the second sub-problem was designed to make corrections to the voltage deviation and reactive power violation in an efficient reactive power planning of multi static var compensator (SVC). Numerical results on three test systems IEEE 30-bus, IEEE 118-bus and 15 generation units with prohibited zones were presented and compared with results of stochastic search algorithms, enhanced GA, ant colony optimization, and GA-fuzzy system approach.

4.2 Particle swarm optimization based approach

A novel particle swarm optimization approach based on multi-agent systems (MAPSO) was presented [35] to solve OPF problems. Each agent, representing a particle to PSO, in MAPSO competes and cooperates with its neighbors. Experiment results prove that the proposed MAPSO approach can reach better solutions much faster than the mature approaches. A multi-objective PSO technique was developed to deal with the highly nonlinear and non-convex multi-objective OPF problem [36]. In addition to the conventional objective generation cost, another conflicting objective environmental pollution is formulated and minimized simultaneously. A fuzzy based hybrid PSO approach for solving OPF problem considering the forecasting uncertainties of wind speed and load demand in power systems was proposed in [37]. A comprehensive learning PSO (CLPSO) was developed to reactive power dispatch to reduce grid congestions [38]. A new multi-objective PSO (MOPSO) technique for solving OPF problem was proposed in [39]. The proposed MOPSO methodology is formulated via the redefinition of global best and local best individuals in multi-objective optimization domain. Reference [40] presented a hybrid particle swarm optimization algorithm (HPSO) to solve the discrete OPF problem. Newton-Raphson algorithm for the minimization of the mismatch of the power flow equations was integrated to the proposed HPSO algorithm. PSO technique was applied for the transient-stability constrained OPF (TSCOPF) problem modeled as an extended OPF with additional rotor angle inequality constraints [41]. Reference [42] proposed a PSO algorithm with reconstruction operators to solve the OPF problem with embedded security constraints (OPF-SC), represented by a mixture of continuous and discrete control variables. The major objective is to minimize the total operating cost, taking into account both operating security constraints and system capacity requirements. The reconstruction operators guarantee searching the optimal solution within the feasible space, reducing the computation time and improving the quality of the solution. An improved PSO algorithm was developed for multi-objective OPF problem. The improved PSO that profits from chaos queues and self-adaptive concepts was used to improve the quality of the solution, particularly to avoid being trapped in local optima. In addition, a new mutation strategy combining different mutant rules was proposed to increase the search ability of the proposed algorithm. The proposed multi-objective OPF considers the fuel cost, loss, voltage stability and emission impacts involved in the objective functions. A fuzzy decision-based mechanism was used to select the best compromise solution of Pareto set obtained by the proposed PSO. In [43], PSO and group search optimizer (GSO) were used to solve the OPF problem with distributed generator failures in power networks. An OPF problem considering controllable and uncontrollable distributed generators was formulated, and cases with single and multiple generator failures were addressed.

4.3 Differential evolution based approach

A multi-agent based differential evolution (MADE) based on multi-agent systems was developed for dealing with OPF problem with non-smooth and non-convex generator fuel cost curves in [44]. A novel robust differential evolution algorithm (RDEA) with new recombination operator was introduced to solve multi-objective OPF problem including two objective functions of generation cost and voltage stability margin, for OPF problem [45]. Similarly, DE was used to solve OPF problem with multiple and competing objectives [46]. The OPF problem was divided into two sub-problems, i.e, active power dispatch and reactive power dispatch were considered. A DE-based approach to solve the OPF problem was developed in [47]. In their formulation, different objective functions that reflect fuel cost minimization, voltage profile improvement, and voltage stability enhancement were examined. Non-smooth pricewise quadratic cost function was also been considered. Reference [48] proposed a similar formulation of OPF with non-smooth and non-convex generator fuel cost curves. They employed a modified DE with a more exploitative mutation strategy and a random mutant factor. For testing purpose, the authors adopted a six-bus and the IEEE 30 bus test systems with three different types of generator cost curves. Comparisons were made among EP, PSO, typical DE, and results were in favor of the proposed modified DE. In [49], DE was comprehensively studied in terms of concept, mechanism, and parameter setting for solving OPF problems. The effectiveness of parallel computing technology for speeding up the computation was also analyzed. It has been concluded that DE requires relatively large populations to avoid premature convergence for medium-size test systems. In order to overcome this disadvantage, a decomposition and coordination method was proposed by the same authors based on the cooperative co-evolutionary architecture and the voltage-var sensitivity-based power system decomposition technique and incorporated with DE in [50]. The framework was implemented with a three-level parallel computing topology. Basu has used DE to minimize the generator fuel cost in OPF with flexible AC transmission systems (FACTS) devices including thyristor-controlled series capacitor (TCSC) and thyristor-controlled phase shifter (TCPS) [51]. Comparisons among DE, EP, and GA were conducted, indicating that the DE approach can obtain better solution and less computational complexities. Considering the transient stability constraints in OPF, In [52], DE was used to find the optimal setting for power system operation. To deal with the large-scale system and speed up the computation, DE was parallelized and implemented on a Beowulf PC-cluster. Reference [53] proposed a hybrid algorithm combining sequential quadratic programming (SQP) and DE for the solution of the OPF problem. In this hybrid method, SQP was used to generate an individual, which is a member of an initial population, for DE algorithm. This manipulation made DE more effectively to reach the optimal solution.

4.4 Evolutionary programming

In [54] and [55], an efficient and reliable EP algorithm was developed to solve OPF problem using the gradient information. The proposed algorithm has been successfully tested on IEEE 30-bus system with different highly nonlinear curves of generator performance. An improved EP and its hybrid version combined with the nonlinear interior point technique were proposed in [56] to solve the optimal reactive power dispatch problems, indicating the superiority of computational efficiency and optimality. The common practices in regulating reactive power were integrated in modifying the mutation direction of control variables of EP to improve its speed. The interior point method was applied to reach a fast initial solution which assisted the initial population of the improved EP method. An improved EP with multiple subpopulations and parallel search for solving OPF with non-smooth and non-convex generator fuel cost curves was proposed in [57]. Gaussian and Cauchy mutation operators have been included in different subpopulations to improve the search diversity and avoid the local optimum. In [58], EP was applied for solving security constrained optimal power flow (SCOPF) problem, where contingency-case security constraints were involved in the optimization of the defined objective function. The EP based OPF in deregulated electric market environment was used and validated in [59]. In [60], EP algorithm was proposed to solve the OPF problem of generator units with ramp rate limits and non-smooth fuel functions such as quadratic, piece-wise, valve point loading and combined cycle cogeneration plants.

4.5 Other techniques

In [61], artificial neural networks (ANNs) have been employed to model stability and security constraints in OPF to formulate the system security boundary (SB). The key novelties of the proposed algorithm include that a NN is trained to derive the SB model and a differentiable mapping function obtained from the NN is used as a constraint in the formulation of OPF problem. This approach ensures that the operating points resulting from the OPF solution process are within a feasible and secure region, comparing with typical security-constrained OPF models. A new ANN memory model-based algorithm was proposed to online implement for solving the unified OPF [62]. The ANN memory model was used to store the load patterns and their related optimal schedules. The proposed algorithm maximizes voltage stability margin while minimizing two other objectives generation cost and transmission loss.

Different ant colony optimization (ACO) algorithms were proposed to handle optimal reactive power dispatch problem in [63], including Ant system (AS), elitist ant system (EAS), rank-based ant system (ASrank) and max-min ant system (MMAS). The problem was modeled as a combinatorial optimization problem involving nonlinear objective function with multiple local minima. The proposed ACO algorithms have been compared to conventional mathematical methods, i.e. genetic algorithm, evolutionary programming, and particle swarm optimization to demonstrate the effectiveness and efficiency. An ant colony system (ACS) method for constrained load flow problem was proposed in [64]. The proposed ACS is a distributed algorithm consisting of a set of cooperating ants to collaboratively search an optimum solution of the constrained load flow problem. In addition, the ACS algorithm was also applied to the reactive power control problem with network operating constraints to minimize real power losses.

Simulated annealing (SA) technique was proposed for solving OPF in [65]. It has been demonstrated that SA is able to solve the OPF problem as well as the load flow and the economic dispatch problem simultaneously. In [66], a novel HOA algorithm, called biogeography based optimization (BBO) was employed to solve constrained OPF problems in power systems with the consideration of valve

point nonlinearities of generators. The simulation results of the proposed approach have been compared with EP, GA, PSO, mixed-integer particle swarm optimization (MIPSO) and sequential quadratic programming (SQP) to indicate its effectiveness for the global optimization of multi-constraint OPF problems. In [67], a quantum-inspired evolutionary algorithm (QEA) based on quantum computation was developed for bid-based active and reactive OPF problems. In [68], artificial bee colony (ABC) algorithm based on the intelligent foraging behavior of honeybee swarm was proposed for optimal reactive power flow (ORPF) problem.

4.6 Hybrid approach

A hybrid tabu search and simulated annealing (TS/SA) approach was proposed in [69] to deal with OPF control with FACTS devices including two types TCSC and TCPS. Test results on IEEE 30-bus system demonstrate that the proposed hybrid TS/SA approach can perform better than GA, SA, or TS alone. A hybrid approach integrating fuzzy systems with GA and PSO algorithm was proposed for the application for OPF problem [70]. A hybrid algorithm of DE and EP (DEEP) was proposed for solving ORPF problem [71]. The proposed DEEP algorithm reduces the required population size by using the advantages of DE and EP. In order to overcome the limits of DE and artificial bee colony (ABC), a hybrid DE and ABC technique (DE-ABC) was developed for solving the ORPF problem [72]. Numerical tests indicate the robustness of the DE-ABC approach. A hybrid evolving ant direction differential evolution (EADDE) algorithm was developed to deal with the OPF problem with non-smooth and non-convex generator fuel cost characteristics in [73].

5 Discussion

HOAs are typically very versatile with respect to OPF problem format. In addition, most HOAs are able to escape local optima which is critical for solving OPF problem.

However, all the HOAs discussed tend to be computationally intensive, yielding impractically long execution times for OPF problems involving large scale systems. To overcome this, parallel processing was executed by most of the reviewed population based HOAs, therefore, the computational time can be significantly reduced.

Furthermore, the reviewed HOAs possess several parameters which must be tuned to ensure good performance. Consideration of the tuning of pre-determined parameters of HOAs and the time limitation of online OPF operation will make these algorithms less robust. Hence, one of the most challenging aspects for HOAs lies in how to consistently converge to a feasible solution that provides an acceptable objective value within a limited function evaluations.

In the reviewed literatures, on the premise of a proper pre-defined parameter choices, almost every HOA method is claimed as being more robust or can converge to a better solution compared with other HOAs. However, comparisons between different HOAs are difficult, as the selection of pre-defined parameters for each HOAs dominates the results. Moreover, according to the no-free-lunch (NFL) theorem, there cannot exist any algorithm for solving all problems that is generally superior to its peers [74]. Therefore, the authors suggest that HOAs should be designed with respect to solving a specific aspect or formulation of OPF problem, i.e. MINLP formulation, so as the comparisons being made.

6 Conclusions

This paper presents relevant research work applying HOAs for solving the OPF problem. It highlights the difficulties of deterministic methods facing i.e. non-continuity and non-differentiability of the OPF objective function. The reviewed articles are organized according to different categories of HOAs. A brief discussion of the limitations and trends of HOA applications with respect to OPF problem is presented in addition to the corresponding literature reviews.

Acknowledgment This work was partially supported by Hong Kong RGC Theme Based Research Scheme Grants No. T23-407/13 N and T23-701/14 N. The work of Ming NIU was supported by a Ph.D. Research Studentship. The work of Can WAN was supported by a Hong Kong Ph.D. Fellowship.

References

[1] Almeida KC, Salgado R (2000) Optimal power flow solutions under variable load conditions. IEEE Trans Power Syst 15(4):1204–1211

[2] Wood AJ, Wollenberg BF (2012) Power generation, operation, and control, 3rd edn. Wiley, New York

[3] Boyd S, Vandenberghe L (2009) Convex optimization. Cambridge University Press, New York

[4] Lima FGM, Galiana FD, Kockar I et al (2003) Phase shifter placement in large-scale systems via mixed integer linear programming. IEEE Trans Power Syst 18(3):1029–1034

[5] Holland JH (1975) Adaptation in natural and artificial systems: an introductory analysis with applications to biology, control, and artificial intelligence. University of Michigan Press, Ann Arbor

[6] Goldberg DE (1989) Genetic algorithms in search, optimization and machine learning. Addison-Wesley Longman Publishing Co, Reading

[7] Kennedy J, Eberhart R (1995) Particle swarm optimization. In: Proceedings of the IEEE international conference on neural networks, vol 4. Perth, WA, USA, 27 Nov–1 Dec 1995, pp 1942–1948

[8] Eberhart R, Kennedy J (1995) A new optimizer using particle swarm theory. In: Proceedings of the 6th international symposium on micro machine and human science (MHS '95), Nagoya, Japan, 4–6 Oct 1995, pp 39–43

[9] Storn R, Price K (1995) Differential evolution: a simple and efficient adaptive scheme for global optimization over continuous spaces. TR-95-012, International Computer Science Institut (ICSI), Berkeley, CA, USA

[10] Lakshminarasimman L, Subramanian S (2006) Short-term scheduling of hydrothermal power system with cascaded reservoirs by using modified differential evolution. IEE P-Gener Transm Distrib 153(6):693–700

[11] Wang Z, Chung CY, Wong KP et al (2008) Robust power system stabiliser design under multi-operating conditions using differential evolution. IET Gener Transm Distrib 2(5):690–700

[12] Vakula VS, Sudha KR (2012) Design of differential evolution algorithm-based robust fuzzy logic power system stabiliser using minimum rule base. IET Gener Transm Distrib 6(2):121–132

[13] Fogel LJ, Burgin GH (1969) Competitive goal-seeking through evolutionary programming. Air Force Cambridge Research Laboratories, Cambridge

[14] Back T, Fogel DB, Michalewicz Z (1997) Handbook of evolutionary computation. IOP Publishing Ltd, Bristol

[15] Dorigo M, Birattari M (2010) Ant colony optimization. In: Sammut C (ed) Encyclopedia of machine learning. Springer Science, Boston, pp p36–p39

[16] Hagan MT, Demuth HB, Beale MH (1996) Neural network design, vol 1. PWS Publishing Company, Boston

[17] Van Laarhoven PJM, Aarts EHL (1987) Simulated annealing: theory and applications. Reidel, Publishing Co, Dordrecht

[18] Glover F, Laguna M (1999) Tabu search. Kluwer Academic Publishers, Boston

[19] Han KH, Kim JH (2002) Quantum-inspired evolutionary algorithm for a class of combinatorial optimization. IEEE Trans Evolut Comput 6(6):580–593

[20] Han KH, Kim JH (2004) Quantum-inspired evolutionary algorithms with a new termination criterion, H_ε, gate, and two-phase scheme. IEEE Trans Evolut Comput 8(2):156–169

[21] Karaboga D, Basturk B (2008) On the performance of artificial bee colony (ABC) algorithm. Appl Soft Comput 8(1):687–697

[22] Iba K (1994) Reactive power optimization by genetic algorithm. IEEE Trans Power Syst 9(2):685–692

[23] Lai LL, Ma JT, Yokoyama R et al (1997) Improved genetic algorithms for optimal power flow under both normal and contingent operation states. Int J Electr Power Energy Syst 19(5):287–292

[24] Numnonda T, Annakkage UD (1999) Optimal power dispatch in multinode electricity market using genetic algorithm. Electr Power Syst Res 49(3):211–220

[25] Subbaraj P, Rajnarayanan PN (2009) Optimal reactive power dispatch using self-adaptive real coded genetic algorithm. Electr Power Syst Res 79(2):374–381

[26] Sailaja Kumari M, Maheswarapu S (2010) Enhanced genetic algorithm based computation technique for multi-objective optimal power flow solution. Int J Electr Power Energy Syst 32(6):736–742

[27] Wu QH, Cao YJ, Wen JY (1998) Optimal reactive power dispatch using an adaptive genetic algorithm. Int J Electr Power Energy Syst 20(8):563–569

[28] Paranjothi SR, Anburaja K (2002) Optimal power flow using refined genetic algorithm. Electr Power Compon Syst 30(10):1055–1063

[29] Devaraj D, Yegnanarayana B (2005) Genetic-algorithm-based optimal power flow for security enhancement. IEE P-Gener Transm Distrib 152(6):899–905

[30] Bakirtzis AG, Biskas PN, Zoumas CE et al (2002) Optimal power flow by enhanced genetic algorithm. IEEE Trans Power Syst 17(2):229–236

[31] Gaing ZL, Chang RF (2006) Security-constrained optimal power flow by mixed-integer genetic algorithm with arithmetic operators. In: Proceedings of the Power Engineering Society general meeting. Montreal, Canada, 18–22 Jun 2006, 8 pp

[32] Yan W, Liu F, Chung CY et al (2006) A hybrid genetic algorithm-interior point method for optimal reactive power flow. IEEE Trans Power Syst 21(3):1163–1169

[33] Wankhade CM, Vaidya AP (2014) Optimal power flow using genetic algorithm: parametric studies for selection of control and states variables. Br J Appl Sci Technol 4(2):279–301

[34] Mahdad B, Srairi K, Bouktir T (2010) Optimal power flow for large-scale power system with shunt FACTS using efficient parallel GA. Int J Electr Power Energy Syst 32(5):507–517

[35] Zhao B, Guo CX, Cao YJ (2005) A multiagent-based particle swarm optimization approach for optimal reactive power dispatch. IEEE Trans Power Syst 20(2):1070–1078

[36] Hazra J, Sinha AK (2011) A multi-objective optimal power flow using particle swarm optimization. Eur Trans Electr Power 21(1):1028–1045

[37] Liang RH, Tsai SR, Chen YT et al (2011) Optimal power flow by a fuzzy based hybrid particle swarm optimization approach. Electr Power Syst Res 81(7):1466–1474

[38] Mahadevan K, Kannan PS (2010) Comprehensive learning particle swarm optimization for reactive power dispatch. Appl Soft Comput 10(2):641–652

[39] Abido MA (2011) Multiobjective particle swarm optimization for optimal power flow problem. In: Panigrahi B, Shi Y, Lim MH et al (eds) Handbook of swarm intelligence, vol 8. Springer, Berlin, pp p241–p268

[40] AlRashidi MR, El-Hawary ME (2007) Hybrid particle swarm optimization approach for solving the discrete OPF problem considering the valve loading effects. IEEE Trans Power Syst 22(4):2030–2038

[41] Mo N, Zou ZY, Chan KW et al (2007) Transient stability constrained optimal power flow using particle swarm optimisation. IET Gener Transm Distrib 1(3):476–483

[42] Onate Yumbla PE, Ramirez JM, Coello Coello CA (2008) Optimal power flow subject to security constraints solved with a particle swarm optimizer. IEEE Trans Power Syst 23(1):33–40

[43] Kang Q, Zhou MC, An J et al (2013) Swarm intelligence approaches to optimal power flow problem with distributed generator failures in power networks. IEEE Trans Automat Sci Eng 10(2):343–353

[44] Sivasubramani S, Swarup KS (2012) Multiagent based differential evolution approach to optimal power flow. Appl Soft Comput 12(2):735–740

[45] Amjady N, Sharifzadeh H (2011) Security constrained optimal power flow considering detailed generator model by a new robust differential evolution algorithm. Electr Power Syst Res 81(2):740–749

[46] Varadarajan M, Swarup KS (2008) Solving multi-objective optimal power flow using differential evolution. IET Gener Transm Distrib 2(5):720–730

[47] Abou El Ela AA, Abido MA, Spea SR (2010) Optimal power flow using differential evolution algorithm. Electr Power Syst Res 80(7):878–885

[48] Sayah S, Zehar K (2008) Modified differential evolution algorithm for optimal power flow with non-smooth cost functions. Energy Convers Manage 49(11):3036–3042

[49] Liang CH, Chung CY, Wong KP et al (2007) Study of differential evolution for optimal reactive power flow. IET Gener Transm Distrib 1(2):253–260

[50] Liang CH, Chung CY, Wong KP et al (2007) Parallel optimal reactive power flow based on cooperative co-evolutionary differential evolution and power system decomposition. IEEE Trans Power Syst 22(1):249–257

[51] Basu M (2008) Optimal power flow with FACTS devices using differential evolution. Int J Electr Power Energy Syst 30(2):150–156

[52] Cai HR, Chung CY, Wong KP (2008) Application of differential evolution algorithm for transient stability constrained optimal power flow. IEEE Trans Power Syst 23(2):719–728

[53] Sivasubramani S, Swarup KS (2011) Sequential quadratic programming based differential evolution algorithm for optimal power flow problem. IET Gener Transm Distrib 5(11): 1149–1154

[54] Yuryevich J, Wong KP (1999) Evolutionary programming based optimal power flow algorithm. IEEE Trans Power Syst 14(4):1245–1250

[55] Wong KP, Yuryevich J (1999) Optimal power flow method using evolutionary programming. In: Simulated evolution and learning, Proceedings of the 2nd Asia-Pacific conference on simulated evolution and learning (SEAL'98), LACS 1585, Canberra, Australia, 24–27 Nov 1998, pp 405–412

[56] Yan W, Lu S, Yu DC (2004) A novel optimal reactive power dispatch method based on an improved hybrid evolutionary programming technique. IEEE Trans Power Syst 19(2):913–918

[57] Ongsakul W, Tantimaporn T (2006) Optimal power flow by improved evolutionary programming. Electr Power Compon Syst 34(1):79–95

[58] Somasundaram P, Kuppusamy K, Kumudini Devi RP (2004) Evolutionary programming based security constrained optimal power flow. Electr Power Syst Res 72(2):137–145

[59] Sood YR (2007) Evolutionary programming based optimal power flow and its validation for deregulated power system analysis. Int J Electr Power Energy Syst 29(1):65–75

[60] Gnanadass R, Venkatesh P, Padhy NP (2004) Evolutionary programming based optimal power flow for units with non-smooth fuel cost functions. Electr Power Compon Syst 33(3):349–361

[61] Gutierrez-Martinez VJ, Cañizares CA, Fuerte-Esquivel CR et al (2011) Neural-network security-boundary constrained optimal power flow. IEEE Trans Power Syst 26(1):63–72

[62] Venkatesh B (2003) Online ANN memory model-based method for unified OPF and voltage stability margin maximization. Electr Power Compon Syst 31(5):453–465

[63] Abbasy A, Hosseini SH (2007) Ant colony optimization-based approach to optimal reactive power dispatch: a comparison of various ant systems. In: Proceedings of the Power Engineering Society conference and exposition in Africa (PowerAfrica'07), Johannesburg, South Africa, 16–20 Jul 2007, 8 pp

[64] Vlachogiannis JG, Hatziargyriou ND, Lee KY (2005) Ant colony system-based algorithm for constrained load flow problem. IEEE Trans Power Syst 20(3):1241–1249

[65] Roa-Sepulveda CA, Pavez-Lazo BJ (2003) A solution to the optimal power flow using simulated annealing. Int J Electr Power Energy Syst 25(1):47–57

[66] Roy PK, Ghoshal SP, Thakur SS (2010) Biogeography based optimization for multi-constraint optimal power flow with emission and non-smooth cost function. Expert Syst Appl 37(12):8221–8228

[67] Vlachogiannis JG, Lee KY (2008) Quantum-inspired evolutionary algorithm for real and reactive power dispatch. IEEE Trans Power Syst 23(4):1627–1636

[68] Ayan K, Kılıç U (2012) Artificial bee colony algorithm solution for optimal reactive power flow. Appl Soft Comput 12(5):1477–1482

[69] Ongsakul W, Bhasaputra P (2002) Optimal power flow with FACTS devices by hybrid TS/SA approach. Int J Electr Power Energy Syst 24(10):851–857

[70] Kumar S, Chaturvedi DK (2013) Optimal power flow solution using fuzzy evolutionary and swarm optimization. Int J Electr Power Energy Syst 47:416–423

[71] Chung CY, Liang CH, Wong KP et al (2010) Hybrid algorithm of differential evolution and evolutionary programming for optimal reactive power flow. IET Gener Transm Distrib 4(1):84–93

[72] Li Y, Wang Y, Li B (2013) A hybrid artificial bee colony assisted differential evolution algorithm for optimal reactive power flow. Int J Electr Power Energy Syst 52:25–33

[73] Vaisakh K, Srinivas LR (2011) Evolving ant direction differential evolution for OPF with non-smooth cost functions. Eng Appl Artif Intell 24(3):426–436

[74] Wolpert DH, Macready WG (1997) No free lunch theorems for optimization. IEEE Trans Evolut Comput 1(1):67–82

Ming NIU received the B.Eng. degree from Dalian University of technology, Dalian, China, in 2010, the MSc degree from The Hong Kong Polytechnic University, Hong Kong, in 2012. He is currently working toward the Ph.D. degree from The Hong Kong Polytechnic University. His research interests include power system planning and operation, computational intelligence, and optimization techniques applied in system networks.

Can WAN received the B.Eng. degree from Zhejiang University, Hangzhou, China, in 2008. He is currently pursuing his Ph.D. degree from The Hong Kong Polytechnic University. He was previously a Visiting Ph.D. with Centre for Electric Power and Energy, Technical University of Denmark, and State Key Laboratory of Power Systems, Department of Electrical Engineering, Tsinghua University. His research interests include power system uncertainty analysis and operation, grid integration of renewable energies, smart grids, machine learning and system engineering and control.

Zhao XU received the B.Eng. degree from Zhejiang University, Hangzhou, China, in 1996, the M.Eng. degree from National University of Singapore, Singapore, in 2002, and the Ph.D. degree from The University of Queensland, Australia, in 2006. He is now with The Hong Kong Polytechnic University. He was previously an Associate Professor with Centre for Electric Power and Energy, Technical University of Denmark. His research interests include demand side, grid integration of renewable energies and EVs, electricity market planning and management, and AI applications in power engineering. He is an editor of Electric Power Components and Systems.

11

Robust current control design of a three phase voltage source converter

Wenming GONG, Shuju HU (✉), Martin SHAN,
Honghua XU

Abstract In this paper, a robust design method for current control is proposed to improve the performance of a three phase voltage source converter (VSC) with an inductor-capacitor-inductor (LCL) filter. The presence of the LCL filter complicates the dynamics of the control system and limits the achievable control bandwidth (and the overall performance), particularly when the uncertainty of the parameters is considered. To solve this problem, the advanced H∞ control theory is employed to design a robust current controller in stationary coordinates. Both control of the fundamental frequency current and suppression of the potential LC resonance are considered. The design procedure and the selection of the weight functions are presented in detail. A conventional proportional-resonant PR controller is also designed for comparison. Analysis showed that the proposed H∞ current controller achieved a good frequency response with explicit robustness. The conclusion was verified on a 5 kW VSC that had a LCL filter.

Keywords Current control, H∞ control, Robust control, Inductor-capacitor-inductor (LCL) filter, Voltage source converter (VSC)

W. GONG, University of Chinese Academy of Sciences, Beijing 100190, China
e-mail: gwm@mail.iee.ac.cn
S. HU, H. XU, Institute of Electrical Engineering, Chinese Academy of Sciences, Beijing 100190, China
(✉) e-mail: hushuju@mail.iee.ac.cn
H. XU
e-mail: hxu@mail.iee.ac.cn
M. SHAN, Control Engineering and Energy Storage Systems, Fraunhofer IWES, 34119 Kassel, Germany
e-mail: martin.shan@iwes.fraunhofer.de

1 Introduction

With the use of a pulse width modulation (PWM) technique, the voltage source converter (VSC) had excellent P&Q operation characteristics, which is shown in Fig. 1. VSCs have been widely used in renewable energy production, smart grids, uninterruptable power supplies (UPSs), active power filtering (APF), and electrical drives [1]. With the development of self-commutated power electronic devices, the capacity of VSCs have been sustained with their growth. The state of the art of this technology is its application in VSC-HVDC systems [2, 3].

In high power applications, a LCL filter is usually employed to filter out the PWM switching frequency components. Compared with the L filter of a single inductor, the LCL filter is much smaller in weight and size, but has a similar low frequency performance. Therefore, the LCL filter is a better choice in this context because it allows for the use of a lower switching frequency that meets the harmonic limits and further decreases the switching losses [4, 5].

A critical innate problem that the LCL filters have is a low damped LC resonance. Resistors can be simply used to provide passive damping to the resonance. However, this has the disadvantage of an increased power loss [6]. Several active damping control strategies have been suggested, such as lead-lag compensation [7] and virtual resistance [8, 9]. However, these control methods usually limit the control bandwidth. There is a tradeoff between the closed loop performance and the system stability. In addition, the uncertainty in the grid environment worsens this problem. Grid operation, cable overload, saturation, and temperature effects can cause variations in the interfacing impedance and the resonance frequency, which makes the control design much more difficult [10].

To enhance the performance of a VSC with all these restrictions, the H∞ control theory was applied to design a robust current controller for the VSC system in stationary coordinates in this paper. The H∞ control theory dealing with the model and disturbance uncertainty. This is a relatively new method for the controller design of VSCs. However, there are still several successful examples. In [11], the authors developed a tuning strategy for designing multiple-loop lag-lead compensators for uninterruptable power supplies (UPSs) based on the H∞ robust control theory. The H∞ tuned compensator was asymptotically stable with guaranteed performance. In [12], a robust control scheme with an outer H∞ voltage control loop and an inner current control loop was designed for a dynamic voltage restorer (DVR). The H∞ voltage controller was effective in both balanced and unbalanced sag compensation as well as load disturbance rejection. In [13], a H∞ current controller was designed for a grid connected to a power factor correction (PFC). It was possible to explicitly specify the degree of robustness with variations in the parameters using the proposed method. A mixed-sensitivity method was used to synthesize the H∞ controller. A high pass weight function was used to guarantee a robust performance and to avoid possible oscillations caused by the capacitor connected in parallel. However, none of the previous works focused on VSCs with an LCL filter. In this paper the H∞ method is further developed; the ability to achieve LC resonance attenuation is enhanced. Alternatively, from the previous work, the LC resonance in this paper is modeled as a noise channel and the corresponding weight functions are proposed to provide a robust performance.

The remainder of this paper is organized as follows: in Section 2, the system model is introduced and the problem with the LC resonance is described. In Section 3, the design procedure of the proposed H∞ current controller and the selection of weight functions are presented. A conventional PR controller is also designed for comparison. The performance is then analyzed in detail. In Section 4, the experimental results are explained. Finally, conclusions are given in Section 5.

2 System modelling

2.1 Mathematical model of the VSC system

As shown in Fig. 1, the studied VSC was connected to the grid via the LCL interface. R_c and L_c are the resistance and inductance for the inductor of the converter side, respectively; R_g and L_g represent the equivalent resistance and inductance for the inductor of the grid side and the AC grid line impedance, respectively; C_f is the filter capacitor; C_{dc} is the DC capacitor; u_g is the grid voltage; u_f is the filter capacitor voltage; i_g is the injected grid current; i_c is the converter output current; u_c is the converter output voltage; u_{dc} is the DC capacitor; and i_{dc} is the DC load current.

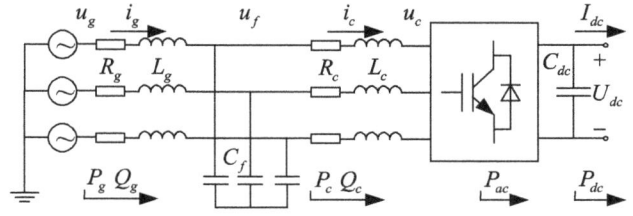

Fig. 1 The grid connected voltage source converter

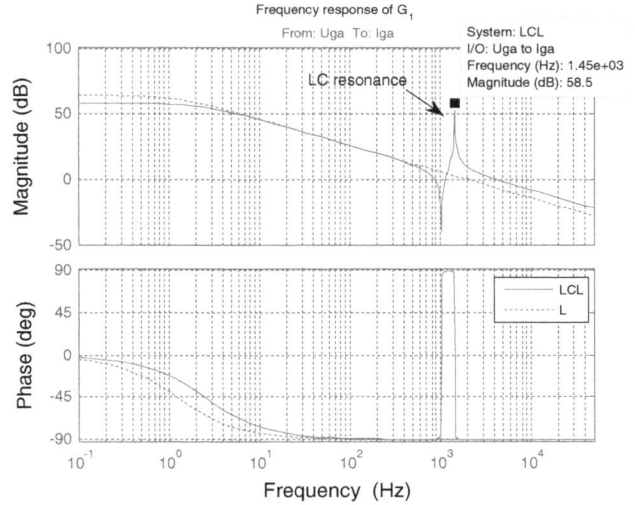

Fig. 2 Frequency response of the open loop system of a VSC

In stationary α-β coordinates, the per-phase mathematic model of the AC side can be written as:

$$\begin{cases} L_g \frac{di_g}{dt} = -R_g i_g + u_g - u_f \\ C_f \frac{du_f}{dt} = i_g - i_c \\ L_c \frac{di_c}{dt} = -R_c i_c + u_f - u_c. \end{cases} \quad (1)$$

On the DC side the model can be written as,

$$C_{dc} \frac{dU_{dc}}{dt} = \frac{P_{ac}}{U_{dc}} - I_{dc} = \frac{3(u_{c\alpha}i_{c\alpha} + u_{c\beta}i_{c\beta})}{2U_{dc}} - I_{dc}. \quad (2)$$

The values for the parameters are shown in Appendix A. The impedance of the AC grid line was uncertain. In addition, taking into consideration of the measurement error and nonlinear characteristics, ±50% uncertainty in all of the parameters were reasonable for robust analysis.

2.2 LC resonance

From (1), the transfer function of the AC side was derived as:

$$i_g = G_1 u_g + G_2 u_c. \quad (3)$$

The frequency response of G_1 is depicted in Fig. 2. The response is similar to that of G_2. For comparison, a system that used a common single L ($L_g + L_c$) filter is also

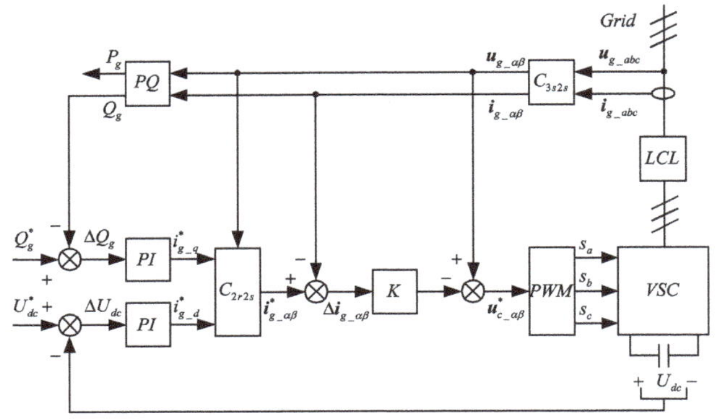

Fig. 3 Multiloop vector control scheme of a VSC

depicted. In the low frequency region, the two systems were similar to each other, which implied that there was a similar control design for the fundamental components. In the high frequency region, there was a single resonance peak with the LCL filter, which can be modeled as noise from the feedback system. The resonance frequency was about 1.45 kHz, which was calculated using:

$$f_{res} = \frac{\omega_{res}}{2\pi} = \frac{1}{2\pi} \sqrt{\frac{L_g + L_c}{L_g L_c C_f}}. \tag{4}$$

The LC resonance, which was caused by the LCL filter, needs to be dampened properly, otherwise the system will become unstable. Uncertainties in the system parameters can make the resonance frequency difficult to identify. In the next section, both the control design of the fundamental components and the suppression of the LC resonance are discussed.

3 Control design

3.1 Multi-loop control scheme

In most applications, a multi-loop vector control scheme, either in stationary or rotating coordinates, is adopted to improve the transient and steady-state performances [1]. The inner current loop is responsible for the overall stability of the system and for attenuating the LC resonance introduced by the filter. The outer loop is usually much slower than the inner loop.

In this paper, the outer loop was used to control the DC voltage and the output reactive power, as shown in Fig. 3. Meanwhile, the inner current loop was designed in stationary coordinates. The injected grid current was fed back into the system for ease of controlling the reactive power.

For coordinate transformations, the C_{3s2s} and C_{2r2s} blocks are:

$$C_{3s2s} : \begin{bmatrix} u_{g\_\alpha} \\ u_{g\_\beta} \end{bmatrix} = \begin{bmatrix} 1 & -\frac{1}{2} & -\frac{1}{2} \\ 0 & \frac{\sqrt{3}}{2} & -\frac{\sqrt{3}}{2} \end{bmatrix} \begin{bmatrix} u_{g\_a} \\ u_{g\_b} \\ u_{g\_c} \end{bmatrix} \tag{5}$$

$$\begin{bmatrix} i_{g\_\alpha} \\ i_{g\_\beta} \end{bmatrix} = \begin{bmatrix} 1 & -\frac{1}{2} & -\frac{1}{2} \\ 0 & \frac{\sqrt{3}}{2} & -\frac{\sqrt{3}}{2} \end{bmatrix} \begin{bmatrix} i_{g\_a} \\ i_{g\_b} \\ i_{g\_c} \end{bmatrix},$$

$$C_{2r2s} : \begin{bmatrix} i^*_{g\_\alpha} \\ i^*_{g\_\beta} \end{bmatrix} = \frac{1}{u^2_{g\_\alpha} + u^2_{g\_\beta}} \begin{bmatrix} u_{g\_\alpha} & -u_{g\_\beta} \\ u_{g\_\beta} & u_{g\_\alpha} \end{bmatrix} \begin{bmatrix} i^*_{g\_d} \\ i^*_{g\_q} \end{bmatrix}. \tag{6}$$

The PQ block was used to calculate active and reactive power:

$$PQ : \begin{bmatrix} P_g \\ Q_g \end{bmatrix} = \begin{bmatrix} u_{g\_\alpha} & u_{g\_\beta} \\ u_{g\_\beta} & -u_{g\_\alpha} \end{bmatrix} \begin{bmatrix} i_{g\_\alpha} \\ i_{g\_\beta} \end{bmatrix}. \tag{7}$$

3.2 PR current control design

In stationary coordinates, a PR controller is usually used to track the sinusoidal signals. In this paper, a PR controller was used for comparison.

$$K = \left(k_p + \frac{2k_i \omega_c s}{s^2 + 2\omega_c s + \omega_n^2} \right) \begin{bmatrix} 1 & 0 \\ 0 & 1 \end{bmatrix}, \tag{8}$$

where k_p and k_i are gain constants; ω_c is the cut-off frequency and $\omega_n = 100\pi$ for a 50 Hz system. More details of the control design can be found in [14].

3.3 H∞ current control design

The PR controller has been successfully used in many VSC systems. However, further improvements in its performance are limited by the presence of the LC resonance.

The advanced H∞ control theory was employed in this section in the current control design. The overall performance was specified by choosing proper weight functions. The controller was then calculated by solving the algebraic Riccati functions [15]. The order of the synthesized controller was equal to the sum of the controlled plant and the weight functions.

A diagram of the control design is shown in Fig. 4, where G_1 and G_2 are the same as in (3). W_p, W_u and W_n are the performance, control and noise weight functions, respectively. As can be seen in Fig. 2, L and LCL had similar frequency responses near the fundamental frequency. Therefore, an L filter model could be used to control the fundamental frequency components by reducing the order of the final controller. The LC resonance, which was caused by the LCL filter, was considered as noise. This is a new development for the design of a VSC control and it is sensible and effective, according to the experimental results.

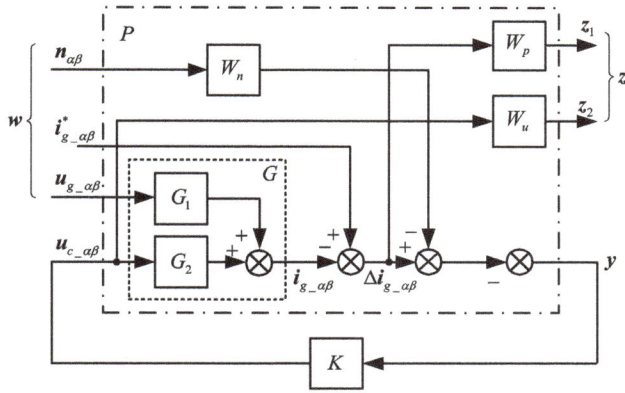

Fig. 4 Diagram of the H∞ current control design

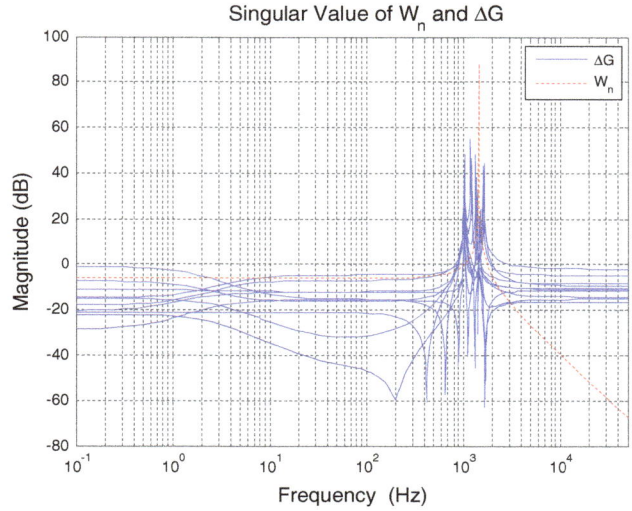

Fig. 5 Frequency response of the open loop system

In this mixed H∞ problem, the control design aimed to obtain a stable controller that minimized the H∞ norm of the generalized transfer function (P) from w to z as shown in Fig. 4.

$$\left\| \begin{array}{ccc} W_p S_o & -W_p S_o G_1 & W_p S_o G_2 K W_n \\ W_u S_i K & -W_u S_i K G_1 & -W_u S_i K W_n \end{array} \right\| < \gamma, \qquad (9)$$

where $S_i = (I + KG_2)^{-1}$ and $S_o = (I + G_2 K)^{-1}$ are the input and output sensitivity transfer functions, respectively, and γ is the peak value of the H∞ norm. For $\gamma < 1$, the system was robust stable within the desired uncertainty range.

From (6), the closed loop performance of the system was largely dependent on the shape of the weight functions. To prevent the control signal from being too large, $W_u = 1$ was chosen for the scaled signals. The other two weight functions were related to the tracking performance and the robust stability.

3.3.1 Performance weight function

The sensitivity function, S, is an indication of the tracking performance of the system. As such, the weight function (W_p) can be used to specify the control performance [15]. As shown in Fig. 3, the current reference is a sinusoidal signal in α-β coordinates. Thus, S should be as small as possible at the fundamental frequency. For the sinusoidal signal, a second order resonant filter was adopted:

$$W_p = \frac{k_n \omega_n^2}{s^2 + 2\zeta_n \omega_n s + \omega_n^2} \begin{bmatrix} 1 & 0 \\ 0 & 1 \end{bmatrix}, \qquad (10)$$

where ω_n, k_n and ζ_n are the fundamental frequency, the gain constant and the damping ratio, respectively.

3.3.2 Noise weight function

To improve the robust stability and suppress the LC resonance, a proper noise weight function needs to be

defined. The variations in the model can be measured using a singular input-output value [16].

$$\Delta(G) = \sigma_{\max} \left(\frac{G_P - G_N}{G_N} \right), \qquad (11)$$

where G_N is the nominal plant and G_P is the disturbed plant.

From Fig. 5, the variation in the circuit parameters caused a large uncertainty near the resonance frequency. There was only one resonance peak in the LCL filter. Thus, a second order resonant filter can be used to model the LC resonance a noise:

$$W_n = \frac{k_{res} \omega_{res}^2}{s^2 + 2\zeta_{res} \omega_{res} s + \omega_{res}^2} \begin{bmatrix} 1 & 0 \\ 0 & 1 \end{bmatrix}, \qquad (12)$$

where ω_{res}, k_{res} and ζ_{res} are the resonant frequency, the gain constant and the damping ratio, respectively.

3.4 Design results

Here, one PR controller and two types of H∞ controllers are compared. PR is a conventional control, for comparison. H∞(I) is a basic H∞ control. Only W_u and W_p were used (W_n was removed). For H∞(II): W_u, W_p, and W_n were used at the same time to guarantee that the overall performance was robust.

An important step in the design of the H∞ control is loop-shaping, which is similar to the traditional Bode's response method, but can be applied to Multiple-Input Multiple-Output systems. First, the weight functions were designed. The controller was then synthesized by iterations, to reduce the H∞ norm. The frequency response could then be depicted. The frequency response of the weight functions covered the critical frequency, allowing

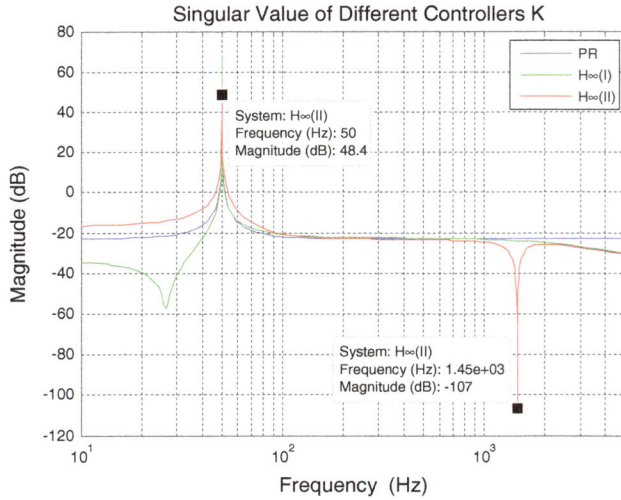

Fig. 6 The singular value of the controllers

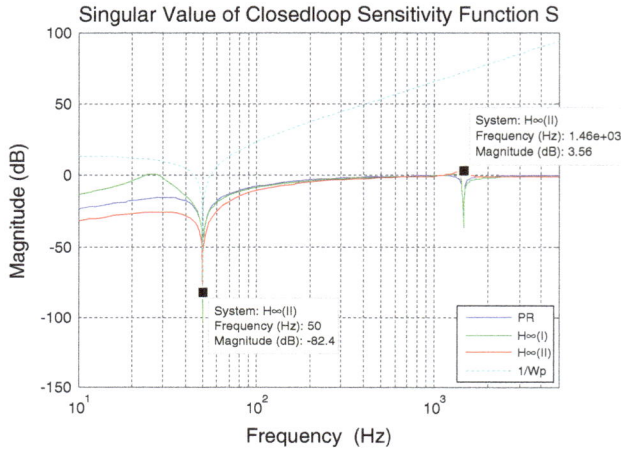

Fig. 7 The singular value of the sensitivity function, S

for a satisfactory stability and performance to be obtained. By checking the frequency response, weight functions can be redesigned or the parameters modified to obtain a more ideal shape.

In this paper, the parameters of the weight functions were tuned by loop-shaping of the T-S response. The circuit parameters and the tuned weight function parameters can be found in Appendix A. The H∞ controllers were synthesized with the robust control toolbox included in MATLAB. In all cases, a H∞ norm of $\gamma = 1.0046$ was achieved, indicating the robust stability of the designed uncertainty.

The singular values of the controllers are depicted in Fig. 6. For comparison, the controllers were tuned to have similar responses in the low frequency range.

The singular value of the closed loop sensitivity function (S) and complementary sensitivity function (T) are shown

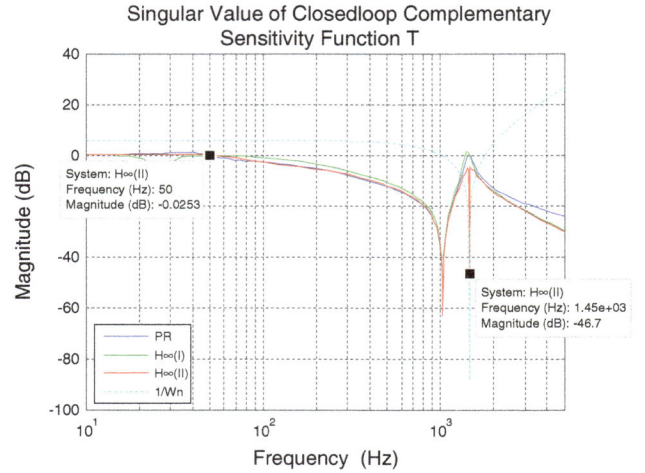

Fig. 8 The singular value of the complementary sensitivity function, T

Table 1 THDs (%) of the injected grid currents

Power (kVA)	1	2	3	4	5
PR	4.8	5.1	4.8	4.3	2.4
H∞(I)	4.7	4.7	4.2	3.7	2.6
H∞(II)	4.0	3.8	3.5	2.4	1.7
H∞(II$_2$)	4.2	3.9	3.6	2.6	1.7

in Figs. 7 and 8, respectively. In Fig. 7 the ∥S∥ value was small around $f = 50$ Hz for all three controllers, which indicated that there was a small tracking error in the fundamental components. Thus, the three controllers were all capable for fundamental control of the current.

In Fig. 8, the H∞ controllers had a faster roll off speed than the PR controller in the high frequency range, which indicated that they had a better harmonic rejection ability. Because of the shaping effect of W_n, H∞(II) had a deep subsidence around the LC resonance frequency; while the other frequencies were not affected. This feature made the system much more stable and insensitive to LC resonance.

4 Experimental results

Experiments were carried out with all three of the controllers designed in Section 3.4. The experimental system included a 5 kW VSC with a LCL filter. The circuit parameters can be found in Appendix A. An additional type II controller with an incorrect resonance frequency (W_{n2}) was also designed to verify the robust performance of the system. System stability was guaranteed during the design of the controller.

The overall performance was evaluated by the total harmonic distortion (THD) of i_g, which is shown in

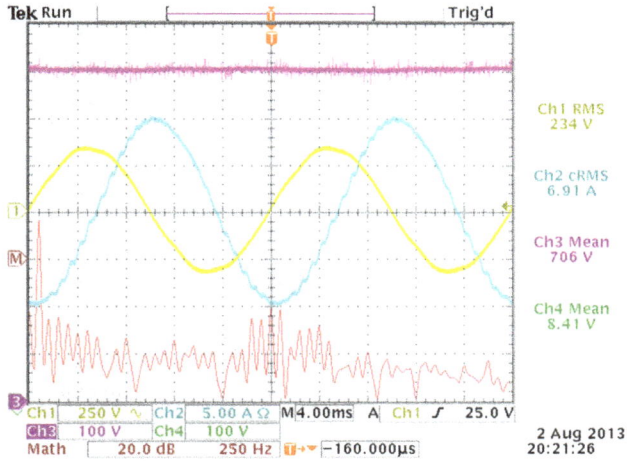

Fig. 9 Experiment results with PR controller

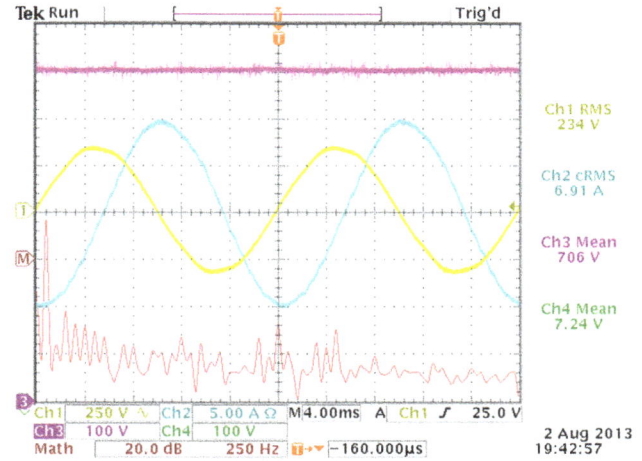

Fig. 11 Experiment results with H∞(II) controller

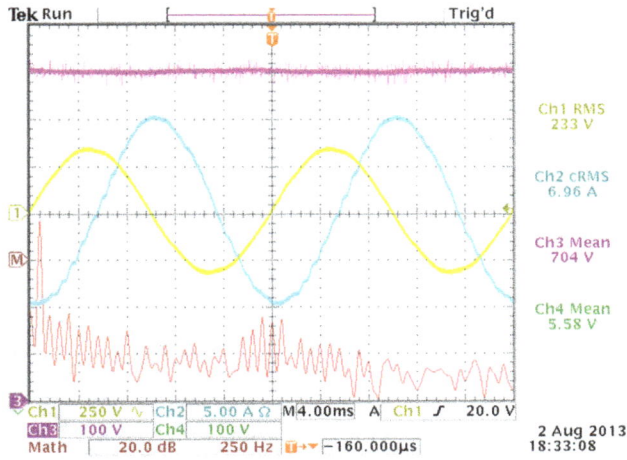

Fig. 10 Experiment results with H∞(I) controller

voltage (U_{dc}); Ch4 was not used; and ChM is the FFT analysis of Ch2. From these figures, the current with the H∞(II) controller (Fig. 11) was much smoother than those with the PR (Fig. 9) and H∞(I) (Fig. 10) controllers, and the FFT curve was less volatile around the LC resonance frequency.

The LC resonance made the control bandwidth of the PR controller difficult to increase. The H∞(II) controller successfully suppressed the LC resonance and was less sensitive to variations in the parameters than the PR controller, i.e. the H∞(II) controller became more robust in terms of its stability and performance.

5 Conclusion

In this paper, a robust current control design method was proposed to enhance the performance of VSC systems with LCL filters, and the details on the design procedure were explained.

Based on the theoretical analysis and the experimental results, it can be concluded that the proposed H∞ current controller was capable of fundamental control of the frequency current and it potentially suppressed the LC resonance. The overall performance of the proposed H∞ current controller was better than that of a conventional PR controller. The H∞ controller was stable with certain parameter uncertainties and it was not sensitive to the LC resonance frequency. The unified H∞ control tools have proven to be effective for multivariable and multi-objective control designs in VSC applications.

Acknowledgments This research was supported by the CAS Fraunhofer Joint Doctoral Promotion Program (DPP) and the National High Technology Research and Development Program of China (863 program) (No. 2011AA050204).

Table 1. The THD of the grid voltage was about 1.2%–1.4% during all of the experiments. A smaller THD for the output current resulted in a better performance of the controller.

From the measured THD, it was verified that all three controllers were capable of fulfilling the control requirements for the fundamental frequency components, in which the PR and H∞(I) controllers had similar performances owing to their similar frequency responses, as shown in Section 3.4. The H∞(II) controller had the best performance. Even with the incorrect parameters, the H∞(II) controller still achieved a relatively good result because of the deep subsidence around the LC resonance frequency, as shown in the singular response plots in Figs. 6, 7 and 8.

Three oscilloscope screenshots are given in Figs. 9, 10, and 11, in which, Ch1 is the grid voltage of phase a (u_{g_a}); Ch2 is the grid current of phase a (i_{g_a}); Ch3 is the DC side

Appendix A

System parameters:

$P_n = 5$ kW; $U_{line\_n} = 400$ V; $U_{DC\_n} = 700$ V; $f_n = 50$ Hz;

$R_g = R_c = 18 \times (1 \pm 50\%)$ mΩ; $L_g = L_c = 1.2 \times (1 \pm 50\%)$ mH;

and

$C_f = 20 \times (1 \pm 50\%)$ μF; $C_{dc} = 500$ μF.

PR controller:

$k_p = 0.07$; $k_i = 100$ and $\omega_c = 0.1$ rad/s.

Performance weight function:

W_p: $\omega_n = 50 \times 2\pi$ rad/s; $k_n = 0.06$ and $\zeta_n = 0.00001$.

Noise weight function:

W_n: $\omega_{res} = 1453 \times 2\pi$ rad/s; $k_{res} = 0.01$; $\zeta_{res} = 0.00001$;
W_{n2}: $\omega_{res} = 1250 \times 2\pi$ rad/s; $k_{res} = 0.01$; and $\zeta_{res} = 0.00001$.

References

[1] Wu B (2006) Hing-power converters and AC drives. Wiley, Hoboken

[2] Xu SK, Liang YY, Guo ZY (2011) Field commissioning test of CNOOC Wenchang VSC-HVDC transmission system. South Power Syst Technol 5(4):1–4 (in Chinese)

[3] Qiao WD, Mao YK (2011) Overview of Shanghai flexible HVDC transmission demonstration project. East China Electr Power 39(7):1137–1140 (in Chinese)

[4] Park SY, Chen CL, Lai JS et al (2008) Admittance compensation in current loop control for a grid-tie LCL fuel cell inverter. IEEE Trans Power Electron 23(4):1716–1723

[5] Jovcic D, Zhang L, Hajian M (2013) LCL VSC converter for high-power applications. IEEE Trans Power Deliver 28(1):137–144

[6] Peña-Alzola R, Liserre M, Blaabjerg F et al (2013) Analysis of the passive damping losses in LCL-filter-based grid converters. IEEE Trans Power Electron 28(6):2642–2646

[7] Blasko V, Kaura V (1997) A novel control to actively damp resonance in input LC filter of a three-phase voltage source converter. IEEE Trans Ind Appl 33(2):542–550

[8] Dahono PA (2003) A method to damp oscillations on the input LC filter of current-type AC-DC PWM converters by using a virtual resistor. In: Proceedings of the 25th international telecommunications energy conference (INTELEC'03), Yokohama, Japan, 19–23 Oct 2003, pp 757–761

[9] Wessels C, Dannehl J, Fuchs FW (2008) Active damping of LCL-filter resonance based on virtual resistor for PWM rectifiers—stability analysis with different filter parameters. In: Proceedings of the IEEE power electronics specialists conference (PESC'08), Rhodes, Greece, 15–19 Jun 2008, pp 3532–3538

[10] Mohamed YA-RI, A-Rahman M, Seethapathy R (2012) Robust line-voltage sensorless control and synchronization of LCL-filtered distributed generation inverters for high power quality grid connection. IEEE Trans Power Electron 27(1):87–98

[11] Willmann G, Coutinho DF, Pereira LFA et al (2007) Multiple-loop H-infinity control design for uninterruptible power supplies. IEEE Trans Ind Electron 54(3):1591–1602

[12] Li YW, Vilathgamuwa DM, Blaabjerg F et al (2007) A robust control scheme for medium-voltage-level DVR implementation. IEEE Trans Ind Electron 54(4):2249–2261

[13] Li YW, Vilathgamuwa DM, Loh PC (2007) Robust control scheme for a microgrid with PFC capacitor connected. IEEE Trans Ind Appl 43(5):1172–1182

[14] Teodorescu R, Blaabjerg F, Liserre M et al (2006) Proportional-resonant controllers and filters for grid-connected voltage-source converters. IEEE Xplore-Electr Power Appl 153(5):50–762

[15] Skogestad S, Postlethwaite I (2005) Multivariable feedback control: analysis and design. Wiley, Chichester

[16] Aten M, Werner H (2003) Robust multivariable control design for HVDC back to back schemes. IEEE Xplore-Gener Transm Distrib 150(6):761–767

Wenming GONG is now a Ph.D. candidate at University of Chinese Academy of Sciences. His research interests include the control of power electronics in wind energy and VSC-HVDC systems.

Shuju HU is an associate researcher at Institute of Electrical Engineering, Chinese Academy of Sciences. His research interests includes the control and test technologies of wind power systems and PV systems

Martin SHAN is head of Department Control Engineering, Fraunhofer IWES (Kassel). His research interests include modeling, simulation and control for wind turbines.

Honghua XU is a professor and research fellow at Institute of Electrical Engineering, Chinese Academy of Sciences. His research interests include wind power, photovoltaic and hybrid power system technologies.

12

"Charge while driving" for electric vehicles: road traffic modeling and energy assessment

Francesco Paolo DEFLORIO (✉), **Luca CASTELLO**,
Ivano PINNA, Paolo GUGLIELMI

Abstract The aim of this research study is to present a method for analyzing the performance of the wireless inductive charge-while-driving (CWD) electric vehicles, from both traffic and energy points of view. To accurately quantify the electric power required from an energy supplier for the proper management of the charging system, a traffic simulation model is implemented. This model is based on a mesoscopic approach, and it is applied to a freight distribution scenario. Lane changing and positioning are managed according to a cooperative system among vehicles and supported by advanced driver assistance systems (ADAS). From the energy point of view, the analyses indicate that the traffic may have the following effects on the energy of the system: in a low traffic level scenario, the maximum power that should be supplied for the entire road is simulated at approximately 9 MW; and in a high level traffic scenario with lower average speeds, the maximum power required by the vehicles in the charging lane increases by more than 50 %.

Keywords Wireless charging, Cooperative driving, Traffic simulation, Mesoscopic, Energy estimation

1 Introduction

Electric vehicles that provide zero local emissions and high energy efficiencies are becoming a real alternative for future motorized mobility. However, their acceptance in the market is limited by the following disadvantages when compared with diffused classical internal combustion engine vehicles: autonomy, lack of recharging infrastructures with public access, the time consuming charging process, limited battery life, battery cost and compliant masses. Charge-while-driving (CWD) technology could represent an interesting opportunity to support the deployment of electric vehicles as a possible solution.

The majority of fully electric vehicles (FEVs) currently satisfy the electric energy requirements for motion with an on-board battery. Reference [1] analyzed the problems related to battery charging management, the uncertainty surrounding the monitoring of the state of charge (SOC), the limited availability of charging infrastructure and the long time required to recharge; problems that have generated range anxiety. Extensive research has claimed that the challenges of battery inefficiency and the large and wasted space in the FEVs can be overcome by the wireless power transfer (WPT) technology. This technology electrically conducts energy from a source to an electric device without any interconnecting mediums [2]. The maglev system, developed in the late 1970s, utilises the high speed of a travelling vehicle to generate electricity using a linear generator [3]. Reference [4] proposed a design methodology for loosely coupled inductive power transfer systems. Such systems were used for non-contact power transfer, normally, over large air-gaps to the moving loads. Reference [5] explored the integrated pricing of electricity and roads enabled with wireless power transfer technology. The on-line electric vehicle (OLEV) system [6] and its non-contact power transfer mechanism were developed by the Korea Advanced Institute of Science and Technology (KAIST) and presented in 2009. The OLEV is an electric transport system in which the vehicles absorb the power from power lines underneath the surface of the road. The

F. P. DEFLORIO, L. CASTELLO, I. PINNA, P. GUGLIELMI,
Politecnico di Torino, 10129 Turin, Italy
(✉) e-mail: francesco.deflorio@polito.it

aim of this research study is to present a method for analyzing the performance of the CWD system, from both traffic and energy points of view. Beginning with an electric vehicle supply equipment (EVSE) layout defined and analyzed in a previous study [7] and using the system requirements defined in the eCo-FEV project [8], a model for the traffic flow simulation is implemented to quantify and describe the time-dependent traffic parameters along the charging lane and the electric power that should be provided by an energy supplier for proper management of the charging system. The results of this analysis confirm the influence of different traffic conditions and system requirements on the quality of the charging service.

2 Simulation model for the EVSE management

The model developed in this paper could be applied to a freight distribution service. The FEV traffic flow simulated here represents a fleet of light vans that could be generated by, or directed to, a logistics centre for a distribution service. The fleet management in this case could include the CWD usage in the common route segment to allow vehicles to cover greater distances, avoiding wasted time for a stationary recharge and to control the mass of the batteries.

The analysis is applied to a 20 km roadway with multiple lanes scenario. The right-hand lane is reserved for the charging activities. In an actual road infrastructure example, this solution could be applied by allocating the slowest lane to CWD operations or by using the emergency lane with dynamic lane management. Figure 1 shows a CWD lane scheme, with two charging zones (CZs) represented. The EVSE includes inductive coils placed under the pavement surface, at a relative distance, which generate a high frequency alternating magnetic field to which the coil on the car couples and power is transferred to charge the battery.

A proper design procedure should consider both the service provider's need to minimize the installation and maintenance costs and the users' acceptance of the time required for a proper recharge in the CWD lane. Taking into account the results obtained in previous studies [7] performed for an electric light van, with a power provided per unit of length (Pcz) of 50 kW/m in the CZs and adopting a system efficiency η_s of 85 % (from energy grid distribution to EV battery), the identified CWD system can be described by the following technical requirements: ①

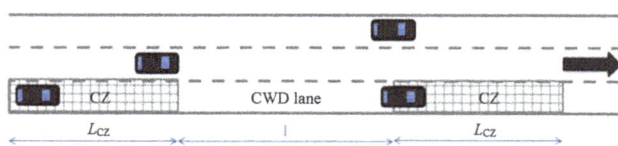

Fig. 1 Scenario layout for CWD in a road with three lanes

Length of the charging zones (L_{CZ}) = 20 m; ② Inter-distance (I) = 30 m; ③ Longitudinal dimension of the on-board charging device (L_{CD}) = 1 m.

In this layout, the energy equilibrium is possible at 60 km/h, whereas at lower speeds the SOC gain is positive. The two following operational speeds are defined for CWD: the highest speed (60 km/h) should allow the vehicle to maintain its entering SOC, whereas the lowest speed (30 km/h) should be a compromise between the recharge needs of vehicles with a low SOC and a minimum speed that can be accepted by the users. In this layout, by driving at the lowest speed, after 20 km in the CWD lane, the SOC increases by more than 7 kWh. This last case has been defined as "emer" status because this refers to a strategy applicable to emergency situations. The other charging vehicles have been classified with the "charge" status.

2.1 Models for energy estimation

In the CWD lane, a balance between the energy consumed for vehicle motion and the energy provided by the CZs should be established to monitor the SOC of the vehicle batteries during the observation period. The vehicle type included in the traffic flow is relevant because the mass and the aerodynamic parameters affect the energy consumption. After estimating the total average resistance force to motion R_{tot} between two consecutive nodes, the average power consumed is calculated according to the following relationship, based on simple mechanical concepts:

$$P_{electric} = \frac{R_{tot} \cdot s}{\eta_d} + P_{aux} \tag{1}$$

where η_d is the average driveline efficiency, which is assumed here constant for any average speed s of the vehicle along the section; P_{aux} is the auxiliary power that includes all consumption not related to the vehicle motion, such as the on-board electrical devices (e.g., lights and air conditioning). Finally, the energy consumed by the vehicle over time is obtained by multiplying the power consumed by the duration. In our scenarios, for sake of simplicity, the average slope will be assumed to equal zero.

The energy that the vehicle receives from the coils in the CWD lane $E_{received}$ can be calculated (as in (2)) by the electric power (P) received by any CZ, the number of CZ n_{CZ} and the occupancy time t_{CZ}. This can then be related to the system element dimensions (CZs and on-board devices-LCD) and P_{cz}, according to the following:

$$E_{received} = P \cdot n_{CZ} \cdot t_{CZ}$$
$$= (P_{CZ} \cdot L_{CD} \cdot \eta_s) \cdot \left(\frac{L_{roadsection}}{L_{CZ} + I}\right) \cdot \left(\frac{L_{CZeff}}{s}\right) \tag{2}$$

In (2), L_{CZeff} is the CZ length in which the vehicles effectively recharge, considering the initial and final partial overlaps of the on-board device. When the vehicle crosses a transmitting coil, it receives the energy according to system efficiency η_s that depends on the distance between the coil(s) of the on-board device and the coil(s) of the CZ installed in the road pavement. Each CZ is subdivided into coils that are excited only if a receiving (and authorized) vehicle is above them. In this way, only the coils that are under the vehicle work, thus maintaining the emitted power inside a shielded zone, correspond to the vehicle occupancy.

2.2 Traffic modelling

The choice of the traffic modelling is derived from the specific requirements of the CWD system [8] as synthesized below.

1) It has been assumed as installed only along the right-hand lane of the motorway because that lane is generally used by slower vehicles. Consequently, the model considers the lane disaggregation of traffic data.

2) The charging lane can be used for two different charging needs ("emer" or "charge") corresponding to two different vehicle speeds. Consequently, the model must consider different classes of vehicles.

One possible approach to effectively model this type of problem (multilane and multiclass) could be microsimulation, in which single vehicle trajectories are modelled with a small time step resolution and with their interaction on the road. An extensive review of traffic modelling approaches can be found in [9], whereas a microsimulation model application example is reported by [10]. Although the microsimulation approach meets the principal requirements of the traffic model for CWD, it does not model vehicle behavior according to their energy needs. The current SOC level of the vehicles and the fleet operators' eventual SOC target requirements influence drivers' decisions concerning lane changing, i.e., vehicles try to enter into the charging lane or to exit according to their needs. Therefore, specific rules must be defined and implemented to obtain realistic results from the traffic model. In addition, the detailed rules implemented in a micro-simulation model usually require an accurate calibration process, aimed at replicating the actual driving in traffic. However, the calibration process can be compromised in a CWD scenario whenever various ADAS are available on-board because they affect driving and traffic. Consequently, a mesoscopic approach would be more accurate than a microscopic one, because the latter is too detailed for this preliminary stage of CWD technology. Further comments on this issue will be reported in Sect. 3. A framework of mesoscopic traffic models can be found in [11], whereas a recent

application of this type of model was proposed in [12].

The developed model represents single vehicle trajectories without introducing a detailed time resolution of the driving activities. It assumes that the CWD lane conditions can be described knowing only the data related to consecutive points. The point spacing, typically hundreds of meters, can be set based on the analysis required. For this reason, detailed traffic information has been updated only at these defined points, defined as "detection points" or "nodes", where it is interesting to know the time series of traffic parameters and the energy provided for the entire vehicle set detected in the related time period. The road segment between the consecutive nodes will be defined as "road section" or "section". Aggregated traffic information, such as average headways, delays and the number of overtake maneuvers, can be estimated along the CWD lane for any road section.

The logic scheme adopted for two consecutive nodes of the traffic model is depicted in Fig. 2.

In the traffic model, the arrival time of a vehicle at the node i is first estimated based on its arrival time at the node (i-1) and its desired speed. It is then adjusted, in a second step, according to the feasible headway for vehicles in the lane. Because of safety and possible technical reasons, headway less than a threshold value between two vehicles in the charging lane may not be allowed. If two vehicles detected at a certain node are too close, in terms of headway, the following one has to slow down until its headway is equal to the threshold.

The headway verification and correction is therefore performed only at discrete space steps, according to the mesoscopic modeling of traffic. In an actual scenario, it can be managed by drivers or by the cooperative system

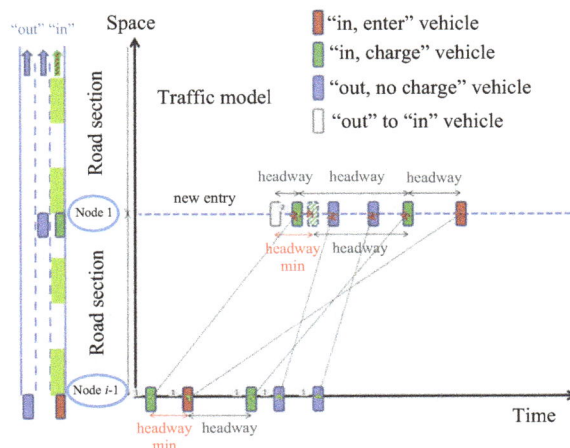

Fig. 2 Several trajectories in the time–space diagram to trace the arrival times of different vehicle types at consecutive nodes

adapting the vehicle speed along the entire section before the node where the headway adjustment is performed.

The battery SOC, monitored along the road at each node, plays a crucial role because it influences drivers' decisions to use the CWD service or not. It is also the parameter used to divide the vehicles into different speed classes. In the model, the CWD lane entries are managed according to the following cooperative behavior: each vehicle requiring recharge moves into the CWD lane at the node, creating the necessary gap in the vehicle flow by slowing down the following vehicles. A block diagram reported in Fig. 3 describes the logic of the procedures and the various functions applied at every simulation step. More details on all the functions can be found in [7].

The proposed scenario refers to a freight distribution service. The decision to charge may be simplified because it depends not only on drivers and their final destinations, but primarily on the fleet operator. To restart the delivery operations in the second part of the day, all of the vehicles in the fleet may require an energy level adequate for their operation.

The analysis considers even the overtaking cases: a cooperative overtaking model at constant speed is implemented and the vehicle does not recharge while it is outside the charging lane. The traffic simulator has been implemented in Microsoft Excel platform using Visual Basic programming language, and more details on the model can be found in [13].

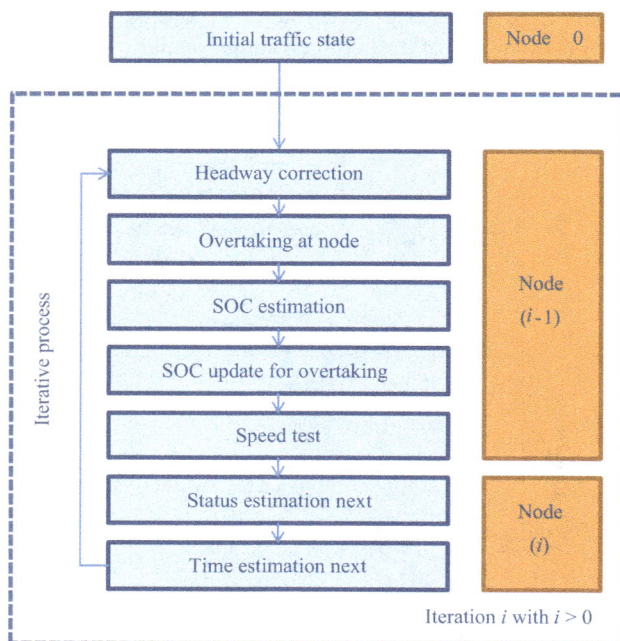

Fig. 3 Logic and procedures of the model

3 Verification and validation process

This chapter explains the model approach chosen, clarifying the reasons for the simplified assumptions and introducing a short discussion on verification and validation issues. Currently, the CWD system has been installed only in small test sites and, unfortunately, there are no opportunities to observe driver behavior in large-scale systems. Furthermore, even fully cooperative driving systems are not completely deployed. An actual traffic scenario, similar to that simulated, can be observed in long road tunnels in which vehicle spacing or headway greater than a predefined threshold should be maintained and all vehicles travel in a predefined speed range for safety reasons (e.g., the Mont Blanc tunnel).

Another important issue that should be considered is that the CWD technological environment will expand in the future. Therefore, it will involve another generation of vehicles, in which vehicle-to-vehicle communications will be used and many cooperative functions will be activated to facilitate the drive. In such a system, the observation of the current driving features is not relevant to model the traffic because vehicle motions and interactions depend more on the settings of the ADAS systems than on drivers' decisions.

For these reasons and considering the current stage of CWD technology development, calibration and validation operations based on empirical and on field observations are not possible. However, an extensive verification process can be performed by analyzing, testing and reviewing activities, according to the concepts defined in the ECSS standards [14]. In particular, a technical verification of the model response can be performed based on the following three consecutive test cases, each one aimed to verify different aspects: ① single vehicle motion and the relationship between its behavior and its energy needs; ② uniform vehicle flow without overtakes to verify if the model is able to correctly manage the headways between vehicles; ③ complex traffic interaction with overtaking maneuvers to assess the global interaction between vehicles, introducing overtaking maneuvers.

In the third stage of the model verification process, traffic results may be controlled by the following relevant parameters affecting traffic behavior.

1) Input traffic distribution (average headway, standard deviation and minimum value);
2) Vehicle speed for the two CWD classes (in the CWD lane where the speed is controlled and in the other lanes where the speed is derived from the density-speed relationship);
3) Overtake management (duration, event detection, event activation, event recovery and multiple overtakes);

4) Vehicle energy parameters (initial SOC, target SOC, SOC thresholds and energy consumption);
5) CWD parameters (system layout and power).

At this stage of the model development, the presented model has been validated by checking the satisfaction of the established technical requirements, based on the system engineering approach [15]. The main functional requirements for the model are the following: ① the model shall estimate the number of vehicles in the CWD for any detection point; ② the model shall consider possible random effects of input flow; ③ the model shall represent the traffic flow at any detection point and reveal if concentration of traffic and congestion occur along the lane; ④ the model shall take into account different values of the minimum headway allowed in the CWD to estimate possible effects on traffic and energy for the various CZs over time.

In the following sections, the model testing results are reported in an "ideal case", in which all of the subsystems and applications involved, such as the CWD booking and authorization functions, or the cooperative ADAS, which enables the vehicle cruise control or the cooperative overtaking, work properly. In this scenario, all the related system information, such as the vehicle position and its SOC, is accurately known. This validation approach could be considered as a "best-case" testing, and it is consistent with the test-case-design methods applied to test software, such as boundary value analysis [16] or distributed real time systems [17].

4 Experiments for model testing and first results

After defining the CWD model, it is necessary to estimate its capability to determine the quality level assessment for the charging service. The electrical power distribution type that should be supplied at each node is an interesting result from this preliminary stage of CWD development. The traffic and energy results will be reported in the two following sections, and two operational testing scenarios will be analyzed.

4.1 Parameter setting for the simulated scenarios

The Reference scenario represents a compatible flow of light vans generated by a logistics centre for multiple deliveries. A second scenario (Alternative) will be explored to analyze how the system performance could be affected by the increase of both the FEV traffic and the minimum allowed technical headway in the CWD lane. In the Alternative scenario, vehicles are generated closer than those in the Reference scenario, but they cannot stay too close while charging, thus creating a delay phenomenon with vehicle platoons in queue.

Table 1 Data related to traffic

Traffic	
Average density for input traffic flow	10 (20) veh/km/lane
Critical density (at max capacity)	30 veh/km/lane
Number of simulated vehicles	500 veh
Coefficient of variation of the headway	0.3
Minimum traffic headway	1.5 s

Table 2 Data related to infrastructure

Infrastructure	
Total length of the road	20 km
Average slope	0 %
Sections length	1 km
Length of the charging zones (L_{CZ})	20 m
Interdistance (I)	30 m
Transition coefficient (T_{rk})	1
System efficiency (η_s)	0.85
Power per unit of length (P_{CZ})	50 kW/m
Minimum headway in CWD lane	1.5 (3) s

In Table 1 and Table 2, basic traffic feature data and infrastructure layout parameters are reported for the Reference scenario. The data between brackets indicates the variations introduced in the Alternative scenario.

A critical density value of 30 veh/km/lane has been assumed based on the generally adopted values for freeways under basic conditions [18]. Minimum headway values between 1.5 and 2.5 s have been chosen to consider the use of ADAS [19].

In Table 3, the vehicle data specifies motion performance, energy consumption and energy needs. At each node of the modelled road, the SOC of every vehicle is assessed.

Although some car manufacturers use the currently available adaptive cruise control (ACC) to give the drivers the opportunity to manually choose the minimal headway, they set the absolute minimum headway at 0.9 s [20]. In this study, a more prudent value of 1.5 s has been assumed. According to the analysis reported in a previous study [7], a vehicle with a SOC less than 30 % of its target is assumed in an emergency situation (state = "emer") and its desired speed along the CWD lane is set to 30 km/h; if the charging level is between 30 % and 60 % of the target value, then the vehicle is assumed to be charged in the CWD lane to preserve its SOC (state = "charge") and its desired speed is set to 60 km/h. Vehicles with a current charge level greater than 60 % of the target SOC are assumed "out" of the CWD lane because they do not need to

Table 3 Data related to vehicles features

Vehicles	
Average starting SOC	10 kWh
Standard deviation of SOC	4.5
SOC target	20 kWh
Length of charging the device (L_{CD})	1 m
SOC limit for "charge" vehicles	60 %
SOC limit for "emer" vehicles	30 %
Desired speed of "charge" vehicles on CWD	60 km/h
Desired speed of "emer" vehicles on CWD	30 km/h
Max free flow speed on other lanes	110 km/h
Average acceleration	0 m/s²
Overtake duration	10 s
Mass (m)	2500 kg
Cross sectional area (A)	4.9 m²
C_x	0.38
f_0	0.12 m/s²
f_2	0.000005 m⁻¹
Driveline efficiency (η_d)	0.75
Auxiliary power (P_{aux})	0.8 kW

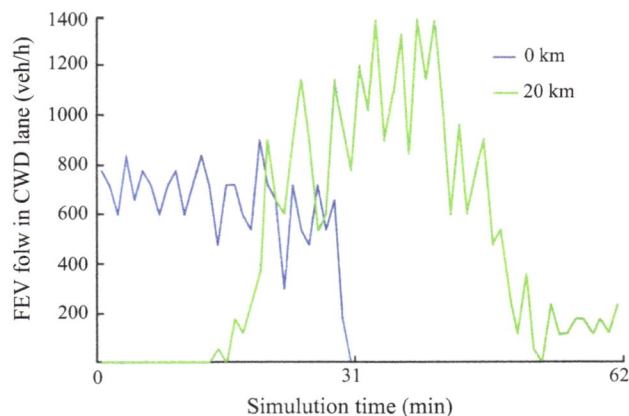

Fig. 4 Traffic flow into the CWD lane at the entrance (0 km) and the exit node (20 km) for the Reference scenario

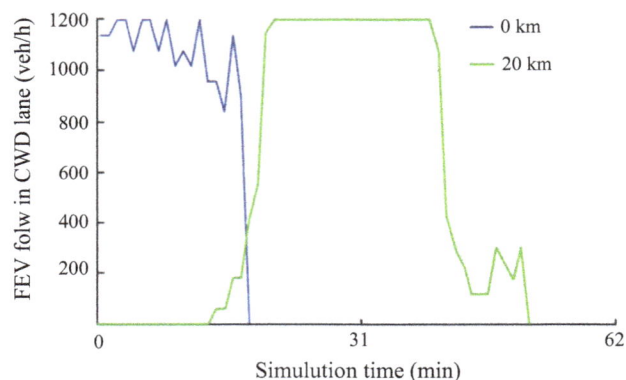

Fig. 5 Traffic flow in the CWD lane at the entrance (0 km) and the exit node (20 km) for the Alternative scenario

recharge. Their speed is then set according to the feasible speed in the other lanes, which depend on the estimated traffic density.

4.2 Primary traffic results

In this section, a comparison of selected principal traffic results in the Reference and Alternative scenarios is reported. Because all results depend on the random variables generated at the initial traffic and energy states, multiple replications of this experiment should be examined to observe, using statistical analysis, how the random effects influence the simulation results. However, to better show the traffic and energy performance of the implemented simulation model, through the reading of the calculated variables in identical conditions, the following results will focus on one selected replication that is close to the average value.

The first parameter analyzed is the FEV traffic flow in the CWD lane. Fig. 4 and Fig. 5 compare the traffic flows at the entrance and at the exit of CWD lane, respectively for the Reference and the Alternative scenarios. Fig. 4 shows that in the Reference scenario, the traffic flow in the CWD lane increases along the lane, with concentration phenomena at the exit section, although never reaching the maximum value of 2400 veh/h related to the minimum technical headway (1.5 s). In particular, based on the values set for the parameters, an "emer" vehicle increases its SOC and, after reaching the SOC threshold, it increases its

speed according to the "charge" vehicles desired speed, whereas a "charge" vehicle maintains a constant SOC over time. Consequently, no vehicle leaves the CWD lane, whereas "out" vehicles can enter into the CWD lane during the simulation.

An identical effect can also be observed for the Alternative scenario, in which the higher minimum technical headway value (equal to 3 s) defines a lower maximum admissible flow of 1200 veh/h in the CWD lane. Therefore, traffic conditions at 0 km approximate the maximum allowable flow. At 20 km, the limit conditions occur for the majority of the simulation time, as illustrated by the plateau in Fig. 5, which is caused by vehicle platoon conditions. In this case, an entrance into the CWD lane or an overtake maneuver may cause a relevant disturbance in the traffic flow, resulting in a sensible reduction in the average speeds of the following vehicles.

Figures 6 and 7, for the Reference and the Alternative scenario, respectively, report the vehicle counts that are detected at each kilometer (at each node), along the CWD

Time (min)

Fig. 6 FEV counts along the CWD lane over time in the Reference scenario

Node	Counts over time (min 0–62)	Total
0	13 12 10 14 11 13 12 10 12 13 10 12 14 12 8 12 12 10 9 15 12 11 5 12 9 8 12 9 11 3	326
1	10 10 9 17 10 13 11 9 14 13 10 15 12 11 8 13 14 9 9 11 16 11 4 8 13 9 11 8 9 7 2	326
2	1 10 8 7 16 13 13 11 8 13 15 10 15 15 12 7 14 15 11 8 11 13 17 4 9 13 13 8 8 5 6 2	340
3	2 9 9 5 16 11 16 11 7 13 16 11 17 15 15 4 15 16 13 12 10 13 15 8 9 9 9 17 9 7 4 4 6 2	355
4	1 2 10 10 5 13 11 14 14 7 14 15 13 17 16 15 7 13 15 14 14 12 13 16 4 13 9 10 15 14 8 3 3 5 5 2	372
5	1 3 10 10 7 13 10 12 13 10 13 15 12 17 18 15 7 16 16 13 14 14 15 16 5 9 12 10 15 10 12 4 2 4 4 5 2	384
6	1 3 11 10 7 14 10 10 11 10 17 15 11 16 18 19 5 16 20 12 16 13 18 17 5 9 9 14 15 9 9 7 3 3 4 6 1	397
7	2 4 11 10 7 14 11 11 8 9 18 19 12 14 17 20 11 14 20 16 16 14 16 19 7 10 9 12 17 8 8 4 6 4 2 3 6 4 1	414
8	1 3 4 11 10 7 15 12 12 8 6 17 18 15 17 14 21 13 15 19 15 20 12 17 16 9 14 8 13 15 10 7 3 5 5 5 3 2 6 3 4 1	426
9	1 4 5 14 10 7 15 12 12 8 6 14 16 15 20 16 19 13 16 19 13 16 21 21 17 17 19 7 17 9 12 15 8 9 3 3 5 3 5 3 3 4 1	441
10	1 1 4 5 14 10 8 15 13 13 8 7 14 14 13 21 19 22 10 16 22 14 19 17 20 17 8 14 12 13 14 8 7 5 3 4 2 3 6 3 2 3 4 1	449
11	1 2 4 5 14 10 9 15 13 13 8 7 14 14 12 20 19 23 12 14 21 15 20 16 20 20 7 14 9 16 15 7 7 5 3 5 5 5 2 3 2 5	453
12	1 2 4 5 14 10 9 15 13 14 8 8 14 14 12 19 18 23 13 16 19 15 22 16 26 20 11 18 11 13 9 13 18 8 6 5 3 5 1 5 5 2 2 3 2 5	458
13	2 2 4 5 14 11 9 15 14 14 8 9 14 15 12 20 16 24 13 17 21 13 21 18 21 18 10 16 8 13 16 10 7 4 3 6 1 3 4 3 2 3 2 4	469
14	2 2 4 5 14 11 9 15 14 14 9 15 15 12 20 16 22 13 18 21 14 21 14 21 18 22 18 10 16 10 13 15 8 9 5 2 6 1 4 2 2 3 3 2 3 2 4	472
15	1 3 2 4 6 15 11 10 15 19 15 9 12 17 23 15 18 22 14 23 19 23 18 10 16 10 13 15 8 9 5 2 6 1 4 2 2 3 3 2 3 2 4	500
16	1 3 2 4 6 15 11 10 15 19 15 9 10 19 16 13 20 17 23 15 18 22 14 23 19 23 18 10 16 10 13 15 8 9 5 2 6 1 4 2 2 3 3 2 3 2 4	500
17	1 3 2 4 6 15 11 10 15 19 15 9 10 19 16 13 20 17 23 15 18 22 14 23 19 23 18 10 16 10 13 15 8 9 5 2 6 1 4 2 2 3 3 2 3 2 4	500
18	1 3 2 4 6 15 11 10 15 19 15 9 10 19 16 13 20 17 23 15 18 22 14 23 19 23 18 10 16 10 13 15 8 9 5 2 6 1 4 2 2 3 3 2 3 2 4	500
19	1 3 2 4 6 15 11 10 15 19 15 9 10 19 16 13 20 17 23 15 18 22 14 23 19 23 18 10 16 10 13 15 8 9 5 2 6 1 4 2 2 3 3 2 3 2 4	500
20	1 3 2 4 6 15 11 10 15 19 15 9 10 19 16 13 20 17 23 15 18 22 14 23 19 23 18 10 16 10 13 15 8 9 5 2 6 1 4 2 2 3 3 2 3 2 4	500
Total	13 23 33 43 58 70 89 96 112 129 138 152 175 183 203 214 232 241 251 273 277 293 291 306 301 303 300 307 308 303 294 292 275 262 245 231 215 194 184 161 146 121 108 96 84 71 62 45 39 29 26 24 22 19 23 17 19 17 14 11 9 6 4	9082

Time (min)

Fig. 7 FEV counts along the CWD lane over time in the Alternative scenario

Node	Counts over time (min 0–62)	Total
0	19 19 20 20 18 20 20 18 20 17 18 17 20 16 16 14 19 15	326
1	15 12 20 20 20 19 20 20 20 17 20 17 18 19 13 18 17 12 9	326
2	1 15 9 16 19 20 19 20 20 20 20 20 20 19 19 16 15 20 12 6 9	335
3	2 14 11 14 14 20 19 20 20 20 20 20 20 20 20 18 17 4 6 9	348
4	2 15 12 13 11 17 19 20 20 20 20 20 20 20 20 20 13 4 7 8	361
5	1 3 15 13 14 11 13 17 20 19 20 20 20 20 20 20 20 16 11 5 6 8	372
6	2 3 15 16 14 11 13 13 18 19 20 20 20 20 20 20 20 10 11 4 7 7	383
7	2 4 15 16 16 11 14 13 14 17 20 20 20 20 20 20 20 16 8 10 5 7 6	394
8	2 4 15 17 16 12 18 14 14 13 20 20 20 20 20 20 20 7 8 9 6 5 6	406
9	1 4 4 15 18 16 12 18 14 14 15 18 20 20 20 20 20 20 14 5 9 7 5 5 6	418
10	2 4 4 15 19 18 12 19 14 15 15 18 14 19 20 20 20 20 19 5 8 5 7 5 4 6	427
11	2 6 6 15 19 18 12 19 14 16 16 18 14 19 20 20 20 20 20 5 9 4 4 7 4 4 6	437
12	1 2 6 7 16 19 20 13 19 14 17 18 14 19 20 20 20 20 20 20 20 7 7 5 4 4 5 5 4 5	448
13	1 2 6 7 17 19 20 13 20 14 17 17 18 14 19 20 20 20 20 20 20 20 10 7 5 4 2 2 6 5 3 5	450
14	2 2 6 7 17 19 20 13 20 15 18 17 18 14 19 20 20 18 19 20 20 20 20 20 12 7 5 4 2 2 2 5 4 3 5	455
15	2 6 7 17 19 20 15 20 15 18 17 19 14 19 20 20 18 19 20 20 20 20 20 12 7 5 4 2 2 2 5 4 3 5	458
16	3 2 6 8 17 19 20 17 20 16 19 17 20 14 19 20 20 19 19 20 20 20 20 20 12 7 5 4 2 2 2 5 4 3 5	466
17	3 2 6 8 18 19 20 18 20 16 19 17 20 15 19 20 20 20 20 20 20 20 20 20 12 7 5 4 2 2 2 5 4 3 5	471
18	3 2 6 8 19 19 20 19 20 16 19 18 20 16 19 20 20 20 20 20 20 20 20 20 12 7 5 4 2 2 2 5 4 3 5	475
19	1 1 3 3 7 9 19 20 20 20 20 20 20 20 20 20 20 20 20 20 20 18 7 5 4 2 2 2 5 4 3 5	500
20	1 1 3 3 7 9 19 20 20 20 20 20 20 20 20 20 20 20 20 20 20 20 18 7 5 4 2 2 2 5 4 3 5	500
Total	19 35 49 66 85 105 121 143 163 181 201 223 248 264 286 295 325 338 344 344 341 334 322 313 307 292 287 272 260 252 239 226 205 192 173 159 139 122 99 83 68 50 30 29 22 23 21 19 17 12 8 5 0 0 0 0 0 0 0 0 0 0 0	8756

lane over time. In the grey scale, the higher values are represented with a darker color.

In the Reference scenario, different color areas can be noted, indicating a certain variability of the traffic flow over time. The initial high traffic conditions of the Alternative scenario cause a uniform distribution of the vehicles, highlighted by a flatter coloration. As expected, the CWD lane flow reaches the maximum value allowed by the degraded system of the Alternative scenario (20 veh/min), confirming the previous platoon considerations.

In the Reference scenario, the number of charging vehicles increases more rapidly because the speed in the other lanes is higher as a result of better traffic conditions, thus increasing vehicle energy consumption. Before the final node (20 km), all generated vehicles must be recharged so they enter the CWD lane.

Because of the battery capacity limitations, all vehicles driving in the unequipped lanes reduce their SOC and reach the "charge" threshold within the last sections. This phenomenon, which is consistent with assumptions, causes the final increase in the vehicle count in the CWD lane.

The second parameter analyzed is the space mean speed. Figures 8 and 9 for the Reference and the Alternative scenario, respectively, report the values on the sections before each node along the CWD lane over time, considering both "charge" and "emer" vehicles. The darker color refers to lower values and therefore the worst traffic condition cases.

The zones in the time–space diagram in which congestion occurs are consistent with the data from the scenarios. Values exceeding the speed limit in the CWD lane (60 km/h) are caused by the entries into the CWD lane from the other lanes, where the speeds are higher, because they are related to the established traffic density.

As expected, the lowest speeds for the first sections are presented at the end of the simulation time because only the "emer" and slow vehicles are presented. Any possible "charge" and fast vehicles have previously crossed this section. This concentration of the slower vehicles at the end of the simulation occurs only for the first sections, because after node 14, all "emer" vehicles have increased their SOC over the "charge" threshold, changing their status. In both analyzed scenarios, the average speeds of the traffic flow exceed 30 km/h.

Finally, delay is the last traffic parameter reported. It is analyzed separately for "charge" and "emer" vehicles. In the Reference scenario, the delay is negligible: considering all simulation time along the CWD lane, it reaches the

Time (min)

Section before	1	2	3	4	5	6	7	8	9	10	11	12	13	14	15	16	17	18	19	20	21	22	23	24	25	26	27	28	29	30	31
1	60	46	38	44	46	46	44	39	42	46	43	47	55	47	40	46	49	54	49	51	44	51	40	44	39	45	51	44	41	32	30
2	104	60	63	42	46	41	50	44	40	43	43	43	47	47	61	42	48	47	47	53	51	53	46	40	53	45	39	50	50	44	33
3		76	60	54	72	49	41	42	44	38	46	48	39	47	47	47	48	48	49	47	51	55	52	56	34	49	51	45	42	45	47
4		104	60	63	63	60	62	45	44	40	38	48	47	45	46	45	47	38	56	50	48	47	53	56	58	52	41	49	54	52	41
5			60	70	60	60	68	60	65	48	43	37	49	47	45	46	47	51	38	48	59	52	51	47	54	56	54	54	42	50	54
6				60	60	62	60	60	62	60	63	47	41	44	54	51	48	47	48	43	48	50	62	56	57	50	53	50	54	56	44
7					76	67	60	60	60	60	62	62	60	46	49	45	53	52	49	49	45	52	50	48	63	52	58	50	48	57	54
8					104	69	60	60	60	60	62	62	62	60	60	52	47	45	52	56	51	47	50	54	50	50	60	54	56	42	55
9						60	66	66	66	60	60	60	60	60	60	60	60	50	45	48	55	57	50	48	52	56	53	53	63	56	55
10						104	60	60	60	60	60	63	60	62	62	60	64	60	62	49	48	50	56	60	50	53	52	57	51	52	60
11							60	76	60	60	60	60	63	60	60	60	60	60	60	60	62	54	52	53	51	60	55	50	55	56	56
12								60	60	60	60	60	60	60	60	60	62	60	63	60	60	60	61	54	53	49	53	60	54	52	56
13									76	60	60	60	60	62	60	60	62	60	60	63	60	62	60	61	60	56	52	54	58	62	55
14										60	60	60	60	60	60	60	60	60	60	60	60	62	60	60	60	60	60	52	54	57	56
15									104		70	60	60	65	62	60	63	62	66	62	63	63	66	62	62	60	61	61	64	60	61
16										60		60	60	60	60	60	60	60	60	60	60	60	60	60	60	60	60	60	60	60	60
17											60		60	60	60	60	60	60	60	60	60	60	60	60	60	60	60	60	60	60	60
18												60		60	60	60	60	60	60	60	60	60	60	60	60	60	60	60	60	60	60
19													60		60	60	60	60	60	60	60	60	60	60	60	60	60	60	60	60	60
20														60		60	60	60	60	60	60	60	60	60	60	60	60	60	60	60	60
Total	62	55	52	50	54	52	54	50	53	51	52	52	53	53	54	52	54	53	54	54	54	55	55	55	54	55	54	55	55	54	55

Section before	32	33	34	35	36	37	38	39	40	41	42	43	44	45	46	47	48	49	50	51	52	53	54	55	56	57	58	59	60	61	62	Total
1																																45
2	30	30																														46
3	34	30	30	30																												46
4	44	36	30	33	30	30																										47
5	46	38	34	30	34	30	30	30																								48
6	58	49	45	32	30	36	30	30	33	30																						49
7	55	51	53	48	34	30	34	30	30	36	30	30																				50
8	60	57	56	46	52	36	37	33	30	30	40	30	30	30																		51
9	49	54	60	56	53	45	45	45	37	30	30	43	30	30	30	30																52
10	48	56	45	56	60	53	52	37	45	48	30	30	40	36	30	30	30	30														53
11	52	60	56	54	48	56	60	52	50	45	50		30	40	34	30	30	30	30	30												54
12	57	57	49	60	54	56	49	53	60	50	45	50	30		43	37	30	30	36	30	33											54
13	54	58	57	55	54	60	56	56	46	52	60	45	51		30	60	40	30	30	36	30	36	30									55
14	60	55	55	57	54	58	50	60	60	53	45	50	60	51	30		48	60	30	30	36	30	36	30	30							55
15	60	62	61	61	60	60	60	60	60	60	60	60	60	60	60	60		60	60	60	60	60	60	60	60	60						61
16	60	60	60	60	60	60	60	60	60	60	60	60	60	60	60	60	60		60	60	60	60	60	60	60	60	60					60
17	60	60	60	60	60	60	60	60	60	60	60	60	60	60	60	60	60	60		60	60	60	60	60	60	60	60	60				60
18	60	60	60	60	60	60	60	60	60	60	60	60	60	60	60	60	60	60	60		60	60	60	60	60	60	60	60	60			60
19	60	60	60	60	60	60	60	60	60	60	60	60	60	60	60	60	60	60	60	60		60	60	60	60	60	60	60	60	60		60
20	60	60	60	60	60	60	60	60	60	60	60	60	60	60	60	60	60	60	60	60	60		60	60	60	60	60	60	60	60	60	60
Total	54	55	54	54	54	54	53	54	53	53	52	53	51	53	50	50	47	46	43	45	44	50	48	54	50	60	60	60	60	60	60	54

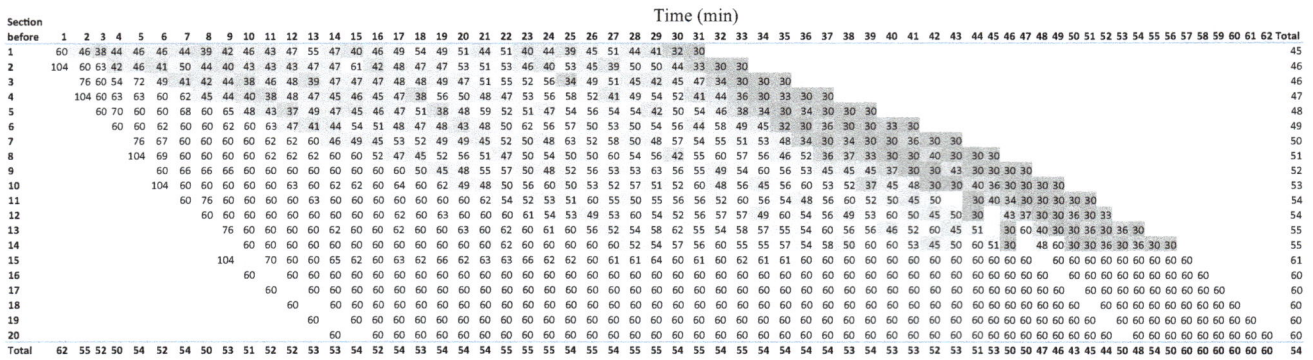

Fig. 8 Space mean speed of FEVs along the CWD lane over time in the Reference scenario

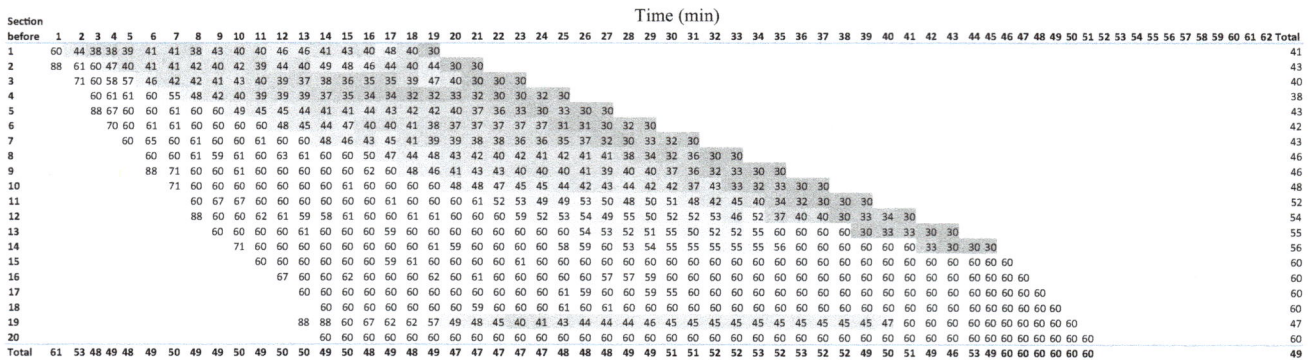

Time (min)

Section before	1	2	3	4	5	6	7	8	9	10	11	12	13	14	15	16	17	18	19	20	21	22	23	24	25	26	27	28	29	30	31
1	60	44	38	38	39	41	41	38	43	40	40	46	46	41	43	40	48	40	30												
2	88	61	60	47	40	41	41	42	40	42	39	44	40	49	48	46	44	40	44	30	30										
3		71	60	58	57	46	42	42	41	43	40	39	37	38	36	35	35	39	47	40	30	30	30								
4			60	61	61	60	55	48	42	40	39	39	39	37	35	34	34	32	32	33	32	30	30	32	30						
5			88	67	60	60	61	60	60	49	45	45	44	41	41	44	43	42	42	40	37	36	33	30	33	30	30				
6				70	60	61	61	60	60	60	60	48	45	44	47	40	40	41	38	37	37	37	37	37	31	31	30	32	30		
7					60	65	60	61	60	60	61	60	60	48	46	43	45	41	39	39	38	38	36	36	35	37	32	30	33	32	30
8						60	60	61	59	61	60	63	61	60	60	50	47	44	48	43	42	40	42	41	42	41	41	38	34	32	36
9						88	71	60	60	61	60	60	60	60	60	62	60	48	46	41	43	43	40	40	40	41	39	40	40	37	36
10							71	60	60	60	60	60	60	60	61	60	60	60	60	48	48	47	45	45	44	42	43	44	42	42	37
11								60	67	67	60	60	60	60	60	60	61	60	60	60	61	52	53	49	49	53	50	48	50	51	48
12								88	60	60	62	61	59	58	61	60	60	61	61	60	60	60	59	52	53	54	49	55	50	52	52
13									60	60	60	60	61	60	60	60	59	60	60	60	60	60	60	60	60	54	53	52	51	55	50
14										71	60	60	60	60	60	60	60	60	61	59	60	60	60	60	58	59	60	53	54	55	55
15											60	60	60	60	60	60	59	61	60	60	60	60	61	60	60	60	60	60	60	60	60
16												67	60	60	62	60	60	60	62	60	61	60	60	60	60	60	57	57	59	60	60
17													60	60	60	60	60	60	60	60	60	60	60	60	61	59	60	60	59	55	60
18														60	60	60	60	60	60	60	59	60	60	60	61	60	61	60	60	60	60
19													88	88	60	67	62	62	57	49	48	45	40	41	43	44	44	44	46	45	45
20														60	60	60	60	60	60	60	60	60	60	60	60	60	60	60	60	60	60
Total	61	53	48	49	48	49	50	49	49	50	49	50	50	49	50	48	49	48	49	47	47	47	47	47	48	48	48	49	49	51	51

Section before	32	33	34	35	36	37	38	39	40	41	42	43	44	45	46	47	48	49	50	51	52	53	54	55	56	57	58	59	60	61	62	Total
1																																41
2																																43
3																																40
4																																38
5																																43
6																																42
7																																43
8	30	30																														46
9	32	33	30	30																												46
10	43	33	32	33	30	30																										48
11	42	45	40	34	32	30	30	30																								52
12	53	46	52	37	40	40	30	33	34	30																						54
13	52	52	55	60	60	60	60	30	33	33	30	30																				55
14	55	55	55	56	60	60	60	60	60	60	33	30	30	30																		56
15	60	60	60	60	60	60	60	60	60	60	60	60	60	60	60																	60
16	60	60	60	60	60	60	60	60	60	60	60	60	60	60	60	60																60
17	60	60	60	60	60	60	60	60	60	60	60	60	60	60	60	60	60															60
18	60	60	60	60	60	60	60	60	60	60	60	60	60	60	60	60	60	60														60
19	45	45	45	45	45	45	45	45	47	60	60	60	60	60	60	60	60	60	60													47
20	60	60	60	60	60	60	60	60	60	60	60	60	60	60	60	60	60	60	60	60												60
Total	52	52	53	52	53	52	52	49	50	51	49	46	53	49	60	60	60	60	60	60												49

Fig. 9 Space mean speed of FEVs along the CWD lane over time in the Alternative scenario

maximum value of 47 s for "charge" vehicles at node 13. This indicates that a delay of 47 s is assessed by considering all 429 "charge" vehicles in 40 minutes of simulation. Consequently, no space–time table will be reported, which is consistent with the data assumed for this scenario.

The delay for the Alternative scenario is relevant. Figures 10 and 11 report the time-dependent average delay for "charge" and "emer" vehicles over time, respectively, along the CWD lane. The mean delay values are similar for "charge" and "emer" vehicles, between 0 and 45 s on average. The Alternative scenario traffic conditions generate queues and consequently delay in the traffic flow, causing a decrease of the average speeds. In Fig. 10, the increased delay at the section before node 19 confirms the entry of the last charging vehicles into the CWD lane, where the traffic flow proceeds with vehicle platoons in queue.

4.3 Energy estimation for CWD

In this section, selected simulation results related to the CWD energy issues are reported. In Figs. 12 and 13, the energy received at each node by FEVs from the single CZ placed on the detection point over time are presented for the Reference and Alternative scenarios, respectively. These results confirm that the simulation can describe the CWD energy dynamics. This analysis confirms that the energy required may vary significantly along the road, and it may change over the time. From the grey scale in Fig. 12, multiple waves travelling ahead with an approximate speed of 30 km/h, which is the speed for emergency vehicles, can be observed for the Reference scenario. The maximum value observed for any CZ at nodes is 0.3 kWh during one minute; in most cases, it does not continue for more than three consecutive minutes. In the Alternative scenario (Fig. 13) the variation is uniform: for example, the value of 0.4 kWh is constant for longer periods (in some cases, approaching 20 minutes). In this scenario, the higher value of 0.5 kWh was detected for CZs at nodes after 3 km, 4 km and 7 km, but only for few minutes. After reporting the simulation results for the energy required by vehicles along the CWD lane at the selected detection points, a global energy analysis is described here.

Cumulative power profiles can be simulated for the Reference and the Alternative scenarios to estimate the power a single energy provider should supply along the entire CWD system. To obtain complete information about

Section before	1	2	3	4	5	6	7	8	9	10	11	12	13	14	15	16	17	18	19	20	21	22	23	24	25	26	27	28	29	30	31	32	33	34	35	36	37	38	39	40	41	42	43	44	45	46	47	48	49	50	51	Total
1	0	1	10	13	5	3	11	16	15	4	13	8	2	12	1	4	1	0																																		7
2	0	0	0	1	5	7	1	1	8	15	17	4	12	5	1	1	1	7	2																																	5
3		0	0	0	0	1	3	3	3	2	12	19	19	19	25	32	29	11	1	2		0																														11
4			0	0	1	0	0	2	6	12	9	12	17	25	28	34	32	41	45	36	32	22		0																												20
5			0	0	0	1	0	0	0	1	3	2	1	8	16	11	12	17	15	23	33	28	24	0	0	0																										11
6		1	0	0	1	0	0	0	0	0	1	3	5	3	13	19	20	23	26	27	30	32	28	8	2		0	0																								13
7			0	0	0	0	1	0	0	0	0	3	3	7	9	18	21	24	23	26	30	30	33	32	2	0	2	2																								13
8				0	0	0	1	1	0	1	0	0	0	2	1	7	9	12	17	21	19	18	19	19	20	11	0	0	1																							9
9				0	0	0	0	0	0	0	0	2	1	6	12	15	20	25	25	24	23	23	21	20	0	1	0	2	0																							10
10					0	0	0	0	2	1	0	0	0	0	0	0	0	2	3	8	16	15	18	21	21	19	17	17	0	1	0	0	0																			8
11					0	0	0	0	0	0	0	0	0	1	0	0	2	1	3	4	8	5	6	9	9	7	7	2	0	0	0	0	0	0																		3
12					0	0	0	0	2	3	0	0	0	0	0	0	1	3	1	2	4	7	3	6	6	6	4	1	0	0	0	0	0	0																		2
13						0	0	0	0	0	0	0	0	2	0	0	0	0	0	0	1	2	7	7	3	6	9	9	5	0	0	0	0	0	0	0																2
14						0	0	0	0	0	0	0	1	0	1	0	0	0	0	3	2	0	2	4	6	6	6	6	4	0	0	0	0	0	0	0	0	0	0													2
15							0	0	0	0	0	1	1	0	0	0	0	0	0	0	0	0	0	0	0	0	0	0	0	0	0	0	0	0	0	0	0	0	0													0
16								0	0	0	1	0	0	0	0	0	0	1	0	1	0	3	3	1	0	0	0	0	0	0	0	0	0	0	0	0	0	0	0	0												0
17									0	0	0	0	1	0	0	0	0	0	1	0	0	0	1	0	3	5	0	0	0	0	0	0	0	0	0	0	0	0	0	0												0
18									0	0	0	0	1	0	0	0	2	0	0	0	0	0	0	0	0	0	0	0	0	0	0	0	0	0	0	0	0	0	0	0	0											0
19									0	0	0	1	0	7	14	18	24	30	29	23	22	23	19	19	19	19	19	19	19	19	19	19	16	0	0	0	0	0	0	0	0	0	0	0								17
20									0	0	0	0	0	0	0	0	0	0	0	0	0	0	0	0	0	0	0	0	0	0	0	0	0	0	0	0	0	0	0	0	0	0									0
Total	0	1	3	3	2	2	2	3	4	3	5	3	4	5	5	6	6	7	7	8	10	10	11	10	9	9	8	7	7	5	4	4	4	3	3	3	3	4	5	5	0	0	0	0	0	0	0	0	0	0	0	6

Fig. 10 Average delay for "charge" FEVs along the CWD lane over time in the Alternative scenario

Section before	1	2	3	4	5	6	7	8	9	10	11	12	13	14	15	16	17	18	19	20	21	22	23	24	25	26	27	28	29	30	31	32	33	34	35	36	37	38	39	40	41	42	43	44	45	46	47	48	49	50	51	Total
1	2	9	14	3	3	9	14	12	5	13	6	3	13	1	4	1	1	0																																		6
2	0	0	0	4	5	2	2	7	15	17	2	11	2	0	2	1	8	1	0	0																																4
3		0	1	2	3	2	3	10	17	18	18	22	31	22	9	1	1	0	0	0																																7
4				1	5	10	9	11	18	25	27	32	32	39	42	32	33	5	0	0	0																															14
5				1	3	0	0	6	15	9	11	17	13	21	30	26	14	0	0	0	0																															7
6				0	1	3	4	11	18	19	21	24	27	30	30	24	7	1	0	0	0																														9
7				2	2	5	8	17	20	23	23	25	29	29	31	17	0	0	0	1	0																													9
8				1	0	4	7	10	17	20	18	18	20	18	18	7	0	0	0	0	0																													6
9				0	7	12	12	18	23	24	24	21	22	21	10	1	1	0	0	0	0																													7
10				1	1	9	15	17	15	18	21	18	16	3	1	0	0	0	0	0																														5
11				0	3	3	6	3	6	6	6	6	5	0	0	0	1	0	0	0	0																													2
12				0	2	3	6	0	5							0	0	0	0	0																														1
13				0	0												0	0	0	0																														0
14																																																			0
Total	1	8	9	3	4	4	5	5	6	10	6	8	9	11	9	10	10	9	11	14	12	11	8	8	6	6	4	4	1	1	0	0	0	0	0	0	0	0	0	0	0	0	0	0	0	0	0	0	0	0	0	7

Fig. 11 Average delay for "Emer" FEVs along the CWD lane over time in the Alternative scenario

Node	1	2	3	4	5	6	7	8	9	10	11	12	13	14	15	16	17	18	19	20	21	22	23	24	25	26	27	28	29	30	31	32	33	34	35	36	37	38	39	40	41	42	43	44	45	46	47	48	49	50	51	52	53	54	55	56	57	58	59	60	61	62	Total
1	0	0	0	0	0	0	0	0.2	0.2	0.1	0.1	0.2	0.2	0.1	0.1	0.1	0.2	0.1	0.1	0.1	0.2	0.1	0.1	0.2	0.1	0.2	0.1	0.1	0.2	0.1																																	4.67
2	0	0	0	0	0	0	0	0.2	0.1	0.2	0.2	0.2	0.2	0.1	0.2	0.2	0.1	0.1	0.2	0.1	0.1	0.2	0.1	0.1	0.2	0.1	0.1	0.2	0.1																																		4.76
3	0	0	0	0	0	0	0	0.2	0.1	0.1	0.2	0.2	0.2	0.2	0.2	0.1	0.2	0.2	0.1	0.1	0.2	0.2	0.1	0.1	0.2	0.1	0.1	0.1	0.1	0.1																																	4.99
4	0	0	0	0	0	0.1	0.2	0.2	0.1	0.2	0.2	0.2	0.2	0.2	0.2	0.2	0.1	0.2	0.2	0.1	0.1	0.2	0.1	0.1	0.1	0.1	0.1	0.1	0																																		5.33
5	0	0	0	0	0.1	0.2	0.2	0.2	0.1	0.2	0.1	0.2	0.2	0.2	0.2	0.1	0.2	0.2	0.2	0.1	0.1	0.2	0.1	0.2	0.1	0.1	0.1	0.1	0.1	0.1																																	5.34
6	0	0	0.1	0.1	0.2	0.1	0.1	0.2	0.2	0.1	0.2	0.2	0.3	0.2	0.2	0.2	0.2	0.1	0.2	0.2	0.2	0.1	0.1	0.1	0.1	0.1	0.1	0																																			5.51
7	0	0.1	0.1	0.2	0.1	0.1	0.1	0.2	0.2	0.1	0.2	0.2	0.2	0.2	0.1	0.3	0.2	0.2	0.2	0.2	0.1	0.1	0.1	0.1	0.1	0.1	0.1	0.1	0.1	0																																	5.6
8	0	0	0.1	0.1	0.2	0.2	0.2	0.1	0.1	0.2	0.2	0.3	0.2	0.2	0.3	0.2	0.2	0.2	0.2	0.1	0.1	0.1	0.1	0.1	0.1	0.1	0.1	0																																				5.79
9	0	0.1	0.2	0.1	0.1	0.2	0.2	0.1	0.1	0.2	0.2	0.3	0.2	0.3	0.2	0.2	0.3	0.1	0.3	0.2	0.2	0.2	0.1	0.1	0.1	0.1	0.1	0.1	0	0																																		6.08
10	0	0.1	0.2	0.1	0.1	0.2	0.2	0.1	0.1	0.2	0.2	0.3	0.2	0.2	0.3	0.2	0.2	0.2	0.1	0.2	0.2	0.1	0.1	0.1	0.1	0.1	0.1	0	0	0																																		6.14
11	0	0.1	0.1	0.2	0.1	0.1	0.2	0.2	0.1	0.1	0.2	0.2	0.3	0.2	0.2	0.3	0.2	0.2	0.2	0.2	0.1	0.1	0.1	0.1	0.1	0	0.1	0	0	0	0																																	6.13
12	0	0.1	0.1	0.2	0.1	0.1	0.2	0.2	0.1	0.1	0.2	0.2	0.3	0.2	0.3	0.2	0.3	0.2	0.2	0.2	0.1	0.1	0.3	0.1	0.1	0	0	0.1	0	0	0	0																																	6.33
13		0	0.1	0.1	0.2	0.1	0.1	0.2	0.2	0.1	0.1	0.2	0.2	0.3	0.2	0.3	0.2	0.3	0.2	0.2	0.2	0.2	0.1	0.1	0.1	0.1	0	0	0	0	0	0																																6.41
14		0	0.1	0.1	0.2	0.1	0.1	0.2	0.2	0.1	0.1	0.2	0.2	0.3	0.2	0.3	0.2	0.3	0.3	0.3	0.3	0.1	0.2	0.2	0.2	0.1	0.2	0.1	0	0	0	0	0	0	0	0																												6.88
15		0	0	0.1	0.1	0.2	0.1	0.1	0.2	0.2	0.1	0.1	0.2	0.2	0.3	0.2	0.3	0.2	0.3	0.2	0.2	0.2	0.2	0.1	0.1	0	0	0	0	0	0	0	0	0																													6.57
16		0		0	0	0.1	0.2	0.1	0.1	0.2	0.3	0.2	0.1	0.1	0.3	0.2	0.2	0.3	0.2	0.2	0.3	0.2	0.3	0.3	0.3	0.1	0.2	0.1	0.2	0.2	0.1	0	0	0	0	0	0																										6.73
17		0	0	0.1	0.1	0.2	0.1	0.1	0.2	0.3	0.2	0.1	0.1	0.3	0.2	0.2	0.3	0.2	0.2	0.3	0.2	0.2	0.3	0.2	0.2	0.1	0.1	0.2	0.2	0.1	0	0	0	0	0	0	0	0																									6.73
18		0		0	0	0.1	0.1	0.2	0.1	0.1	0.3	0.2	0.1	0.1	0.3	0.2	0.2	0.3	0.2	0.2	0.3	0.3	0.3	0.3	0.2	0.1	0.1	0.2	0.1	0.2	0.2	0	0	0	0	0	0	0	0	0																							6.73
19		0		0	0	0.1	0.1	0.2	0.1	0.1	0.2	0.3	0.2	0.1	0.1	0.3	0.2	0.2	0.3	0.3	0.3	0.3	0.3	0.2	0.1	0.2	0.1	0.2	0	0	0	0	0	0	0	0	0	0	0	0																							6.73
20			0		0	0	0.1	0.1	0.2	0.1	0.1	0.2	0.3	0.2	0.1	0.1	0.3	0.2	0.2	0.3	0.2	0.2	0.3	0.3	0.3	0.2	0.1	0.2	0.1	0	0	0	0	0	0	0	0	0	0	0	0																						6.73
Total	0	0	0	1	1	1	1	1.4	1.6	1.8	1.9	2.2	2.3	2.5	2.7	2.9	3	3.2	3.4	3.4	3.7	3.7	3.9	3.8	4	3.8	4	4	4.1	4	3.8	3.6	3.5	3.3	3.1	2.8	2.7	2.4	2.2	1.8	1.6	1.5	1.3	1	1	1	1	1	1	0	0	0	0	0	0	0	0	0	0	0	0	0	120

Fig. 12 Energy received (kWh) by FEVs at nodes along the CWD lane over time in the Reference scenario

all of the CZs, a higher resolution of simulation sections is required. Two additional experiments for the Reference and the Alternative scenarios have been performed.

The analyzed nodes were set at a distance of LCZ + I, equal to 50 m. Fig. 14 and Fig. 15 report the cumulative number of coil on/off switching during the simulation for a 20 s time widow (1500 ∼ 1520 s) respectively for the Reference and Alternative scenarios. In the Reference scenario, there is generally a higher occurrence of switching on compared to the Alternative scenario. This result can be confirmed because of the larger number of vehicles in the CWD lane. The variability of the power provided, as estimated by simulation, is evident in the charts in Fig. 14 and Fig. 15 in which the instantaneous

Time (min)

Node	1	2	3	4	5	6	7	8	9	10	11	12	13	14	15	16	17	18	19	20	21	22	23	24	25	26	27	Total
1	0	0	0	0	0	0	0.4	0.4	0.3	0.4	0.3	0.3	0.4	0.2	0.4	0.3	0.2	0.2										6.36
2	0	0	0	0	0	0	0.4	0.4	0.4	0.4	0.4	0.3	0.3	0.4	0.2	0.2	0.2											6.32
3	0	0	0	0	0	0.4	0.4	0.4	0.4	0.4	0.4	0.5	0.5	0.5	0.4	0.3	0.3	0.1	0.2	0.2								6.97
4	0	0	0	0	0.3	0.4	0.4	0.4	0.4	0.4	0.4	0.4	0.5	0.5	0.5	0.5	0.3	0.1	0.2	0.2								7.72
5	0	0	0	0	0.1	0.2	0.3	0.4	0.3	0.4	0.4	0.4	0.4	0.4	0.4	0.4	0.4	0.3	0.1	0.2	0.2							7.05
6	0	0	0.2	0.1	0.2	0.2	0.3	0.3	0.4	0.4	0.4	0.4	0.4	0.4	0.4	0.4	0.4	0.3	0.1	0.2	0.2							7.33
7	0	0	0.2	0.2	0.1	0.2	0.2	0.2	0.3	0.4	0.4	0.4	0.4	0.4	0.4	0.4	0.5	0.4	0.2	0.3	0.1	0.2	0.2					7.47
8	0	0	0.2	0.2	0.2	0.2	0.2	0.2	0.2	0.3	0.4	0.4	0.4	0.4	0.4	0.4	0.4	0.3	0.4	0.1	0.2	0.1	0.2					7.13
9	0	0	0.1	0.2	0.2	0.2	0.2	0.2	0.2	0.2	0.3	0.4	0.4	0.4	0.4	0.4	0.4	0.4	0.2	0.4	0.1	0.2	0.2	0.1	0.1	0.2		7.4
10	0	0.1	0.1	0.2	0.3	0.2	0.2	0.2	0.2	0.2	0.3	0.3	0.3	0.4	0.4	0.4	0.4	0.4	0.1	0.1	0.1	0.2	0.1	0.1	0.2			7.17
11	0	0.1	0.1	0.2	0.3	0.2	0.3	0.2	0.2	0.2	0.2	0.3	0.3	0.3	0.3	0.3	0.3	0.3	0.1	0.2	0.1	0.1	0.2	0.1	0.1	0.2		6.79
12	0	0.1	0.1	0.2	0.3	0.3	0.3	0.2	0.2	0.2	0.2	0.2	0.3	0.3	0.3	0.3	0.3	0.3	0.1	0.1	0.1	0.1	0.1	0.1	0.1			6.75
13	0	0.1	0.1	0.2	0.3	0.3	0.2	0.2	0.2	0.2	0.2	0.3	0.3	0.3	0.3	0.3	0.3	0.3	0.1	0.1	0.1	0.1	0	0.1	0.1	0.1	0.1	6.64
14	0	0.1	0.1	0.2	0.3	0.2	0.3	0.2	0.2	0.2	0.2	0.3	0.2	0.3	0.3	0.3	0.3	0.3	0.2	0.1	0.1	0	0	0.1	0.1	0.1	0	6.56
15	0	0.1	0.1	0.2	0.3	0.3	0.2	0.3	0.2	0.2	0.2	0.3	0.3	0.3	0.3	0.3	0.3	0.3	0.2	0.1	0.1	0	0	0.1	0.1	0	0	6.16
16	0	0.1	0.1	0.2	0.3	0.3	0.2	0.3	0.2	0.3	0.2	0.3	0.3	0.3	0.3	0.3	0.3	0.3	0.2	0.1	0.1	0	0	0.1	0	0	0	6.28
17	0	0.1	0.1	0.2	0.3	0.3	0.3	0.3	0.2	0.3	0.3	0.3	0.3	0.3	0.3	0.3	0.3	0.3	0.2	0.1	0.1	0.1	0	0	0	0	0	6.37
18	0	0.1	0.1	0.3	0.3	0.3	0.3	0.3	0.2	0.3	0.3	0.3	0.3	0.3	0.3	0.3	0.3	0.3	0.2	0.1	0.1	0.1	0	0	0	0	0	6.39
19	0	0	0.1	0.3	0.3	0.3	0.4	0.4	0.4	0.4	0.4	0.4	0.4	0.4	0.4	0.4	0.4	0.3	0.1	0.1	0.1	0	0	0	0	0	0	8.64
20	0	0	0.1	0.1	0.3	0.3	0.3	0.3	0.3	0.3	0.3	0.3	0.3	0.3	0.3	0.3	0.2	0.1	0.1	0	0	0	0	0	0	0	0	6.73
Total	0	0	1	1	1	2	2	2.3	2.7	3	3.4	3.7	4	4.4	4.6	5.1	5.3	5.7	5.7	5.8	5.7	5.5	5.4	5.3	4.9	4.8	4.6	138

(Total row continues: 4.3 4.1 3.8 3.6 3.2 3 2.6 2.5 2.1 1.9 1.5 1.4 1.1 0.8 0.5 0.5 0.3 0 0 0 0 0 0 0 0 0 0 0 0 0 0 0 0 0 0)

Fig. 13 Energy received (kWh) by FEVs at nodes along the CWD lane over time in the Alternative scenario

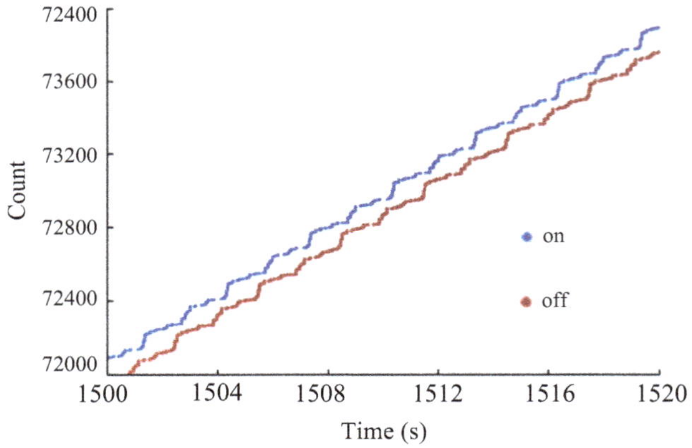

Fig. 14 Cumulative count of on/off switching for all the CZs of the CWD lane during 20 s for the Reference scenario

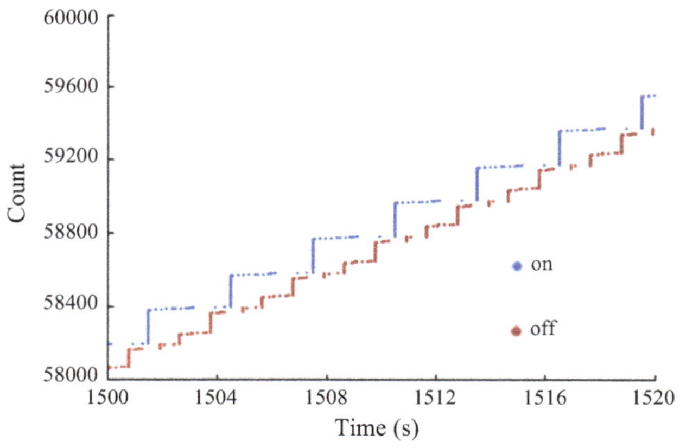

Fig. 15 Cumulative count of on/off switching for all the CZs of the CWD lane during 20 s for the Alternative scenario

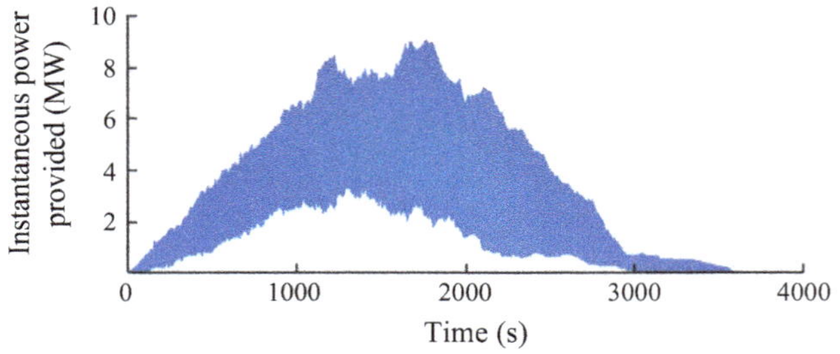

Fig. 16 Instantaneous power provided for the entire 20 km CWD lane in the Reference scenario

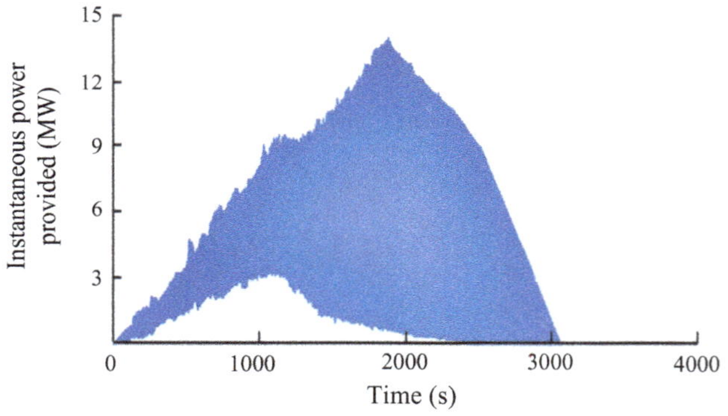

Fig. 17 Instantaneous power provided for the entire 20 km CWD lane in the Alternative scenario

number of CZs in the "ON" state changes in a few seconds for both Reference and Alternative scenarios. The maximum number of CZs simultaneously in the "ON" state is estimated to equal 181 CZs at the simulation time of 1763.2 s for the Reference scenario and 281 CZs at the simulation time of 1876.5 s for the Alternative. To better observe the energy variability, the simulated instantaneous power provided for the entire 20 km CWD lane is also reported in Fig. 16 and Fig. 17. The minimum and maximum power provided can be clearly identified, by multiplying the number of CZs in the "ON" state by the nominal power provided (Pcz), according to LCD. In addition, a detailed chart of the power provided for the entire CWD lane is presented in Figs. 18 and 19 for an identical 20 s time window to show the typical pattern for the two simulated scenarios.

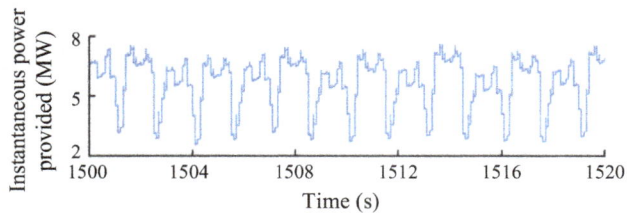

Fig. 18 Instantaneous power provided (MW) for the 20 s time window for the entire 20 km CWD lane in the Reference scenario

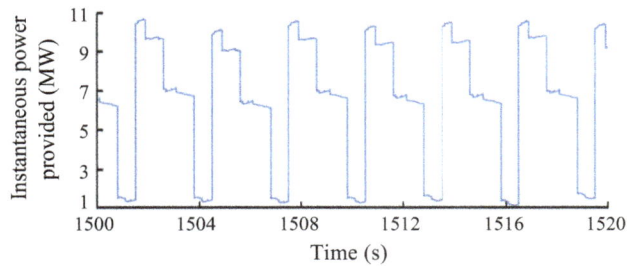

Fig. 19 Instantaneous power provided for the 20 s time window for the entire 20 km CWD lane in the Alternative scenario

5 Conclusions

This study presented a method for assessing the performance of the wireless inductive power transfer used to charge electric vehicles while driving. Assuming the CWD system can operate in a scenario with cooperative behavior, the developed traffic model is able to simulate different traffic conditions. Primary traffic parameters can be estimated for the CWD lane, such as the vehicle count and the average speed that are time dependent and change relevantly along the road. This traffic model can manage even intense traffic conditions by simulating vehicle platoons and delays caused by internal traffic interactions (i.e., different vehicle speeds and new entries into the lane) and technical constraints requiring a minimum headway in the CWD lane. Unlike traditional dynamic traffic models, the vehicle motion in this proposal includes the energy needs and charging opportunities because they influence drivers' decisions and then traffic performance. According to their SOC along the road, vehicles are simulated as inside or outside the charging lane, and their speeds are set according to their charging mode. The model has an approximation consistent with the stage of development of CWD technology and the deployment of cooperative driving. Although simplified, it allows for the prediction of many relevant energy issues and possible operational problems.

From the energy point of view, the analyses presented here for a "best case" scenario demonstrates that the traffic also has a relevant effect on the energy that should be supplied by an energy provider. In the Reference scenario

simulated, characterized by better traffic conditions, the maximum power that should be supplied for the entire road is approximately 9 MW, whereas in the Alternative scenario, in which vehicles proceed slower and are generated closer, the power required by the vehicles on the CWD lane is approximately 14 MW. This result is even more relevant considering that the total switching on number is greater in the Reference scenario, thus indicating a major usage of the CWD lane. However, the slower speeds and the platoon conditions require a larger number of coils to be on simultaneously. This critical traffic condition, characterized by platoons with vehicles at a constant distance, generates high peaks in the power trend; in a few tenths of a second, the power required can change by more than 9 MW. Generally, the required power trend under platoon conditions is more consistent but with higher peaks.

Acknowledgments This study is partially supported by the eCo-FEV project (Grant agreement No. 314411). The authors would like to thank all project partners for their support.

References

[1] Boulanger AG, Chu AC, Maxx S et al (2011) Vehicle electrification: status and issues. P IEEE 99(6):1116–1138
[2] Kotchapansompote P, Wang YF, Imura T et al (2011) Electric vehicle automatic stop using wireless power transfer antennas. In: Proceedings of the 37th annual conference on IEEE Industrial Electronics Society (IECON'11), Melbourne, Australia, 7–10 Nov 2011, pp 3840–3845
[3] Cassat A, Jufer M (2002) MAGLEV projects technology aspects and choices. IEEE Trans Appl Supercon 12(1):915–925
[4] Stielau OH, Boys JT, Covic GA et al (1999) Battery charging using loosely coupled inductive power transfer. In: Proceedings of the 8th European conference on power electronics and applications (EPE'99), Lausanne, Switzerland, 7–9 Sept 1999
[5] He F, Yin Y, Zhou J (2013) Integrated pricing of roads and electricity enabled by wireless power transfer. Transp Res C-Emer 34:1–15
[6] SLee S, Huh J, Park C et al (2010) On-line electric vehicle using inductive power transfer system. In: Proceedings of the 2010 IEEE energy conversion congress and exposition (ECCE'10), Atlanta, GA, USA, 12–16 Sept 2010, pp 1598–1601
[7] Deflorio F, Guglielmi P, Pinna I et al (2013) Modelling and analysis of wireless "charge while driving" operations for fully electric vehicles. In: Proceedings of the convegno annuale e seminario scientifico della Societ Italiana dei Docenti di Trasporti (SIDT'13), Trieste, Italy, 17–18 Oct 2013
[8] Witold K, Lan L, Yuichi K et al (2014) Deliverable D200.1: Use cases and requirements for an efficient cooperative platform. eCo-FEV, Valbonne
[9] Hoogendoorn SP, Bovy PHL (2001) State-of-the-art of vehicular traffic flow modelling. P I Mech Eng I 215(4):283–303

[10] Barceló J (2005) Dynamic network simulation with AIMSUN. In: Kitamura R, MasoKuwahara M (eds) Simulation approaches in transportation analysis, part 1. Springer, Boston, pp 57–98

[11] Cascetta E (2001) Transportation systems engineering: theory and methods. Kluwer Academic Publishers, Dordrecht

[12] Ben-Akiva ME, Gao S, Wei Z et al (2012) A dynamic traffic assignment model for highly congested urban networks. Transp Res C-Emer 24:62–82

[13] Deflorio F, Castello L (2014) Traffic modeling of a cooperative charge while driving system in a freight transport scenario. In: Proceedings of the 4th international symposium of transport simulation and international workshop on traffic data collection and ITS standardisation (ISTS & IWTDCS'14), Ajaccio, France, 1–4 June 2014, pp 7–18

[14] SS-EN 14725:2004. Space engineering – Verification. 2004

[15] Haskins C (2011) Systems engineering handbook V. 3.2.2. INCOSE-TP-2003-002-03.2.2. The International Council on Systems Engineering (INCOSE), San Diego, CA, USA

[16] Myers GJ, Badgett T, Sandler C (2004) The art of software testing. Wiley, New York

[17] Gutiérrez JCP, Gutierrez Garcia JJ, Gonzalez Harbour M (1998) Best-case analysis for improving the worst-case schedulability test for distributed hard real-time systems. In: Proceedings of the 10th Euromicro workshop on real-time systems, Berlin, Germany, 17–19 June 1998, pp 35–44

[18] Daganzo CF (1997) Fundamentals of transportation and traffic operations. Pergamon, Amsterdam

[19] Yannis G, Golias J, Antoniu C (2004) Combining traffic simulation and driving simulator analyses for advanced cruise control system impact identification. In: Proceedings of the 83rd annual meeting of the transportation research board, Washington, DC,USA, 11–15 Jan 2004

[20] Hegeman G, Brookhuis K, Hoogendoorn S (2005) Opportunities of advanced driver assistance systems towards overtaking. Eur J Transp Infrastruct Res 5(4):281–296

Francesco Paolo DEFLORIO received the Ph.D. in Automatics and Information Science in Transportation Systems from Politecnico di Torino, Italy, where he is assistant professor in Transport System Engineering since 2006. His research interests include modeling and applications of traffic and transportation systems, such as dynamic route guidance in road networks, traffic and energy simulation analysis, intelligent transport systems (ITS).

Luca CASTELLO received the M.Sc. in Civil Engineering from Politecnico di Torino, Italy, in 2011, where he worked as a research fellow from 2013 to 2014. Currently, he is working as a consultant for the transportation research center of IVECO. His research interests include traffic simulations and new vehicle technologies, primarily concerning vehicle energy needs and emissions.

Ivano PINNA received the M.Sc. in Mechanical Engineering from Politecnico di Torino, Italy, where he is research contractor and Ph.D student. His research interests include intelligent transport systems, road safety, ADAS, electric and hybrid vehicles, "charge while driving" analysis from transport systems viewpoint.

Paolo GUGLIELMI received the Ph.D. degree in Electrical Engineering from the Politecnico di Torino, Turin, Italy, in 2001. In 1997, he joined the Department of Electrical Engineering, Politecnico di Torino as Research Assistant. Since 2012, he is Associate Professor at the same university. His research interests include power electronics for wireless power transfer, high-performance drives, and computer-aided design of electrical machines.

13

Direct load control by distributed imperialist competitive algorithm

Fengji LUO, Junhua ZHAO (✉), Haiming WANG,
Xiaojiao TONG, Yingying CHEN, Zhao Yang DONG

Abstract Demand side management techniques have drawn significant attentions along with the development of smart grid. This paper proposes a new direct load control (DLC) model for scheduling interruptible air conditioner loads. The model is coordinated with the unit commitment and economic dispatch to minimize the total operation cost over the whole dispatch horizon. The network constraints are also considered in the model. To ensure the thermal comfort of the occupants, we are among the first to incorporate the advanced two-parameter thermal inertia dynamical model of customer houses into the DLC model to calculate the indoor temperature variation. This paper also proposes a distributed imperialist competitive algorithm to effectively solve the model. The simulation studies prove the efficiency of the proposed methodology.

Keywords Direct load control, Imperialist competitive algorithm, Demand side management

1 Introduction

With the increasing penetration of renewable energy sources and ever-increasing load demand, power system is experiencing the transition from a state "where relatively well behaved demand is matched with well predictable generation" to a state where both demand and generation sides are becoming increasingly time-varying and stochastic [1]. This transition encourages the adoption of demand side management to re-shape the load profiles. The most commonly used demand side management approach is known as the direct load control (DLC), which aims to schedule the cycling of the customers' controllable appliances to achieve a certain operation objective.

In the summer days, air-conditioners are large energy consumers. Thus, designing suitable methods to schedule the cycling of air conditioner loads (ACLs) can effectively curtail the peak load in summer days. Since the primary usage of air-conditioners is to provide thermal comfort for customers, the most important constraint of ACL scheduling is obviously minimizing customers' thermal discomfort. Besides, DLC scheduling should be coordinated with the existing power system operation tasks such as unit commitment (UC) and economic dispatch (ED).

In the literature, many efforts have been done in DLC. In [2], the authors integrated interruptible load management (ILM) into DLC to provide ancillary services; in [3], the authors proposed a dynamical optimal power flow model to online select the interruptible loads while incorporating the network constraints; in [4], the authors reported three interruptible load management programs of Taiwan power company; in [5], the authors used the iterative deepening genetic algorithm to perform the DLC scheduling; a DLC based unit commitment model to minimize the system production cost is presented in [6]; the authors in [7] proposed an optimal power flow based framework for the independent system operator (ISO) to real-time select the interruptible load offers; the authors in [8] used probabilistic methods to do distributed interruptible load shedding; the authors in [9] proposed a reliability-constrained unit commitment model by integrating

F. LUO, J. ZHAO, H. WANG, Y. CHEN, Centre for Intelligent Electricity Networks (CIEN), The University of Newcastle, Callaghan, NSW 2308, Australia
(✉) e-mail: andy.zhao@newcastle.edu.au
X. TONG, The Hunan First Normal University, Changsha, China
Z. Y. DONG, The University of Sydney, Sydney, NSW 2006, Australia

probabilistic spinning reserve and interruptible load. Literature [10] adopted a multi-pass dynamic programming to schedule the ACLs such that both of the generating costs and peak load reduction are minimized over the whole dispatch period; literature [11] proposed a unit commitment model incorporating large air conditioner loads, and used the fuzzy dynamic programming method to solve it; the authors in [12] designed a fuzzy logic based controller which recognizes several important customer preferences and desires. Authors in [13] presented a direct load control model in virtual power plant (VPP) operation, and used commercial software to simulate the thermal transition process of different buildings.

By reviewing the literature, we found that most of the previous works are based on the assumption that the scheduled ACL groups are homogeneous, which means that ACL groups are assumed to have the identical building environments and the customers have the same thermal preferences. However, in practice, the heterogeneous building environment will cause different indoor temperature variations, and different people may have different thermal preferences. In order to minimize customers' thermal discomfort, a more accurate model is needed to depict the thermal transition process of the building while considering the customers' different thermal preferences. Furthermore, to the best of the authors' knowledge, the existing DLC methods have not taken the transmission system capability constraints into account. Based on the above considerations, the main contributions of this paper are highlighted as follows.

- Propose a novel DLC model by taking the transmission network constraints into account;
- Among the first to incorporate a comprehensive thermal inertia model into DLC scheduling to satisfy customers' different thermal preferences.
- Propose a new distributed ICA algorithm to effectively solve the proposed DLC model.

This paper is organized as follows. In section 2, proposed DLC model is introduced; in section 3, the thermal inertia model is introduced; in section 4, proposed distributed imperialist competitive algorithm is introduced; approach to solve the model is presented in section 5; case studies are given in section 6; finally, conclusions are drawn in section 7.

2 DLC model formulation

2.1 Objective function

The DLC model aims to minimize the total system operation costs over the dispatch horizon. The objective function is formulated as (1).

$$\min F = \sum_{t=1}^{T} \left(C_{gen}^t + C_{il}^t \right) \tag{1}$$

where t is the time interval index and T is the total number of the time intervals; C_{gen}^t and C_{il}^t represent the generation cost and the ACL interruption cost at time interval t, respectively. We assume that the utility provides the discounted retail price for the customers who participate in the DLC program. Then C_{il}^t can be represented by the profit loss of the utility by promoting the DLC program. C_{gen}^t and C_{il}^t are calculated as (2) and (3).

$$C_{gen}^t = \sum_{G=1}^{G} \left[\left(FC(P_g^t) + SC_g^t \cdot (1 - u_g^{t-1}) \right) \cdot u_g^t \right] \tag{2}$$

$$C_{il}^t = \sum_{n=1}^{N} \left(pr \cdot (1 - \eta) \cdot CP_n \cdot s_n^t + pr \cdot (1 - s_n^t) \cdot CP_n \right) \tag{3}$$

where g and G represent the generator unit index and the total number of the units, respectively; n and N denote the index and the total number of the ACL groups, respectively; u_g^t represents the state of the gth unit at t: 0-OFF and 1-ON; s_n^t represents the nth ACL group at t: 0-OFF and 1-ON; P_g^t is the active power output of the unit g at time t (kW); $FC(P_g^t)$ is the fuel cost function of the units (\$); SC_g^t is the startup cost of the unit g at time t (\$); η is the discount rate of the retailed electricity price provided by the utility (%); CP_n is the capacity of the nth ACL group (kW); pr is the retailing electricity price (\$/kWh). $FC(P_g^t)$ is with the form of (4) and SC_g^t is represented by (5).

$$FC_g^t = a_g \cdot (P_g^t)^2 + b_g \cdot P_g^t + c_g \tag{4}$$

$$SC_g^t = \begin{cases} HSC_g, & \text{if } MDT_g \leq TG_g^{off} \leq MDT_g + CSH_g \\ CSC_g, & \text{if } \quad TG_g^{off} > MDT_g + CSH_g \end{cases} \tag{5}$$

where a_g, b_g, and c_g are fuel cost coefficients of unit g; HSC_g and CSC_g are the hot and cold startup costs of unit g, respectively (\$); TG_g^{on} and TG_g^{off} denote the Duration during which the unit g is continuously ON and OFF (hours); MDT_g is the minimum down time of unit g; CSH_g is the cold startup hour of the unit g (hours).

2.2 Constraints

1) Load balance constraint

$$\sum_{g=1}^{G} P_g^t = PL^t - IL^t \tag{6}$$

$$IL^t = \sum_{n=1}^{N} CP_n \cdot (1 - s_n^t) \tag{7}$$

where PL^t is the forecasted system load at time interval t (kW); IL^t is the total interrupted load capacity at time t (kW).

2) Generator power output constraint

$$P_g^{\min} \leq P_g^t \leq P_g^{\max} \tag{8}$$

where $P_g^{\min}$ and $P_g^{\max}$ are the minimum and maximum power limits (kW) of unit g.

3) Generator ramp rate constraint

$$R_{down,g} \leq P_g^t - P_g^{t-1} \leq R_{up,g} \tag{9}$$

where $R_{down,g}$ and $R_{up,g}$ are ramp up and down rate of unit g.

4) Generator minimum online/offline time constraint

$$\begin{cases} TG_g^{on} \geq GMUT_g \\ TG_g^{off} \geq GMDT_g \end{cases} \tag{10}$$

where $GMUT_g$ and $GMDT_g$ are the minimum offline time (hours) and online time (hours) of unit g.

5) Network power flow constraint

$$\begin{cases} P_i^t = PL_i^t - RL_i^t + PB^t + U_i^t \sum_{j=1}^{I} U_j^t (G_{ij} \cos \theta_{ij}^t + B_{ij} \sin \theta_{ij}^t) \\ \\ Q_i^t = QL_i^t + U_i^t \sum_{j=1}^{I} U_j^t (G_{ij} \cos \theta_{ij}^t - B_{ij} \sin \theta_{ij}^t) \end{cases} \tag{11}$$

where i is the bus index; I is the set of the buses; L is the set of the transmission lines; P_i^t and Q_i^t are the active and reactive power of bus i at time interval t (kW); PL_i^t and QL_i^t are the active and reactive load on bus i at time t (kW); PB_i^t is the payback energy of bus i at time t (kW); RL_i^t is the reduced load of bus i at time t (kW); U_i^t is the voltage magnitude of bus i at time t; θ_{ij}^t is the phase angle deviation of branch ij at time t; G_{ij} and B_{ij} are real part and imaginary part of the nodal admittance matrix.

6) Bus voltage constraint

$$U_i^{\min} \leq U_i^t \leq U_i^{\max} \tag{12}$$

where $U_i^{\min}$ and $U_i^{\max}$ are the minimum and maximum voltage magnitude of bus i.

7) Apparent power constraints for transmission lines

$$S_{ij}^t \leq S_{ij}^{\max} \tag{13}$$

where S_{ij}^t is the apparent power of branch ij at time t (kW); and $S_{ij}^{\max}$ is the maximum apparent power of branch ij (kW).

$S_{ij}^{\max}(t)$ is calculated as

$$S_{ij}^t \sqrt{(P_{ij}^t)^2 + (Q_{ij}^t)^2} \tag{14}$$

$$P_{ij}^t = (U_i^t)^2 G_{ij} - U_i^t U_j^t (G_{ij} \cos \theta_{ij}^t + B_{ij} \sin \theta_{ij}^t) \tag{15}$$

$$Q_{ij}^t = -(U_i^t)^2 B_{ij} - U_i^t U_j^t (G_{ij} \cos \theta_{ij}^t - B_{ij} \sin \theta_{ij}^t) \tag{16}$$

8) ACL minimum online time constraint. To avoid frequently switch on/off and protect the mechanism equipment of the ACLs, the ACL groups are constrained by (18).

$$TA_n^{on} \geq AMUT_n \tag{17}$$

where TA_n^{on} is the duration during which the ACL group n is continuously ON (hours); and $AMUT_n$ is the minimum online time of ACL group n (hours);

9) Indoor temperature dead band. The indoor temperature must be constrained by the temperature dead band, which refects different customers' preferences.

$$TP\_Low_n^{in} \leq TP_n^{in}(t) \leq TP\_Up_n^{in} \tag{18}$$

where $TP_n^{in}(t)$ is the indoor temperature of ACL group n at time t; $TP\_Low_n^{in}(t)$ and $TP\_Low_n^{in}(t)$ are the lower and upper limits of comfort temperature range of ACL group n (°C).

To calculate the indoor temperature profiles of the rooms in each ACL group after switching off the ACs, the thermal inertia model introduced in next section is applied.

3 Thermal appliances modeling

3.1 Thermostatically load modeling

The key to ensure the occupants' thermal comfort is to fully understand and model the dynamic thermal process of the buildings. In the smart house study, the one parameter thermal model has been widely used in many literatures [14–16], which is shown in Fig. 1a. The one-parameter model takes into account of parameters like internal and external temperatures, but only considers the thermal resistance of walls and neglects walls' thermal capacitance. In this paper, a more complex and accurate two-parameter is represented in Fig. 1b. The house is divided into two components, one of which is the internal of the house and the other is the additional thermal mass such as walls with much larger thermal capacitance.

However, the variation of indoor air temperature of a house could be considerably different when taking into account of thermal capacitance of walls [15]. This is due to the fact that the heat gain of a house can be divided into two parts: the relatively steady-state transmission resulting from temperatures differences between the indoor air and outdoor surroundings, and the unsteady-state gain due to the varying intensity of solar radiation on the walls. The unsteady-state heat flow across walls is very complicated

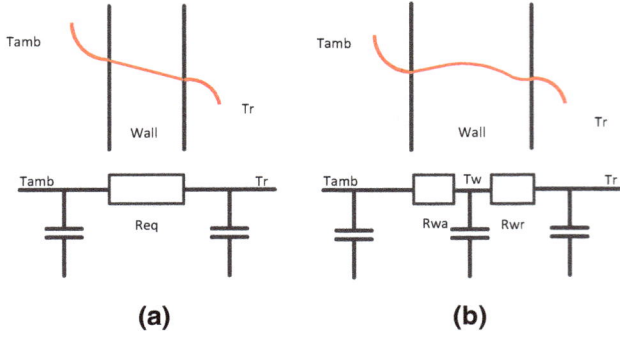

Fig. 1 **a** One parameter model. **b** Two parameter model

as part of heat passing through walls is captured and later released to either the indoor air or outdoor ambient. Therefore, the thermal dynamic model of a two-parameter model could be expressed as below [17].

$$\frac{dT_r(t)}{dt} = \frac{1}{M_a \times Cp_a} \times \left(\frac{dQ_{gain\_a}(t)}{dt} - \frac{dQ_{ex\_w\_r}(t)}{dt} - \frac{dQ_{ac}(t)}{dt} \right) \tag{19}$$

$$\frac{dT_w(t)}{dt} = \frac{1}{M_w \times Cp_w} \times \left(\frac{dQ_{gain\_w}(t)}{dt} + \frac{dQ_{ex\_w\_r}(t)}{dt} \right) \tag{20}$$

$$\frac{dQ_{gain\_a}(t)}{dt} = \frac{T_{amb} - T_r}{R_{eq}} \tag{21}$$

$$\frac{dQ_{ex\_w\_r}(t)}{dt} = \frac{T_w - T_r}{R_{wr}} \tag{22}$$

$$\frac{dQ_{ac}(t)}{dt} = COP \times P_{ac} \tag{23}$$

$$\frac{dQ_{gain\_w}(t)}{dt} = \frac{T_{amb} - T_w}{R_{wa}} \tag{24}$$

where T_r is the room temperature (°C); T_w is the wall temperature (°C); M_a and M_w are the mass of air inside the house and the walls (kg); Cp_a and Cp_w are the heat capacities of the air and the wall (J/kg*k^{-1}); Q_{gain_a} is the heat gain by the indoor air from the ambient (J); Q_{gain_w} is the heat gain by the wall from the ambient (J); Q_{ac} is the cooling energy delivered by air conditioner (J); COP is the coefficient of performance of air conditioner; $Q_{ex\_w\_r}$ is the heat exchange between the wall and indoor air (J); R_{eq} is the equivalent thermal resistance of the house envelope (m^{2}K/W); T_{amb} is the ambient temperature (°C); R_{wr} is the thermal resistance between the wall inner surface and the indoor air (m^{2}K/W); R_{wa} is the thermal resistance between the wall outer surface and the ambient (m^{2}K/W); P_{ac} is the power of the individual air-conditioner (kW).

One of our previous studies [18] has proved that different complexities of models can pose significant impacts on the accuracy of cooling energy calculation, it is therefore decided that the complex thermal model is chosen in order to obtain more accurate results.

3.2 Linearization of the thermal inertia model

The thermal dynamic model in (19)–(24) can be linearized for convenient calculating the indoor temperature variation. For each dispatch time interval, Δt is divided into K steps. Provided that K is large enough, we can assume that the temperatures of the ambient, walls, and the indoor air within any time step are constant. Hence, the change in temperatures can therefore be presented by the temperature difference between two adjacent time steps. Therefore, the thermal dynamic model can be linearized as (25)–(28).

$$TP_n^{in}(k) = \left(1 - \frac{1}{M_a \times Cp_a \times R_{eq}} \right) \times T_{r\_init}$$
$$+ \frac{1}{M_a \times Cp_a \times R_{eq}} \times T_{amb\_init}$$
$$+ \frac{T_{w\_init} - T_{r\_init}}{M_{air} \times Cp_a \times R_{wr}} - S_{ac\_init}$$
$$\times \frac{Q_{ac}}{M_a \times Cp_a}, \ k = 1 \tag{25}$$

$$TP_n^{in}(k) = \left(1 - \frac{1}{M_a \times Cp_a \times R_{eq}} \right) \times TP_n^{in}(k-1)$$
$$+ \frac{1}{M_a \times Cp_a \times R_{eq}}$$
$$\times T_{amb}(k-1) + \frac{T_w(k-1) - T_r(k-1)}{M_a \times Cp_a \times R_{wr}}$$
$$- S_{ac}(k) \times \frac{Q_{ac}(k-1)}{M_a \times Cp_a}, \ \forall k \in [2, K] \tag{26}$$

$$T_w(k) = T_{w\_init} + \frac{T_{amb\_init} - T_{w\_init}}{M_w \times Cp_w \times R_{wa}} + \frac{T_{r\_init} - T_{w\_init}}{M_w \times Cp_w \times R_{wr}},$$
$$k = 1 \tag{27}$$

$$T_w(k) = T_w(k-1) + \frac{T_{amb}(k-1) - T_w(k-1)}{M_w \times Cp_w \times R_{wa}}$$
$$+ \frac{T_r(k-1) - T_w(k-1)}{M_w \times Cp_w \times R_{wr}}, \forall k \in [2, K] \tag{28}$$

where T_{amb_init} is the initial ambient temperature (°C); T_{w_init} is the initial wall temperature (°C); T_{r_init} is the initial room temperature (°C).

4 Distributed imperialist competitive algorithm

The proposed DLC model is a binary, non-convex, high dimension, combinatorial optimization problem, which is hard to be handled by the conventional programming methods. There are many heuristic-based optimization algorithms that can be applied to solve the model, such as genetic algorithm (GA), particle swarm optimization (PSO) algorithm, differential evolutionary (DE) algorithm, etc. Recently, there is a new algorithm called imperialist competitive algorithm (ICA) [19] proposed has been applied to solve many industrial optimization problems [20–22]. A recent paper reported ICA is powerful to solve power system combinatorial problem [23]. As what will be shown in the later section of this paper, the evolution mechanism of ICA makes it be inherently suitable for paralleled implementation and its searching capability can thus be significantly improved. Therefore, in this paper we employ ICA to solve the proposed model. And in order to enhance its searching performance, we propose a distributed processing architecture for ICA.

4.1 Imperialist competitive algorithm

ICA mimics the competition among the imperialists. Each imperialist possess some colonies to form an empire, and tries to expanse its power by possessing the colonies of other empires. During the competition, weak empires collapse and powerful ones take possession of their colonies.

1) Empire initiation

As other heuristic searching algorithms, ICA maintains a population of individuals with the number of N_{pop}. In ICA, each individual is called a *country*. For a N-dimension minimization problem, the ith country is a $1 \times N$ vector with the form of $country_i = [x_i^1, x_i^2, ..., x_i^N]$. Each country represents a solution for a given problem, and its cost c_i can be obtained by evaluating it as $c_i = f(country_i) = f([x_i^1, x_i^2, ..., x_i^N])$.

In the start, N_{imp} of the countries are selected as imperialists. The other $N_{pop} - N_{imp}$ countries act as the colonies and are assigned to the N_{imp} imperialists in proportional to the powers of the imperialists. To calculate the imperialist powers, the cost of an imperialist (denoted as c_n) is firstly normalized,

$$C_n = c_n - \max_i\{c_i\} \tag{29}$$

where C_n is the normalized cost. Then the normalized power of nth imperialist can be calculated by (30).

$$p_n = \left| \frac{C_n}{\sum\limits_{i=1}^{N} C_i} \right| \tag{30}$$

The initial number of the colonies possessed by each imperialist (NC_n) is in proportional to its power,

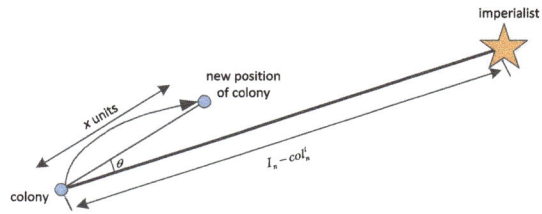

Fig. 2 Movement of the colony [19]

$$NC_n = round\{p_n \cdot N_{col}\} \tag{31}$$

where and N_{col} is the total number of the colonies. Then, NC_n colonies are randomly selected and assigned to nth imperialist.

2) Moving the colonies towards the imperialists

As other heuristic searching algorithms, ICA has an iterative process to mutate the individuals. In each iteration, each colony moves moved toward the relevant imperialist by x units,

$$x = \omega \cdot (I_n - col_n^i) \tag{32}$$

where col_n^i and I_n are the positions of ith colony and imperialist of nth empire, respectively; ω is the weight factor which is usually set as a uniform random number within (0, 2), so as to make the colony move towards the imperialist in both sides. A random angle θ is also added as the deviation.

$$\theta \sim U(-\gamma, \gamma) \tag{33}$$

where γ is the control parameter that adjusts the deviation from the original movement direction. Generally, the movement of the colony can be depicted by Fig. 2.

3) Updating the positions of imperialist and a colony

After the movement, a colony may reach to a position with lower cost than the imperialist. In this case, the colony and the imperialist must exchange positions. The rest colonies of this empire should move forward the new imperialist position.

4) Imperialist competition

In each iteration, all empires compete with each other to try to take possession of colonies of other empires. The competition is based on the empires' powers. The total power of nth empire (denoted as TC_n) is calculated by (34).

$$TC_n = c_n^{imp} + \xi \cdot \frac{\sum\limits_{i=1}^{NC_n} c_n^i}{NC_n} \tag{34}$$

where c_n^{imp} is the cost of the imperialist of nth empire; NC_n is the number nth empire's colonies; c_n^i is the cost of the ith colony of nth empire; ξ is the weight factor. The normalized total power (denoted as NTC_n) and possession probability (denoted as p_{p_n}) of nth empire is represented as below.

Table 1 Procedure of the distributed ICA

```
prepare the input parameters of the optimization model
assign the empire indexes for all computing nodes
for all computing nodes do in parallel
          retrieval model data from the database;
          initialize the imperialist and colonies of the empire;
end parallel
for each iteration
          for all computing nodes do in parallel
                    retrieve the imperialist information;
                    move each colony;
                    evaluate each colony;
                    constraint handling for each colony;
                    send the fitness values to the coordination node;
          end parallel
          do routine on the coordination node
                    calculating the powers of the empires;
                    choose the weakest colony from the weakest
empire;
                    choose a powerful empire to possess that colony;
                    migrate that colony to the computing node which
maintains that empire;
                    if there is empty empire do
                              migrate half of the colonies of the
computing node which has the largest number of colony to the
computing node which maintains the empty empire;
                              eliminate the information of the empty
empire;
                    end if
                    synchronize all the computing nodes;
          end routine
end iteration

do routine on the coordination node
          output the final optimal imperialist and its fitness value;
end routine
```

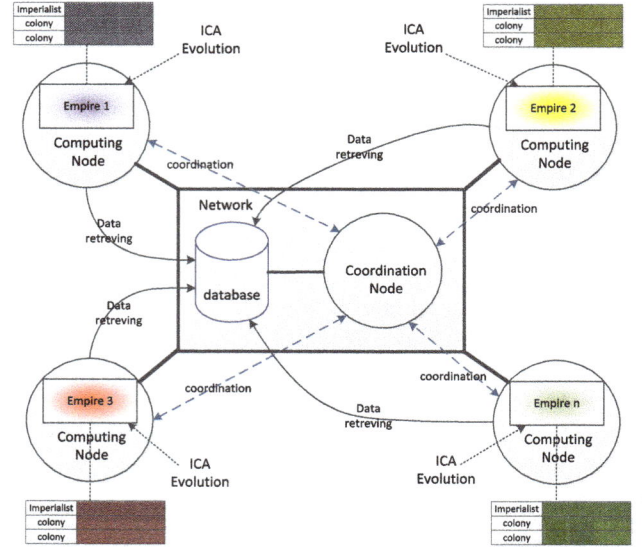

Fig. 3 Distributed processing architecture of the ICA

- All the empires collapse except for the most powerful one, and all the colonies have the same cost with the imperialist.
- The preset maximum iteration number is reached.

4.2 Distributed processing architecture of ICA

In ICA, multiple empires are formed and evolved separately. The communications among the empires occurs at the end of each iteration, where the weakest colony is re-assigned and the empty empire is eliminated. This characteristic makes ICA suitable for parallel and distributed processing in nature. In this paper, we propose a distributed computing architecture for running ICA across networked processors, depicted in Fig. 3.

As depicted by Fig. 3, each computing node is assigned to maintain the countries of $round(N_{emp}/N_{node})$ empires. The coordination node is responsible for coordinating the weakest colony re-assignment, empty empire eliminating, and statistics. A central database is established to store the input data of the optimization problem. The computing nodes retrieve data from the database through the network to perform the empire initialization, fitness evaluation and constraint handling.

With the evolution, the weakest empires are gradually eliminated. To balance the computing load, when $N_{emp} < N_{node}$, the idle computing nodes are assigned to part of the colonies of the empires which have largest number of colony. This is depicted by Fig. 4.

The load balance algorithm is an active research topic in distributed computing. In future, more elaborate load balance strategy for the distributed ICA can be further studied.

$$NTC_n = TC_n - \max_i\{TC_i\} \tag{35}$$

$$p_{p_n} = \left| \frac{NTC_n}{\sum_{i=1}^{N_{imp}} NTC_i} \right| \tag{36}$$

To divide the mentioned colony among empires based on their possession probability, vector P is formed as $P = [p_{p_1}, p_{p_2}, \ldots, p_{p_{imp}}]$. Then the vector R is formed as $R = [r_{r_1}, r_{r_2}, \ldots, r_{r_{imp}}]$, with the elements are uniform distributed random numbers within $[0, 1]$. Then vector D is formed by simply subtracting R from P,

$$D = P - R = [p_{p_1} - r_{r_1}, p_{p_2} - r_{r_2}, \ldots, p_{p_{imp}} - r_{r_{imp}}] \tag{37}$$

The mentioned colony then will be assigned to the empire whose relevant index in D is maximum. After each iteration, ICA checks whether there exists an empire which has lost all the colonies. If so, the empire will collapse and be eliminated.

5) Termination

The algorithm terminates when either of following two conditions is satisfied.

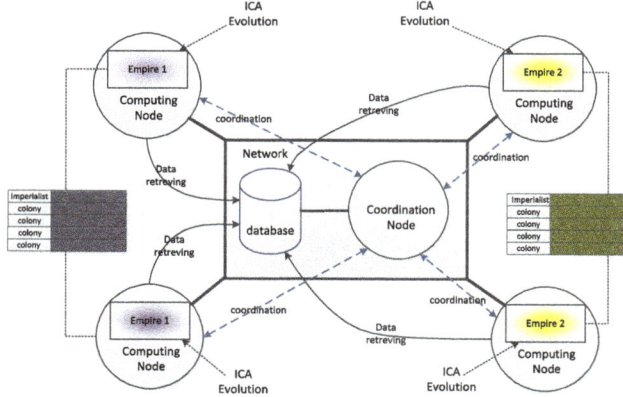

Fig. 4 Re-assignment of the computing nodes

The procedure of the distributed ICA is shown in Table 1. Firstly, the coordination node assigns the different empires to different computing nodes. Then all the computing nodes read the model data and do the initialization works in parallel. After that, all the computing nodes evolve their managed empires simultaneously. At the end of each iteration, the coordination node calculates the powers of all the empires, and mitigates some colonies from the compute node which maintains the most powerful empire to the compute node which maintains the weakest empire. After the evolution, the coordination nodes outputs the optimal solution.

5 Distributed ICA based DLC approach

By employing distributed ICA to solve the proposed model, the states of the generators and the ACL groups are encoded together as a country. The encoding of the population is shown as (38). For each country, the first $T * G$

dimensions represent the states of the generators on each UC time interval and the remaining $T_{dlc} * N$ dimensions represent the states of ACL groups on each DLC time interval. The symbol T_{dlc} denotes the count of the DLC time intervals.

$$\begin{bmatrix} x_1^1 \mid x_1^2 \mid \ldots \mid x_1^{T*G} \mid\mid x_1^{T*G+1} \mid \ldots \mid x_1^{T*G+T_{dlc}*N} \\ x_2^1 \mid x_2^2 \mid \ldots \mid x_2^{T*G} \mid\mid x_2^{T*G+1} \mid \ldots \mid x_2^{T*G+T_{dlc}*N} \\ \vdots \\ x_{N_{pop}}^1 \mid x_{N_{pop}}^2 \mid \ldots \mid x_{N_{pop}}^{T*G} \mid\mid x_{N_{pop}}^{T*G+1} \mid \ldots \mid x_{N_{pop}}^{T*G+T_{dlc}*N} \end{bmatrix} \quad (38)$$

For each country, a certain constraint handling strategy is necessary to make it feasible. For a given country, following ACL group constraint handling procedures are firstly applied.

(1) On each DLC time interval, for each ACL group, if its state is OFF, then check whether the minimum online time constraint is satisfied. If not, then change its state to be ON.

(2) On each DLC time interval, for each ACL group, calculate the indoor temperature variation based on the thermal inertia model. If its state is OFF and the calculated indoor temperature beyond the upper limit, then change its state to be ON.

After handling the constraints of ACL groups, the generator constraint handling algorithm described in [24] is applied for the generators. After that, if the solution is still feasible, then the power flow calculation is performed to check whether the network constraints are satisfied. If so, then the solution is feasible; otherwise, the solution is marked as infeasible. The whole workflow is shown in Fig. 5.

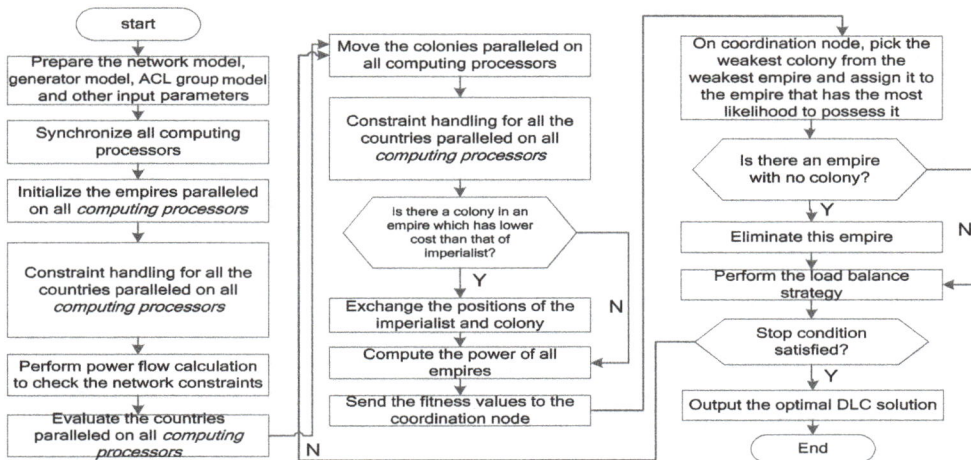

Fig. 5 Workflow of the distributed ICA based DLC

6 Simulation study

6.1 Distributed computing test bench

We aggregate 6 PCs and 1 workstation to establish a distributed computing test bench shown in Fig. 6. All of

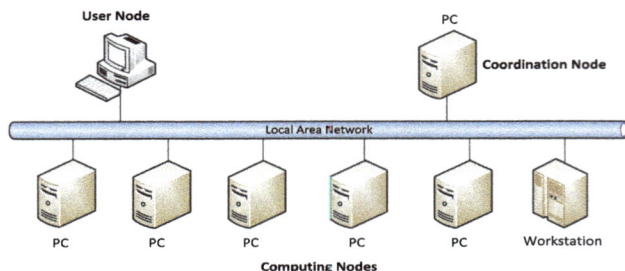

Fig. 6 Topology of the distributed computing test bench

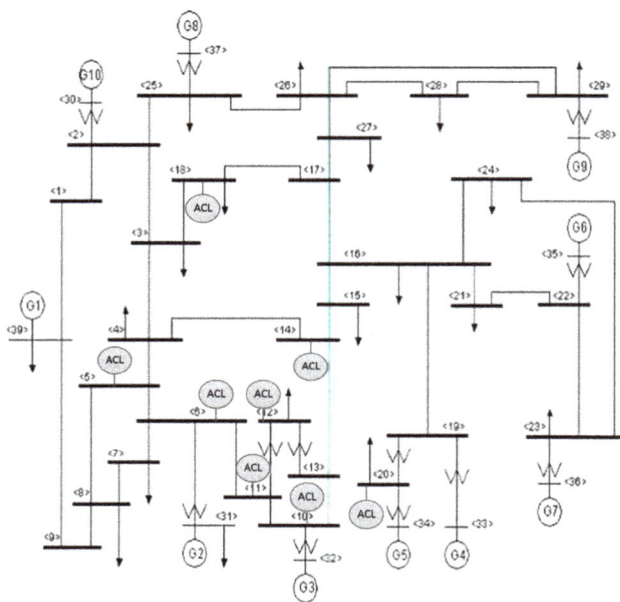

Fig. 7 The modified IEEE 39-bus system

the 6 PCs are DELL 64-bit, 4-core, and with Intel ® Core ™ i5-2400 CPU and 4 Giga-byte RAM; the workstation is DELL 64-bit, 8-core, and with Intel ® Core™ i5-2400 CPU and 8 Giga-byte RAM. The operation systems of the 4 PCs are windows 7 while that of the workstation is windows server 2000. The 7 machines are located in different places and connected by local area network (LAN). One PC is selected as the coordination node and the others act as the computing nodes.

6.2 Programing platform

The MPICH2 [25] is used to implement the distributed ICA, with the C++ programming language. MPICH2 is a high performance and portable implementation of the message passing interface (MPI) standard, and has been applied in many high performance computing applications. The power flow calculation is implemented by the interior point method. The development environment is Microsoft Visual Studio 2010.

6.3 Experiment setup

The IEEE 39-bus benchmark system [26] is used to test the proposed method. The system consists of 10 units and 29 branches. 8 ACL groups are set up on the bus 5, 6, 10, 11, 12, 14, 18, and 20. The ACL groups are assumed to have different building environments and different customer comfort preferences. The system is shown in Fig. 7. The information of the 6 ACL groups is shown in Table 2.

Table 3 ICA parameter setting

Population Size	6000
Maximum Iteration Time	600
N_{imp}	6
γ	$\pi/4$

Table 2 ACL group information

ACL group	Total capacity/ MW	Capacity of each AC/kW	Building material	Building length/m	Building width/m	Building height/m	Wall thickness/ m	Minimum online time/ min	$TP\_Low_n^{in}/$ °C	$TP\_Up_n^{in}/$ °C
1	20	6	Brick	20	10	12	0.24	45	24	27
2	40	4	Brick	18	12	10	0.3	30	24	28
3	20	2	Brick	14	10	10	0.24	30	25	28
4	10	4	Wood	18	10	8	0.3	30	22	26
5	40	4	Wood	16	15	10	0.4	45	24	28
6	20	6	Wood	20	15	12	0.35	45	26	28
7	20	4	Concrete	18	12	11	0.4	30	22	27
8	20	6	Concrete	19	15	12	0.3	30	22	26

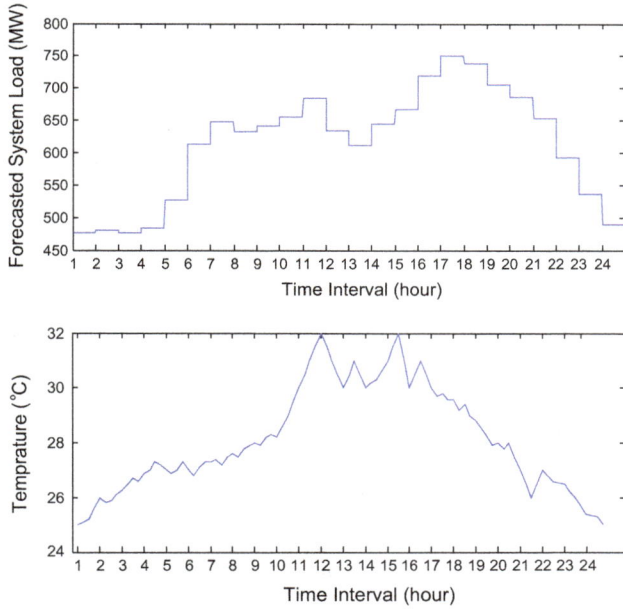

Fig. 8 Forecasted system load (up) and outdoor air temperature profile (down)

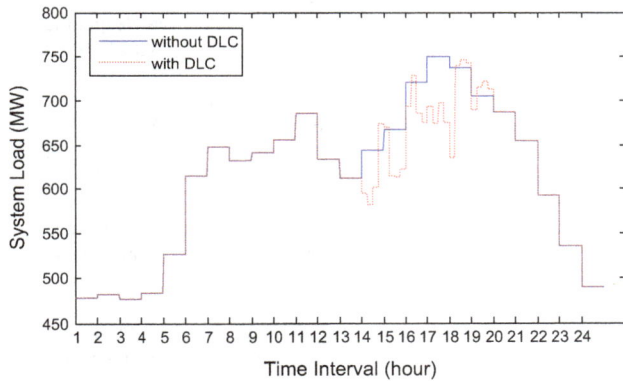

Fig. 9 Load patterns before and after the DLC

Table 4 System characteristics of the case studied

Total system operation cost with DLC ($)	53927
Generation cost ($)	49089
Load shedding cost ($)	4838
Total shed load of the day (MWh)	285.25
Energy saving ratio (of interruptible capacity) (%)	27.43

The generator data follows the literature [25]. The control parameters of the ICA are set as Table 3.

The 24-hour forecasted system load and outdoor air temperature profiles are shown as Fig. 8. It can be seen that the peak load occurs between 14:00 and 20:00, thus the DLC is executed during this period. The dispatch interval

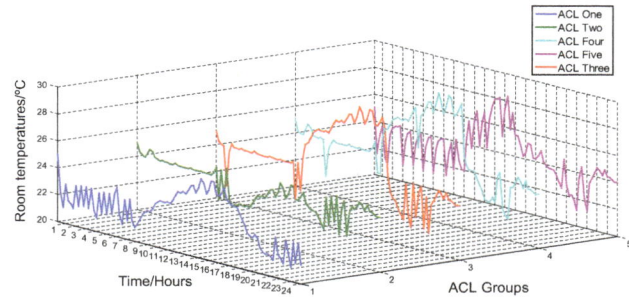

Fig. 10 Indoor temperature profile of the 5 ACL groups

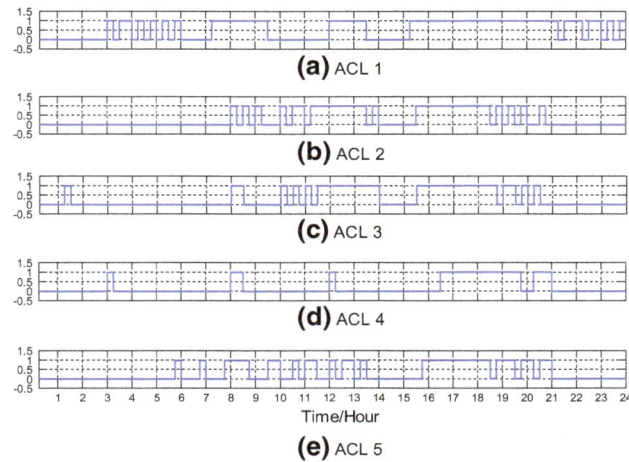

Fig. 11 ON/OFF actions of the 5 ACL groups

is set to be 15 minutes. Thus there are totally 24 dispatch intervals.

6.4 Numerical results

The load patterns before and after the DLC scheduling are shown in Fig. 9. Table 4 reports the system characteristics of the case studied. The results show that the system peak loads are effectively shed by the DLC, with some secondary peak loads. Those secondary peak loads are caused by the resuming of the ACL groups which are switched off in previous time intervals. There are totally 285.25 MWh loads are curtailed during the dispatch horizon (occupied 27.43% of the total ACL capacity), and the total operation cost is $53,927.

The indoor temperature variations of the ACL groups are effectively controlled to ensure the customers' thermal comfort. For example, Fig. 10 shows the indoor temperature profile of 5 randomly selected ACL groups during the dispatch horizon. It can be clearly seen that the temperature is controlled within [25 °C, 28 °C]. Figure 11 shows the scheduled ON/OFF states of the 5 ACL groups.

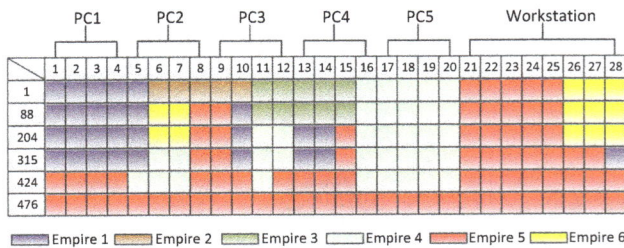

Fig. 12 Empire elimination process of the distributed ICA

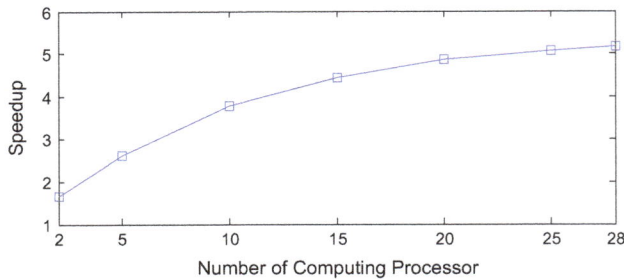

Fig. 13 Speedup of the distributed ICA

Figure 12 shows the empire migrating process of the distributed ICA. Figure 12 clearly shows that along with the elimination of the empires, the coordination node mitigates the colonies of the left empires among the processors to achieve the load balance. Finally, all the colonies of the only left empire (Empire 3) are maintained by all the processors.

Figure 13 reports the speedup achieved by increasing the number of the computing processors, relative to run the same case on a single process serially. In the single process, the total execution time of the simulation is 61,433 seconds. The results show that with the increase of processor number, the distributed ICA can significantly shorten the simulation time.

7 Conclusions

A comprehensive DLC model is proposed in this paper. The objective of the model is to schedule interruptible ACLs and generation units to minimize the total operation costs over the dispatch horizon, while taking the network constraints into account. An advanced two-parameter thermal inertia model is used to represent the thermal transition processes of buildings with heterogeneous characteristics.

A new distributed ICA algorithm is proposed to solve the DLC model. The distributed ICA is based on the ICA algorithm, and employs the process communication model to perform the distributed high performance searching. The proposed method is verified on the IEEE 39-bus benchmark system and a distributed computing test bed.

Acknowledgments This work was supported in part by National Natural Science Foundation of China (Key Project 71331001, General Project 71371065, 11171095, 71071025).

References

[1] Ipakchi A, Albuyeh F (2009) Grid of the future. IEEE Power Energy Mag 7(2):52–62
[2] Huang KY, Huang YC (2004) Integrating direct load control with interruptible load management to provide instantaneous reserve for ancillary services. IEEE Trans Power Syst 19(3): 1626–1634
[3] Majumdar S, Chattopadhyay D, Parikh J (1996) Interruptible load management using optimal power flow analysis. IEEE Trans Power Syst 11(2):715–720
[4] Chen CS, Leu JT (1990) Interruptible load control for Taiwan power company. IEEE Trans Power Syst 5(4):460–465
[5] Yao L, Chang WC, Yen RL (2005) An iterative deepening genetic algorithm for scheduling of direct load control. IEEE Trans Power Syst 20(3):1414–1421
[6] Hsu YY, Su CC (1991) Dispatch of direct load control using dynamic programming. IEEE Trans Power Syst 6(3):1056–1060
[7] Tuan LA, Bhattacharya K (2003) Competitive framework for procurement of interruptible load services. IEEE Trans Power Syst 18(2):889–897
[8] Argiento R, Faranda R, Pievatolo A et al (2012) Distributed interruptible load shedding and micro-generator dispatching to benefit system operations. IEEE Trans Power Syst 27(2): 840–848
[9] Aminifar F, Fotuhi-Firuzabad M, Shahidehpour M (2009) Unit commitment with probabilistic spinning reserve and interruptible load considerations. IEEE Trans Power Syst 24(1):388–397
[10] Wei DC, Chen NM (1995) Air conditioner direct load control by multi-pass dynamic programming. IEEE Trans Power Syst 10(1):307–313
[11] Yang HT, Huang KY (1999) Direct load control using fuzzy dynamic programming. IEE Proc Gener Transm Distrib 146(3):294–300
[12] Salehfar H, Noll PJ, LaMeres BJ et al (1999) Fuzzy logic-based direct load control of residential electric water heaters and air conditioners recognizing customer preferences in a deregulated environment. In: Proceedings of the power engineering society summer meeting, vol 2. Edmonton, Canada, 18–22 Jul 1999, pp 1055–1060
[13] Ruiz N, Cobelo I, Oyazabal J (2009) A direct load control model for virtual power plant management. IEEE Trans Power Syst 24(2):959–966
[14] Hubert T, Grijalva S (2012) Modeling for residential electricity optimization in dynamic pricing environments. IEEE Trans Smart Grid 3(4):2224–2231
[15] Mortensen RE, Haggerty KP (1988) A stochastic computer model for heating and cooling loads. IEEE Trans Power Syst 3(3):1213–1219

[16] Sansawatt T, Ochoa LF, Harrison GP (2012) Smart decentralized control of DG for voltage and thermal constraint management. IEEE Trans Power Syst 27(3):1637–1645

[17] Bălan R, Cooper J, Chao KM et al (2011) Parameter identification and model based predictive control of temperature inside a house. Energy Build 43(2/3):748–758

[18] Wang HM, Meng K, Luo FJ et al (2013) Demand response through smart home energy management using thermal inertia. In: Proceedings of the 2013 Australian universities power engineering conference (AUPEC'13), Hobart, Australia, 29 Sept–3 Oct 2013, 6 pp

[19] Atashpaz-Gargari E, Lucas C (2007) Imperialist competitive algorithm: an algorithm for optimization inspired by imperialistic competition. In: Proceedings of the IEEE congress on evolutionary computation (CEC'07), Singapore, 25–28 Sept 2007, pp 4661–4667

[20] Mohammadi-Ivatloo B, Rabiee A, Soroudi A et al (2012) Imperialist competitive algorithm for solving non-convex dynamic economic power dispatch. Energy 44(1):228–240

[21] Gargari E, Hashemzadeh F (2008) Colonial competitive algorithm: a novel approach for PID controller design in MIMO distillation column process. Int J Intell Comput Cybern 1(3):337–355

[22] Yousefi M, Darus AN, Mohammadi H (2012) An imperialist competitive algorithm for optimal design of plate-fin heat exchangers. Int J Heat Mass Trans 55(11/12):3178–3185

[23] Moghimi Hadji M, Vahidi B (2011) A solution to the unit commitment problem using imperialistic competition algorithm. IEEE Trans Power Syst 27(1):117–124

[24] Chung CY, Yu H, Wong KP (2011) An advanced quantum-inspired evolutionary algorithm for unit commitment. IEEE Trans Power Syst 26(2):847–854

[25] Thakur R, Rabenseifner R, Gropp W (2005) Optimization of collective communication operations in MPICH. Int J High Perform Comput Appl 19(1):49–66

[26] Athay T, Padmore R, Virmani S (1979) A practical method for the direct analysis of transient stability. IEEE Trans Power Apparatus Syst 98(2):573–584

Fengji LUO is a post doctor associate of the Centre for Intelligent Electricity Networks (CIEN), The University of Newcastle, Australia. He obtained his bachelor and master degrees from Chongqing University, and doctor degree from the University of Newcastle. His research interests include smart grid planning, renewable energy dispatch, demand side management, and distributed computing.

Junhua ZHAO received his Ph.D. degree from the University of Queensland, Australia. Currently he is a senior lecturer at the University of Newcastle, Australia. His research interests include power system analysis and computation, smart grid, cyber physical system, electricity market, data mining and its applications.

Haiming WANG is a graduate of the School of Energy Science and Engineering, Central South University, China, and he received his Master of Science degree in New and Renewable Energy from Durham University, UK in 2010. He is now pursuing a PhD in The Centre for Intelligent Electricity Networks, the University of Newcastle. His research interests include thermostatically controlled loads and demand side management.

Xiaojiao TONG received the B.S. and M.S. degrees in mathematics from Wuhan University, Wuhan, China, and the Ph.D. degree in applied mathematics from Hunan University, Hunan, China, in 1983, 1986, and 2000, respectively. She has taught at Changsha University of Science and Technology, Hunan, China, and is now a professor and Vice President at Hunan First Normal University, Hunan, China. Her research interests include nonlinear optimization, stochastic programming and applications, power system analysis and power market.

Yingying CHEN is a research associate at the Centre for Intelligent Electricity Networks (CIEN), The University of Newcastle, Australia. She obtained her bachelor and master degrees from Chongqing University, and doctor degree from the University of Newcastle. Her research interests include wind power dispatch, wind farm planning and distributed computing.

Zhao YANG DONG is now Professor and Head of School of Electrical and Information Engineering, University of Sydney, Australia. He is immediate Ausgrid Chair Professor and Director of the Centre for Intelligent Electricity Networks (CIEN), the University of Newcastle, Australia. He previously held academic and industrial positions with the Hong Kong Polytechnic University, the University of Queensland, Australia and Transend Networks, Australia. His research interest includes smart grid, power system planning, power system security, load modeling, renewable energy systems, electricity market, and computational intelligence and its application in power engineering. He is an editor of IEEE Transactions on Smart Grid, and IEEE Power Engineering Letters.

14

Preheating method of lithium-ion batteries in an electric vehicle

Zhiguo LEI (✉), **Chengning ZHANG,**
Junqiu LI, Guangchong FAN, Zhewei LIN

Abstract To improve the low-temperature charge-discharge performance of lithium-ion battery, low-temperature experiments of the charge-discharge characteristics of 35 Ah high-power lithium-ion batteries have been conducted, and the wide-line metal film method for heating batteries is presented. At −40 °C, heating and charge-discharge experiments have been performed on the battery pack. The results indicate the charge-discharge performance is substantially worse in cold climates, and can be significantly improved by heating the battery pack with a wide-line metal film. Pulse charge-discharge experiments show that at −40 °C ambient temperature, the heated battery pack can charge or discharge at high current and offer almost 80% power.

Keywords Lithium-ion battery, Charge-discharge characteristics, Low-temperature performance, Electric vehicles, Heating method

1 Introduction

In recent years, electric vehicles have developed rapidly. In-depth research work in many countries has improved all aspects of electric vehicle technology so that urban pollution can be reduced, and fruitful results have been achieved. Battery energy storage is one of the key components in electric vehicles, so it receives strong research attention and has developed rapidly as a result. The performance and cost of an electric vehicle depends strongly on the performance and service life of its battery. Currently, battery chemistries used in electric vehicles include lead-acid, nickel cadmium, nickel metal hydride, lithiumion, and supercapacitors [1]. Lithium-ion is gradually replacing other chemistries to become the most common battery technology found in electric vehicles due to its advantages of high power, high energy density, long cycle life, low self-discharge rate, long shelf life and low pollution [2, 3].

With their increased use, the low-temperature performance of lithium-ion batteries begins to attract attention. At low temperature, the charge-discharge performance of lithium-ion batteries significantly reduces [4–6]. Generally, the solid electrolyte interface film, surface charge transfer impedance and Li+ diffusion in the electrode are the main influences on low-temperature performance of lithium-ion batteries [7–10]. So far, it is difficult to improve their performance through innovations in the batteries' materials. Therefore, auxiliary methods to improve the low-temperature performance of lithium-ion batteries become an important research direction, i.e., the AC heating method [11–13], preheating method [14–16], heating plate method [17] and heating bag method [18]. This paper studies the charge-discharge performance of a 35Ah@3.7V LiMn$_2$O$_4$ battery in a 8×8 wheeled electric vehicle from 20 °C to −40 °C. A wide-line metal film is proposed to heat the battery so as to meet the low-temperature operating requirements of the 8×8 wheeled electric vehicle. Experimental results prove that the wide-line metal film heating method can significantly improve the low-temperature performance of the battery.

Z. LEI, School of Mechanical and Electrical Engineering, Fujian Agriculture and Forestry University, Fuzhou 350002, Fujian, China
(✉) e-mail: lzgkkk@163.com
C. ZHANG, J. LI, G. FAN, National Engineering Laboratory for Electric Vehicle, Beijing Institute of Technology, Beijing 100081, China
Z. LIN, Ningde Entry-Exit Inspection and Quarantine Bureau, Fujian 355017, China

2 Test platform for the Li-ion battery

A diagram of the test platform is shown in Fig. 1. Cell-level charge-discharge testing is performed by the HT-V5C200D200 which is manufactured by the LTD company in Guangzhou. Its maximum voltage of charge-discharge is 5 V, and its measurement precision can reach 0.1 mV. Battery-pack testing is performed by the Digatron EVT500-500 which is manufactured by the Digatron Company in Germany. Its maximum current and voltage of charge-discharge are 500 A and 500 V respectively. The function of the thermostatic enclosure is to provide the environment for the tests. The electrochemical workstation produced by the Zahner Company in Germany is used to measure the AC impedance spectra of the battery and the impedance value at a fixed frequency. Its measurement frequency range and AC amplitude are 10 μHz-4 MHz and 1 mV-1 V respectively with frequency accuracy of 0.0025%.The tested battery is a 35Ah@3.7V $LiMn_2O_4$ cell, with cathode of spinel structure $LiMn_2O_4$, anode of artificial graphite, and shell of Al-plastic film. For battery-packtesting, a string of three cells is constructed.

3 AC impedance variations of the battery cell at low temperature

Battery internal resistance is one of the important parameters. To determine the change of battery internal resistance at low temperature, this paper measured the AC impedance of the battery placed in a −40 °C thermostatic enclosure. The DC internal resistance of battery cannot be measured directly because charging and discharging with a DC current will change the state of the battery: its temperature and charge level. Therefore we use the electrochemical workstation to measure the AC impedance of the battery. Figure 2 is an AC impedance spectrum of the Li-ion battery cell at 20 °C, as the frequency is varied from 1 Hz to 100 kHz and the voltage is 5 mV. The magnitude of the AC impedance varies very little at low frequencies (<1 kHz), and its phase is close to zero, thus the AC impedance of battery at low frequencies can be thought as the internal resistance. Because the AC impedance is almost minimum at 260 Hz, this frequency is used to measure the AC impedance of the battery in experiments. Figure 3 is an impedance curve of a Li-ion cell during the eight hours after it was placed in a −40 °C enclosure. It shows that impedance of battery rapidly increased with standing time at first, but impedance of battery is almost constant after 3.5 hours. Therefore, the battery requires 3.5 hours to reach thermal equilibrium with its environment.

4 Influence of low temperature on the discharge performance

To study the effect of low temperature on the discharge performance of the battery, a cell placed in different low-temperature environments was discharged at several different constant currents. Firstly, the cell was charged at 1C/3 at room temperature. Secondly, the cell was placed in the thermostatic enclosure for five hours to reach thermal equilibrium. Finally, the cell was discharged at constant current until its voltage reduced to 3 V, and a discharge

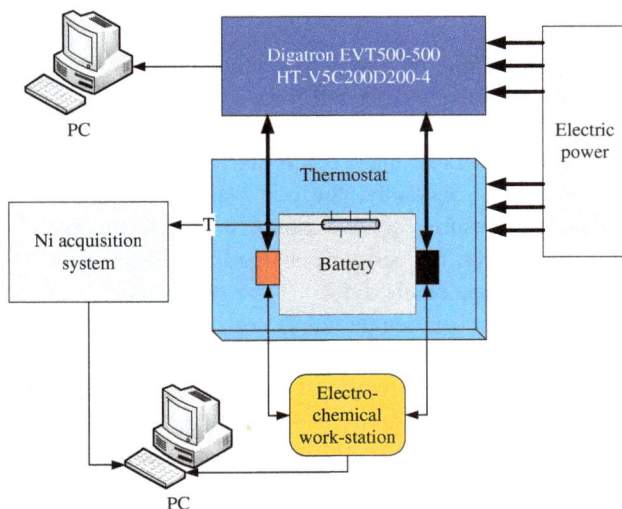

Fig. 1 Diagram of the test platform

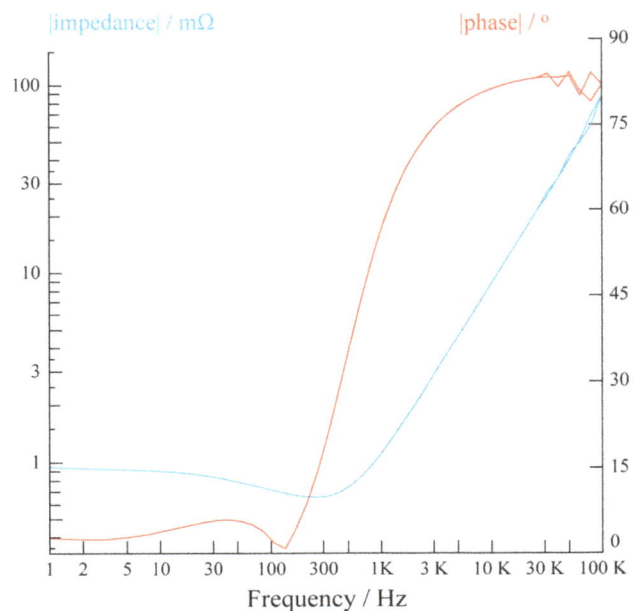

Fig. 2 Impedance spectroscopy of a Li-ion cell at 20 °C

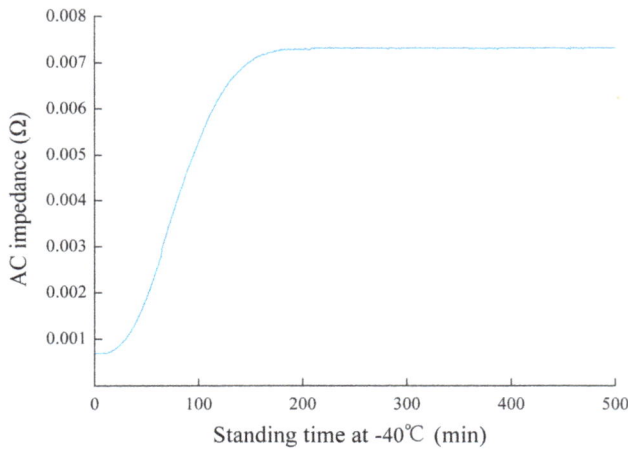

Fig. 3 Impedance convergence of a Li-ion cell at −40 °C

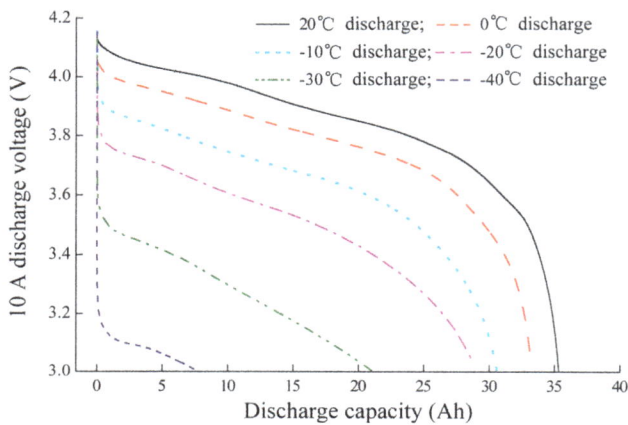

Fig. 5 35 A discharge curves of a Li-ion cell at various temperatures

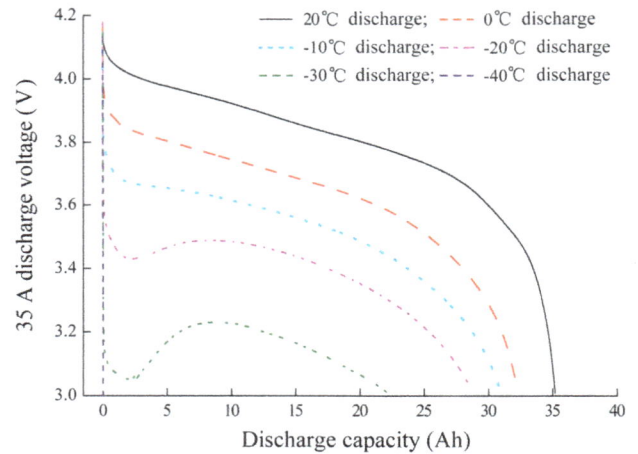

Fig. 4 10 A discharge curves of a Li-ion cell at various temperatures

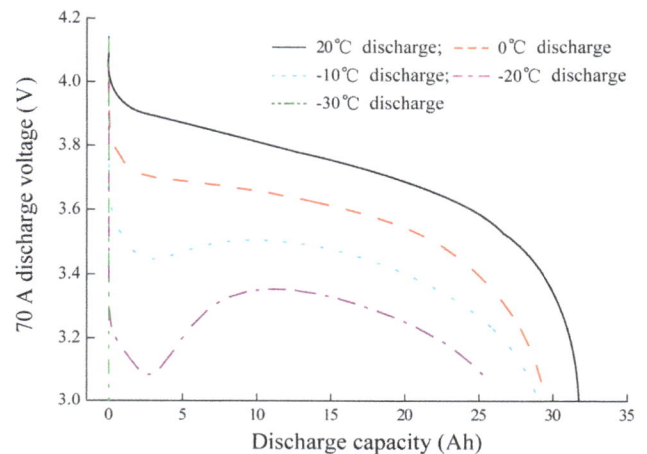

Fig. 6 70 A discharge curves of a Li-ion cell at various temperatures

curve was produced by graphing the voltage and the discharge capacity calculated as current multiplied by elapsed time. The discharge curves of the cell at different low temperatures are shown in Figs. 4, 5 and 6 for 10 A, 35 A, and 70 A discharge currents respectively. The experimental results show that the discharge capacity of the cell declines when its temperature decreases. Moreover, the voltage collapses rapidly at temperatures −10 °C and below, and then recovers for a while before descending below 3 V. In low-temperature environment, the internal resistance of the battery is greatly increased, therefore the voltage of the battery reduces rapidly at the beginning of discharging. Meanwhile, the battery will be heated and the voltage of the battery rise because of the internal resistance during discharging. This collapse-recovery behaviour must give some clue about the physical causes of reduced discharge capacity and should be investigated further.

5 Influence of low temperature on the charge performance

To study the effect of low temperature on the charge performance of the battery, a cell placed in different low-temperature environments was charged at several different constant currents. Firstly, the cell was discharged at 1C/3 at room temperature. Secondly, the cell was placed in the thermostatic enclosure for five hours to reach thermal equilibrium. Finally, the cell was charged at constant current until its voltage reached 4.2 V. The charge curves of the cell at different low temperatures are shown in Figs. 7 and 8 for 10 A and 35 A charge currents respectively. Compared with the low-temperature discharge performance, the charge performance of the cell is even more degraded. Under 0 °C, it can't charge normally even at low current, and at high current the voltage of the cell reaches

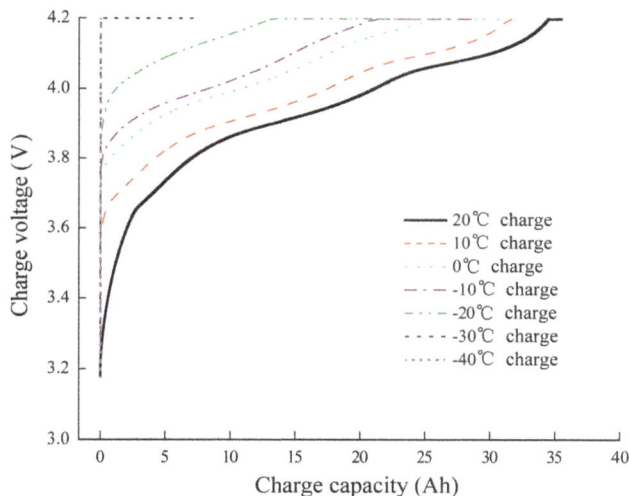

Fig. 7 10 A charge curves of a Li-ion cell at various temperatures

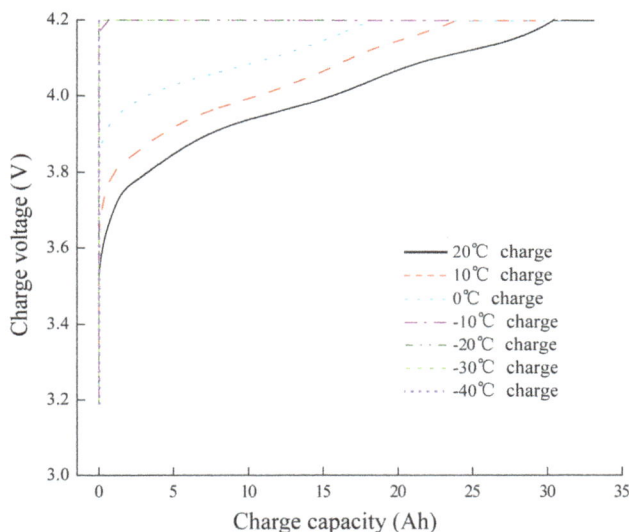

Fig. 8 35 A charge curves of a Li-ion cell at various temperatures

4.2 V immediately. This is why no results are shown for 70 A charging current.

6 Wide-line metal film heating method

The experiments reported above (as shown in Fig. 3) prove that the internal resistance of the battery is greatly increased and the charge-discharge performance is significantly reduced at low temperatures. The performance of an electric vehicle is limited by the low-temperature performance of its batteries, and this is especially for special-purpose electric vehicles that are required to operate under a great variety of temperature conditions. To meet the high reliability requirement of the 8×8 wheeled electric vehicle,

a wide-line metal film heating method is proposed, in which two pieces of wide-line metal film are placed on the two largest surfaces of the battery cell. The wide-line metal film is printed on a FR4 board or aluminum PCB, and its thickness is 1 mm. One side of the wide-line metal film is a complete rectangular copper film, and the other side is a certain width of continuous copper wire. The thickness of copper film and copper wire is 0.035 mm. The copper wire is heated by passing anelectric current through it. The heat is distributed evenly to the battery by the copper film on the other side. The structure of the heating device is simple, and it can be installed conveniently in order to heat batteries without changing the structure of the original battery pack.

7 Experiments on the low-temperature heating

To study the recovery of low-temperature charge and discharge performance of a battery pack that is heated by the wide-line metal film, three 35Ah@3.7V $LiMn_2O_4$ battery cells were connected in series to form a battery pack. The wide-line metal film was installed in four heaters between three cells between two battery cells as shown in Fig. 9.

In order to make the heating experiments conform closely to the conditions of a vehicle battery, the battery pack with the wide-line metal film was put into a battery box. Figure 10 shows the battery box placed in the thermostatic enclosure that is set to −40 °C. The standing time of the battery box was increased from 5 hours to 8 hours to reach thermal equilibrium because the battery box assembly has a higher thermal inertia. The wide-line metal film began to heat the battery pack after 8 hours.

Fig. 9 Photograph of the battery pack and heater

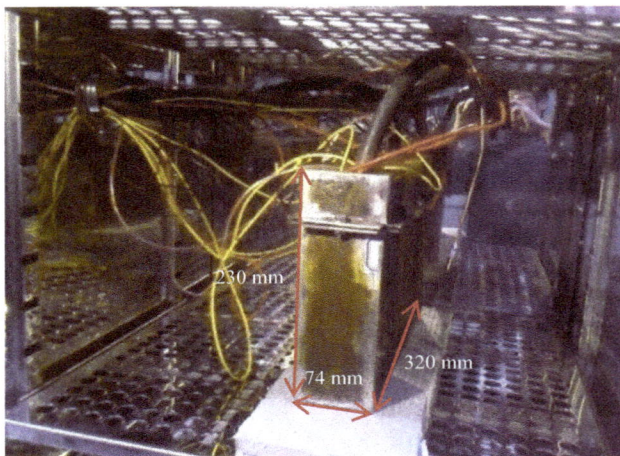

Fig. 10 Photograph of the battery box inside the thermostatic enclosure

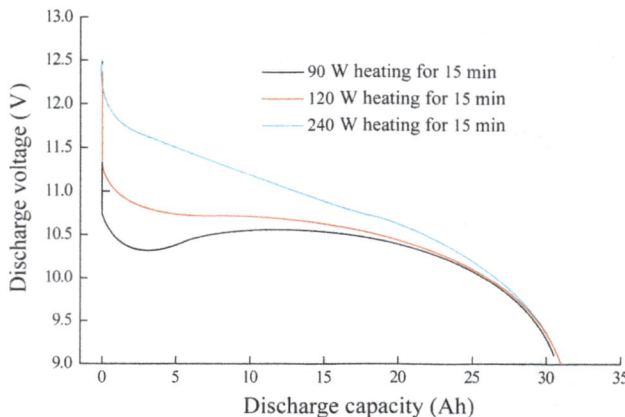

Fig. 12 35 A discharge curves of the unheated cell and heated battery pack

7.1 Discharge performance of battery pack at −40 °C after heating for 15 min

Figure 11 shows three 1C constant current discharge curves of battery pack after it is heated for 15 min by 240 W, 120 W and 90 W power to the wide-line metal film. In all cases, the full discharge capacity of just over 35 Ah is achieved for the series-connected cells, as can be seen by comparing with Fig. 5, while the voltage in the early and middle stages of discharge increases with increasing heating power. The average discharge voltage of the battery is 0.53 V higher when heated by 240 W power compared to 90 W power, and the maximum voltage differential is 1.38 V. Therefore, increasing the heating power above 90 W can improve the discharge voltage and thereby increase the discharge power but does not significantly affect the discharge capacity.

Figure 12 compares three 1C discharge curves of the battery pack at −40 °C after it is heated for 15 min, and three 1C discharge curves of an unheated cell at 0 °C, −10 °C, and −20 °C. The battery pack is composed of three cells, and the discharge voltage of each cell is different, thus the average discharge voltage of the three cells is used for comparison with the discharge voltage of the single unheated cell. After the battery pack at −40 °C is heated for 15 min with 90 W power, its average discharge voltage is close to the discharge voltage of the unheated cell at −20 °C at the beginning of the discharge, and higher than the discharge voltage of the unheated cell at −20 °C at the middle and late stages of the discharge. The discharge capacity of the battery pack at −40 °C heated for 15 min with 90 W power is almost equal to the discharge capacity of the unheated cell at −10 °C. These results suggest that part of the heat generated by the process of battery pack discharging has the effect of heating the battery pack after the external heating stops.

After the battery pack at −40 °C is heated for 15 min with 120 W power, its average discharge voltage is slightly below the discharge voltage of the unheated cell at −10 °C at the beginning of the discharge, and almost equal to the discharge voltage of the unheated cell at −10 °C at the middle and late stages of the discharge. After the battery pack at −40 °C is heated for 15 min with 240 W power, its average discharge voltage is slightly above the discharge voltage of the unheated cell at 0 °C at the beginning of the discharge, and slightly below the discharge voltage of the unheated cell at −0 °C at the middle and late stages of the discharge.

7.2 Charge performance of battery pack after 15 min of heating at −40 °C

Figure 13 compares five 1C constant current charge curves, including one charge curve of the battery pack after

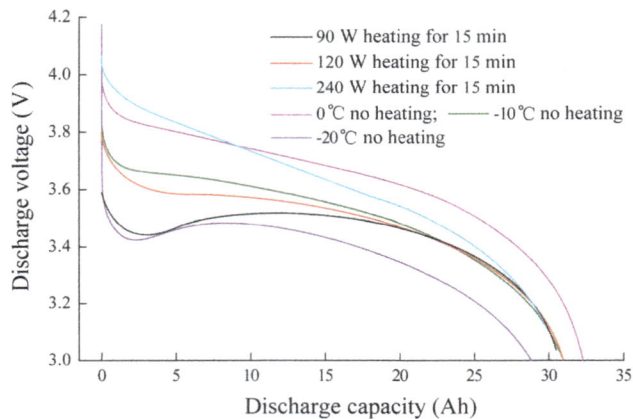

Fig. 11 35 A discharge curves of the battery pack at −40 °C with heating

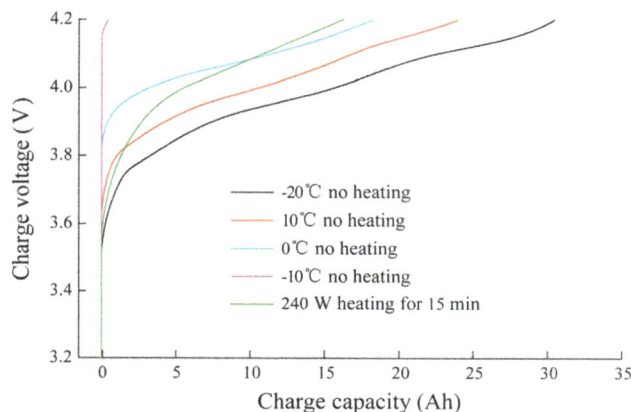

Fig. 13 35 A charge curves of the unheated cell and heated battery pack

it is heated for 15 min and four charge curves of an unheated cell at 10 °C, 0 °C, −10 °C, and −20 °C. The charge performance of the battery pack heated is improved significantly by heating. After the battery pack at −40 °C is heated for 15 min with 240 W power, its charging performance is close to the charging performance of the unheated cell at 0 °C. The main consideration for low-temperature charging performance is the heating time and heating uniformity, which can be controlled when the battery pack is heated by an external power supply.

7.3 Pulse charge-discharge performance of battery pack at −40 °C after 15 min of heating

The result experiments reported above demonstrate that the low-temperature charge-discharge performance of the battery pack heated in low temperature is improved significantly, and can be draw by the above experiments on the low-temperature heating. The full energy storage capacity can be achieved. However, the maximum charge-discharge power of the heated battery pack heated at −40 °C cannot be achieved because the heated battery pack is always charged or discharged at 1C constant current. Poor voltage performance evident in Figs. 6 and 8 prevents the use higher constant currents. Therefore, some experiments were performed using pulse charge-discharge of the heated battery pack at low temperature. Firstly, the battery pack was charged at 1C/3 at room temperature. Secondly, the battery pack was placed at −40 °C for eight hours to reach thermal equilibrium. Thirdly, the battery pack was heated for 15 min with 90 W power, and finally the battery pack was subjected to charge and discharge pulse currents. The pulses had the minimum discharge current of 17.5 A and maximum discharge current of 280 A and minimum charge current of 17.5 A and maximum charge current of 210 A. This pulse profile was designed to discharge the battery pack as quickly as possible.

The charge-discharge curve is shown in Fig. 14 together with the pulse current. In order to show the charge-discharge curve more clearly, the pulse curves at 90% state of

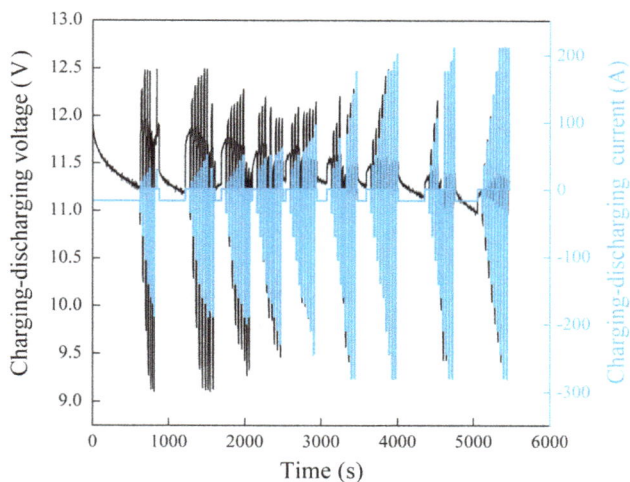

Fig. 14 Charge-discharge pulse curve of battery pack heated at −40 °C

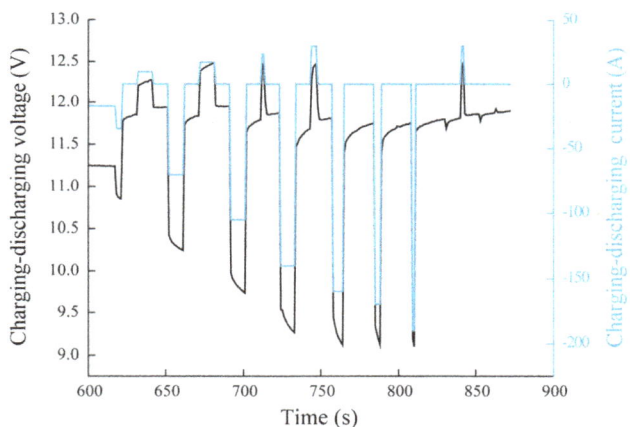

Fig. 15 Charge-discharge pulse curve at SOC 90%

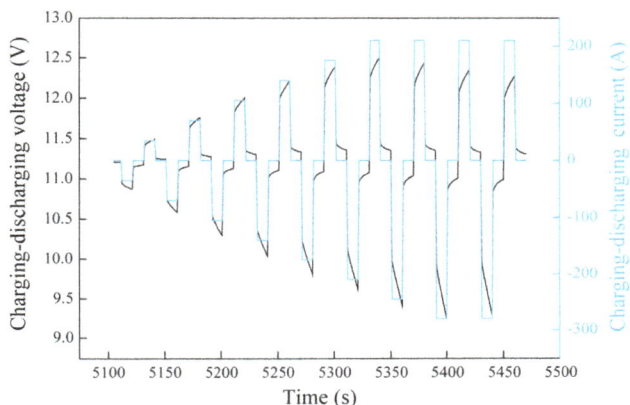

Fig. 16 Charge-discharge pulse curve at SOC 10%

charge (SOC) and 10% SOC are shown in Figs. 15 and 16. At 90% SOC, in Fig. 15, the maximum discharge current of the heated battery pack is near 210 A (6C). At the 10% SOC, in Fig. 16, the maximum discharge current of the heated battery pack is near 280 A (8C). The unheated battery pack at $-40\ ^{\circ}C$ cannot be charged and discharged at such high currents. Therefore, the heating method can effectively improve the discharge performance of the battery pack at low temperatures.

8 Conclusion

This paper reports a series of experiments on the low-temperature charge-discharge of a 35Ah@3.7V $LiMn_2O_4$ battery cell and a battery pack comprising three cells. The results show that the charge-discharge performance of the cell reduces significantly with the decrease of temperature, and the useable capacity of the cell becomes negligible at $-20\ ^{\circ}C$. Therefore, the cell should be heated to improve its low-temperature performance.

A wide-line metal film heating mechanism was introduced to the battery back and improved its low-temperature performance significantly. After heating for 15 min at powers between 90 W and 240 W, the voltage and power performance of the battery pack were improved at 1C charging and discharging rates, and the discharging capacity of the battery pack was restored almost to its room-temperature level, while its charging capacity was restored to approximately half its room-temperature level.

Pulse charge-discharge experiments were performed on the battery pack at $-40\ ^{\circ}C$ after heating for 15 min. High pulse discharge currents were achieved, between 6C and 8C rates, giving an average discharge rate of 4.5C and a discharge capacity of 80% of the room-temperature discharge capacity. Charging with pulse currents is ongoing research. The battery pack at $-40\ ^{\circ}C$ only needs to be preheated at the beginning, because its temperature can be maintained thereafter by the heat that is produced during the charging and discharging process.

Acknowledgement This work was supported by the defense pre-research project (104010108), the Fujian province natural science foundation (2014J01173), the key discipline of mechanical engineering in Fujian province (6112c1600) and the Fujian province department of education (JA12100).

References

[1] Chen QQ, Sun FC, Zhu JG (2002) Modern technology of electric vehicle. Beijing Institute of Technology Press, Beijing (in Chinese)

[2] Huang CK, Sakamoto JS, Wolfenstine J et al (2000) The limits of low temperature performance of Li-ion cells. J Electrochem Soc 147(8):2893–2896

[3] Liu XJ, Xiao CW, Yu B et al (2007) Research on the development of HEVs Li-ion battery. Chin J Power Sour 31(7): 509–514 (in Chinese)

[4] Smart MC, Ratnakumar BV, Surampudi S et al (1999) Irreversible capacities of graphite in low-temperature electrolytes for lithium-ion batteries. J Electrochem Soc 146(11):3963–3969

[5] Shiao HC, Chua D, Lin HP et al (2000) Low temperature electrolytes for Li-ion PVDF cells. J Power Sour 87(1/2): 167–173

[6] Zhang SS, Xu K, Jow TR (2003) The low temperature performance of Li-ion batteries. J Power Sour 115(1):137–140

[7] Smart MC, Ratnakumar BV, Whitcanack LD et al (2003) Improved low-temperature performance of lithium-ion cells with quaternary carbonate-based electrolytes. J Power Sour 119–121: 349–358

[8] Wang CS, Appleby AJ, Little FE (2002) Low temperature characterization of lithium-ion carbon anodes via microperturbation measurement. J Electrochem Soc 149(6):A754–A760

[9] Fan J (2003) On the discharge capability and its limiting factors of commercial 18650 Li-ion cell at low temperatures. J Power Sour 117(1/2):170–178

[10] Fan J (2006) Studies on charging lithium-ion cells at low temperatures. J Electrochem Soc 153(6):A1081–A1092

[11] Hand A, Stuart TA (2002) AC heating for EV/HEV batteries. In: Proceedings of the 2002 conference on power electronics in transportation (PET'02), Auburn Hills, 24–25 Oct 2002, pp 119–124

[12] Hande A (2004) A high frequency inverter for cold temperature battery heating. In: Proceedings of the 2004 IEEE workshop on computers in power electronics (CIPE'04), Urbana, 15–18 Aug 2004, pp 215–222

[13] Hande A, Stuart TA (2004) HEV battery heating using AC currents. J Power Sour 129(2):368–378

[14] Alaoui C, Salameh ZM (2001) Solid state heater cooler: Design and evaluation. In: Proceedings of the 2001 large engineering systems conference on power engineering (LESCOPE'01), Halifax, 11–13 July 2001, pp 139–145

[15] Salameh ZM, Alaoui C (2003) Modeling and simulation of a thermal management system for electric vehicles. In: Proceedings of the 29th annual conference of the IEEE Industrial Electronics Society (IECON'03), vol 1, Roanoke, 2–6 Nov 2003, pp 887–890

[16] Alaoui C, Salameh ZM (2004) A novel thermal management for electric and hybrid vehicles. IEEE Trans Veh Technol 54(2):468–476

[17] Ma X (2014) Research on the thermal characteristics and the thermal management system of electric vehicle power battery. Beijing Institute of Technology, Beijing (in Chinese)

[18] Su ZG (2010) A lithium ion power battery heating device. China Patent, CN101710630A, 19 May 2010

Zhiguo LEI received his Ph.D. degree in Vehicle Engineering from Beijing Institute of Technology, China, in 2013. He is a Lecturer at Fujian Agricultural and Forestry University. His research interests include the thermal management system of the power batteries in Electrical Vehicles and Vehicle Design.

Chengning ZHANG is a Professor and Ph. D. Supervisor at Beijing Institute of Technology. His research interests include the power batteries technology and the motor design and control technology.

Junqiu LI is an Associate Professor and Master Supervisor at Beijing Institute of Technology. His research interests include the power batteries technology and the motor design and control technology.

Guangchong FAN is a Master Candidate at Beijing Institute of Technology. His research interests include the thermal management system of the power batteries in Electrical Vehicle.

Zhewei LIN is now an Engineer with Ningde Entry-Exit Inspection and Quarantine Bureau, Fujian, China. His research interests include the power batteries technology and the motor design and control technology.

15

An optimization strategy of controlled electric vehicle charging considering demand side response and regional wind and photovoltaic

Hong LIU, Pingliang ZENG, Jianyi GUO (✉), Huiyu WU, Shaoyun GE

Abstract Renewable energy, such as wind and photovoltaic (PV), produces intermittent and variable power output. When superimposed on the load curve, it transforms the load curve into a 'load belt', i.e. a range. Furthermore, the large scale development of electric vehicle (EV) will also have a significant impact on power grid in general and load characteristics in particular. This paper aims to develop a controlled EV charging strategy to optimize the peak-valley difference of the grid when considering the regional wind and PV power outputs. The probabilistic model of wind and PV power outputs is developed. Based on the probabilistic model, the method of assessing the peak-valley difference of the stochastic load curve is put forward, and a two-stage peak-valley price model is built for controlled EV charging. On this basis, an optimization model is built, in which genetic algorithms are used to determine the start and end time of the valley price, as well as the peak-valley price. Finally, the effectiveness and rationality of the method are proved by the calculation result of the example.

Keywords Renewable energy, Electric vehicle, Controlled electric vehicle (EV) charging, Demand side response, Peak-valley price

H. LIU, J. GUO, S. GE, Key Laboratory of Smart Grid of Ministry of Education, Tianjin University, Tianjin 300072, China
(✉) e-mail: guojianyigjy@126.com
P. ZENG, Electric Power Research Institute of China, Beijing 100192, China
e-mail: zengpingliang@epri.sgcc.com.cn
H. WU, University of Wisconsin Madison, Madison, WI 53706, USA

1 Introduction

With the environmental degradation around the world, people are increasingly advocating and pursuing the green living concept, and electric vehicles (EVs) are being increasingly seen as eco-friendly vehicles. On the other hand, the oil and other traditional fossil fuel resources are being depleted, wind power and solar power are being widely studied and used as sustainable and clean energy. Wind, solar and other renewable energy naturally have the characteristics of intermittence and volatility, thus renewable energy generation will inevitably produce power fluctuations to the grid. When large-scale renewable energy connects to the grid, how to stabilize the fluctuation and improve the renewable energy integration ability of the grid becomes a serious problem. With the improvement in the EV technology and rapid rollout of EVs, EV charging is likely to have significant impact on the grid [1, 2]. Therefore, it is important to manage and influence EV charging behaviors in a controlled manner so as to reduce their impact. The controlled EV charging strategy can also improve the load characteristics of the grid, smooth the fluctuation of wind and solar power outputs, and optimize peak-valley of the grid load.

For the study of EV controlled charging and related fields, some results have been published. From the perspective of operation benefits of the charging station and by responding to the time-of-use (TOU) price of the grid, [3] uses the control methods of controlled charging to improve the economic benefits of charging stations, but it does not consider smoothing load fluctuations of the grid, it may result in an additional peak at night. Reference [4] proposes an optimized model for TOU price time-period. It uses the controlled charging of EVs to cut the peak and fill the valley of the power grid load curve, but it does not consider the value of peak-valley price and demand side response.

Reference [5] considers reducing the peak-valley difference as the primary goal, taking into account the division of TOU periods and load fluctuations of the local distribution network. It does not consider the influence of the wind, solar and other new energy power on the grid load characteristics. Reference [6] establishes a mathematical model of dispatching EVs and wind power cooperatively, analyses the feasibility of dispatching the EV charging to smooth the load fluctuations and to consume the superfluous wind power at night. However, the paper does not propose specific solutions for controlled charging from the point of demand side response.

This paper develops a probabilistic model of wind and photovoltaic (PV) power outputs, and a method of assessing the peak-valley difference of the stochastic load curve is put forward. A two-stage peak-valley price model is constructed and is used to guide EV charging, an optimization model is built, in which genetic algorithms are used to determine the start and end time of the valley price, as well as the value of the peak-valley price.

2 Controlled EV charging based on demand side response

2.1 Two-stage model of peak-valley price

Firstly, the two-stage peak-valley price model is built for controlled EV charging as follows:

$$f_p(t) = \begin{cases} P_v & t_1 < t < t_2 \\ P_p & \text{else} \end{cases} \qquad (1)$$

where P_v is the price of valley period; P_p is the price of peak period; t_1 is the start time of valley period; t_2 is the end time of valley period.

In this paper, the peak-valley rate β is defined as:

$$\beta = \frac{P_p - P_v}{P_0} \qquad (2)$$

where P_0 is the grid original tariff when TOU pricing policies are not implemented.

2.2 Analysis of user response

With the increasing of β, the user response has three stages, as shown in Fig. 1. In the first stage, β is in the interval $(0, a)$, users choosing to charge in valley period for tariff factor are very few. In the second stage, when β reaches a certain value, with the increasing of β, more and more users choose to charge in valley period for the price factor. In the third stage, the number of users choosing to charge in valley period stops increasing. In the figure,

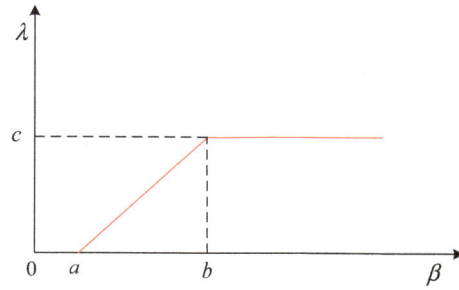

Fig. 1 Schematic diagram of user response

$\lambda = n/N$ is the ratio of the responding EV users, in which n is the number of EVs charged in valley period and N is the total number of EVs.

The EV users' responsiveness to the peak-valley rate β is also affected by the time.

$$T_p + T_y + T_d = 24 \qquad (3)$$

where T_p is the period that users' driving demand for EVs is urgent; T_y is the period that users' driving demand for EVs is common; and T_d is the period that users' driving demand for EVs is low.

In different periods, users' response to peak-valley rate β is different, as shown in Fig. 2.

So, EV users' price response function should be a three-dimensional surface which contains peak-valley rate β and time, as shown in Fig. 3.

The responding users' start time of charging is described by the formula:

$$t_s = \begin{cases} t_1 + \alpha(t_2 - t_c) & 0 \le t_c < t_2 - t_1 \\ t_1 & \text{else} \end{cases} \qquad (4)$$

where t_s is the start time of charging; t_c is the duration of charging; and α is a random number in interval $(0,1)$.

When the charging time is shorter than the valley period, users will choose any time in the valley period that EV can be fully charged, and when the charging time is longer than the valley period, users will choose the start time of valley period to charge.

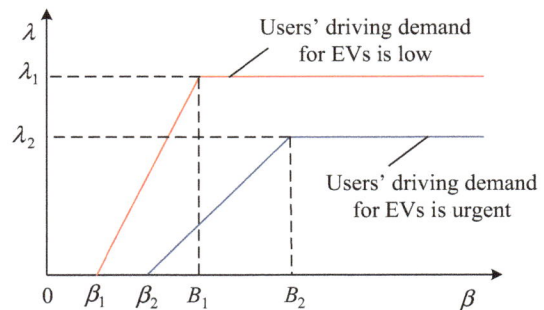

Fig. 2 Responding curve considering driving demand factors

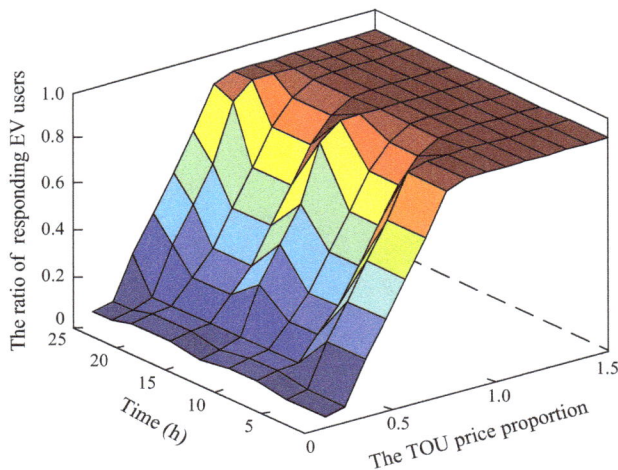

Fig. 3 Response surface considering driving demand factors

3 Probabilistic model of regional wind and PV power

3.1 Load curve transformation when renewable energy is superimposed on

If a area has n wind farms, the number of wind turbines in farm i is N_i ($i = 1, 2, \cdots, n$) and the farm has m PV power stations, the number of PV modules in PV station j is M_j ($j = 1, 2, \cdots, m$), the following assumptions are made.

1) In the same wind farm, the relationship between the outputs of different wind turbines is strongly correlated, the correlation coefficient is 1; and the relationship between the outputs of different wind farms is mutually independent, the correlation coefficient is thus 0.

2) Similarly, in the same PV station, the relationship between the outputs of different PV modules has a correlation coefficient of 1; and the relationship between the outputs of different PV stations is mutually independent with correlation coefficient equal to 0.

3) The relationship between the outputs of PV stations and wind farms is mutually unrelated, the correlation coefficient is 0.

Then for wind farm i, its total output is $P_i = N_i P_w$, in which P_i is the total output of the wind farm i, P_w is the output of a single wind turbine. Similar results could be obtained for PV stations.

According to the methods in [7], when wind speed meets the Weibull distribution, assuming the active power outputs of wind farms are random variables, based on the historical data, we can get their semi-invariants δ_i, $i = 1, 2, \cdots, n$. Similarly, when light intensity meets the Beta distribution, based on the historical data, the semi-invariant of

the active power output of PV station j is ε_j, $j = 1, 2, \cdots, m$.

Since the active power outputs of n wind farms and m PV stations are mutually independent, based on the additivity of the semi-invariant, the semi-invariant of the regional total active power output is:

$$\gamma = \sum_{i=1}^{n} \delta_i + \sum_{j=1}^{m} \varepsilon_j \tag{5}$$

where γ is the semi-invariant of the regional total active power output.

In this paper, we consider that the distribution of wind speed and light intensity are different at different times of the day. Based on the historical hourly data, we can divide one day into 24 parts, therefore, 24 pairs of expectation μ and standard deviation σ for wind speed and light intensity can be obtained, respectively, and we can get 24 Weibull distributions and 24 Beta distributions.

Therefore we can get 24 semi-invariants of the regional total power output:

$$\gamma_k = \sum_{i=1}^{n} \delta_{ki} + \sum_{j=1}^{m} \varepsilon_{kj} \tag{6}$$

where γ_k is the semi-invariant of the regional total active power output of the time k, $k = 1, 2, \cdots, 24$.

In turn, by the Gram–Charlier expanding [8, 9], for the random variables of the regional total active power outputs, their probability density function $f_k(x)$ and cumulative distribution function $F_k(x)$ can be obtained, as shown in Fig. 4 (the semi-invariant of the curve is got from the

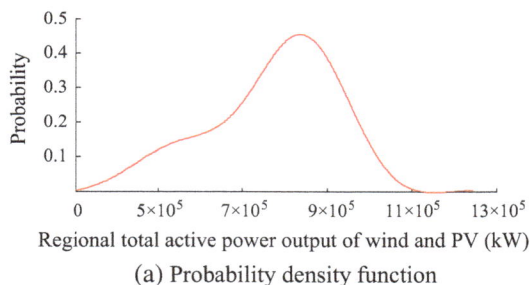

(a) Probability density function

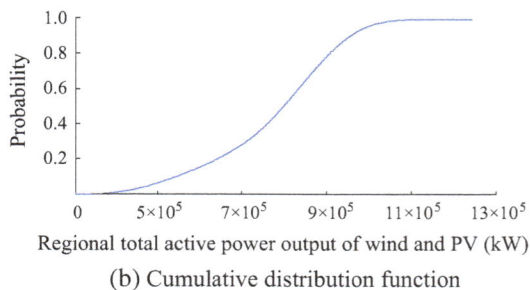

(b) Cumulative distribution function

Fig. 4 Probability density function $f_k(x)$ and cumulative distribution function $F_k(x)$

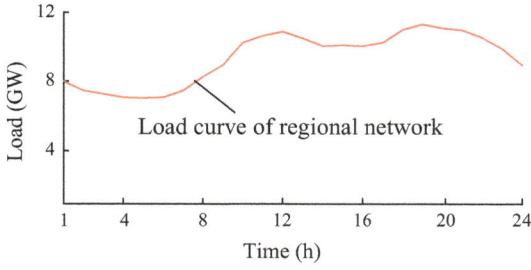

Fig. 5 A typical daily load curve of a certain area

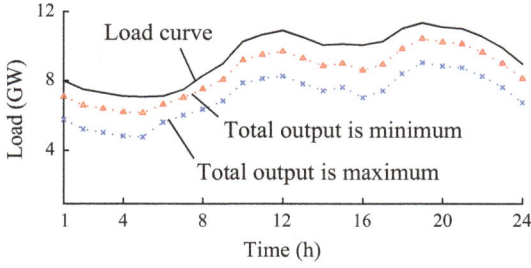

Fig. 6 Changing process of load curve

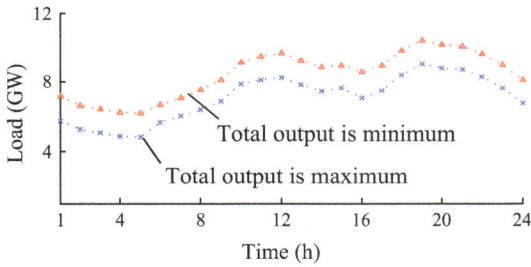

Fig. 7 The 'load belt' of area

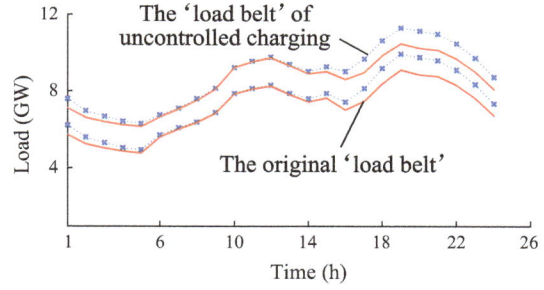

Fig. 8 Impact of large-scale EV uncontrolled charging on original 'load belt'

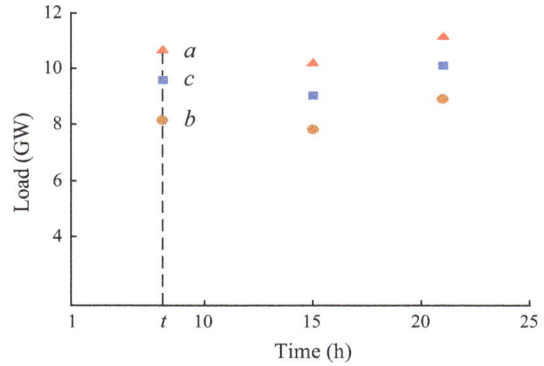

Fig. 9 Range of load fluctuation at time t

historical data of wind speed and light intensity in the certain area).

Generally, the random variable falls inside the interval $(\mu_k - 3\sigma_k, \mu_k + 3\sigma_k)$, which can be considered, for the total active power outputs of wind farms and PV stations, the minimum is $\mu_k - 3\sigma_k$, and the maximum is $\mu_k + 3\sigma_k$, $k = 1, 2, \cdots, 24$.

Figure 5 is a typical daily load curve of a certain area, when regional wind and PV power outputs are superimposed on it, the composite load curve is shown in Fig. 6. So the load curve is no longer deterministic, it is a random curve inside the 'load belt' and meets certain probabilistic rules, as shown in Fig. 7.

3.2 Impact of unordered charging

So when large-scale EVs get charged in an uncontrolled manner, using the Monte Carlo simulation method of [10],

the load curve of uncontrolled charging can be obtained. When there are one million uncontrolled charging EVs in a certain city, its 'load belt' is shown in Fig. 8.

It can be obtained from Fig. 8 that, the charging peak would appear near the peak of the original load, resulting in the lifting of the original 'load belt' peak, and the peak-valley difference gets bigger.

3.3 Probabilistic model of wind and PV power outputs

In every hour, the load is a random value between the minimum and the maximum, and its derivation process of the probability density function is as follows.

In Fig. 9, at time t, $t = 1, 2, \cdots, 24$, if the random variable of load is L_t. L_t is a random value between point b and c. Point a corresponds to the original load of the grid, its value L_{ta} is determined; point b corresponds to the load when total outputs of wind farms and PV stations are maximum, its value is L_{tb}; point c corresponds to the load when total output is minimum, it is L_{tc}. The relations among them are:

$$\begin{cases} L_{tc} = L_{ta} - (\mu_t - 3\sigma_t) \\ L_{tb} = L_{ta} - (\mu_t + 3\sigma_t) \\ L_t = L_{ta} - C_t \end{cases} \quad (7)$$

where C_t is the random variable of the total output of wind farms and PV stations at time t, probability density function of C_t is $f_t(c)$ (the method of calculating $f_t(c)$ has been explained before), the probability density function of random variable L_t is $g_t(l) = f_t(L_{ta} - l)$, in which l is the value of L_t.

In the above process, the probabilistic model of wind and PV power outputs is built, thus, for each of the 24 time points of the 'load belt', the probability density function of the random variable has been obtained [11–13].

4 Method of assessing peak-valley of uncertain load curve

When regional wind and PV power outputs are superimposed on, the load curve is no longer deterministic, it is a random curve inside the 'load belt' [14]. This paper proposes a probabilistic method to assess the peak-valley difference.

If the random variable of load at t_1 is L_{t1}, its probability density function is $g_{t1}(l)$, and at t_2, the probability density function of L_{t2} is $g_{t2}(l)$, assume that $\mu_{t2} > \mu_{t1}$, then:

$$P\left(\frac{L_{t2} - L_{t1}}{L_{t2}} \leq \theta\right) = P((1-\theta)L_{t2} - L_{t1} \leq 0) \tag{8}$$

In this paper, we refer to θ as the peak-valley difference index. Therefore, assume that random variables L_{t1} and L_{t2} are mutually independent, and if $L' = (1-\theta)L_{t2} - L_{t1}$, then by using the additivity of the semi-invariant, the cumulative distribution function of L' can be obtained from $g_{t1}(l)$ and $g_{t2}(l)$, we write it $F_F(l)$, therefore:

$$P((1-\theta)L_{t2} - L_{t1} \leq 0) = P(L' \leq 0) = F_F(0) \tag{9}$$

Firstly, calculate the expectations of the 24 random variables at 24 time points, select the first three maximum points as the possible 'peak' time points and the last three minimum points as the possible 'valley' time points, then use $P((L_{t2} - L_{t1})/L_{t2} \leq \theta)$ to calculate the three pairs of time points above. Choose the minimum of the calculation results as the final value of P_θ, P_θ is the probability that peak-valley difference is less than or equal to the certain peak-valley difference index θ:

$$P_\theta = P\left(\frac{L_{tp} - L_{tv}}{L_{tp}} \leq \theta\right) \tag{10}$$

where L_{tp} is the random variable of load at 'peak' time; L_{tv} is the random variable of load at 'valley' time. As the interval between random variables L_{tp} and L_{tv} is long, they can be considered mutually independent [15, 16].

To illustrate P_θ can effectively reflect the value of the peak-valley difference, for the first graph in Fig. 10,

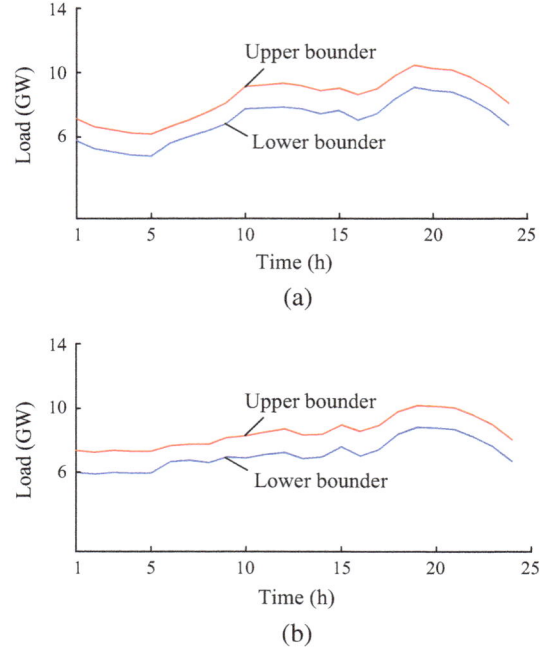

(a)

(b)

Fig. 10 Two different 'load belts'

Table 1 P_θ corresponding to different peak-valley difference indices of two load curves

$\theta/\%$	$P_\theta^{(1)}$	$P_\theta^{(2)}$
25	0	0.120
30	0	0.542
35	0.365	0.750
40	0.721	0.905
50	0.917	1.000

$P_\theta^{(1)} = P((L_{tp} - L_{tv})/L_{tp} \leq \theta)$, and the second graph $P_\theta^{(2)} = P((L_{tp} - L_{tv})/L_{tp} \leq \theta)$, we can get the Table 1.

It can be seen from Table 1 that, for every peak-valley difference indices, $P_\theta^{(1)} < P_\theta^{(2)}$ is right. Therefore it can be concluded that, there is a positive correlation between P_θ and the peak-valley difference of the uncertain load curve, the bigger the value of P_θ is, the smaller the peak-valley difference (peak-valley difference rate) is, and the flatter the load curve is.

5 Optimization model

Based on the ordered charging model, the power supply side guides the charging of EVs by formulating P_p and P_v, and delimiting the start and end time of valley period t_1 and t_2.

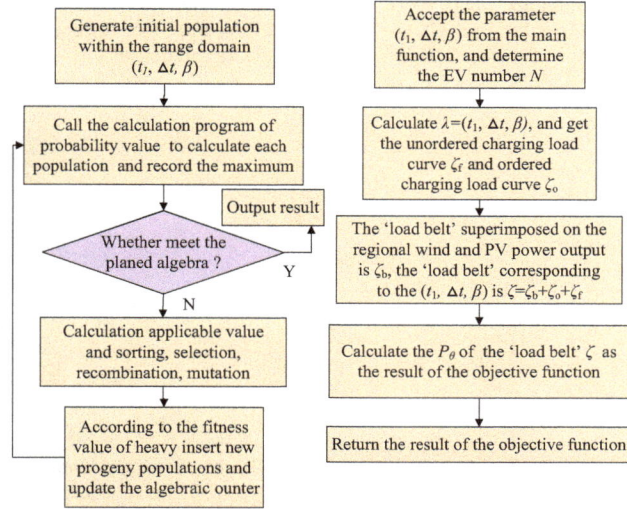

Fig. 11 Flow chart of genetic algorithm optimization and objective function

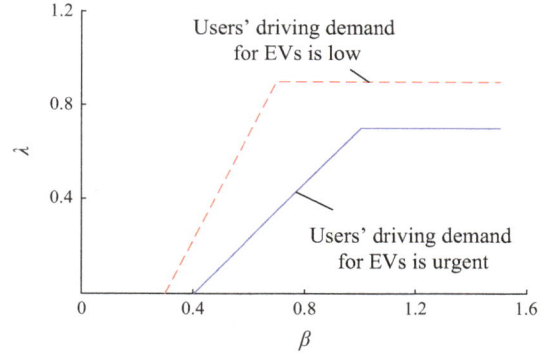

Fig. 12 Response curves in different driving demands

Table 2 Optimization results

$\theta/\%$	t_1	$\Delta t/h$	β	$\lambda/\%$	$P_\theta/\%$
40	01:34	7.5465	1.0914	80.1	89.8
35	00:55	6.4787	1.1822	70.1	79.6
30	01:04	7.0562	1.1093	80.0	65.6

If the total number of EVs is N, corresponding to every (P_p, P_v, t_1, t_2), there are λN EVs choose ordered charging, its load curve is ζ_o; $(1 - \lambda)N$ EVs choose unordered charging, its load curve is ζ_f; the 'load belt' superimposed on the regional wind and PV power outputs is ζ_b, then corresponding to the (P_p, P_v, t_1, t_2), the 'load belt' is $\zeta = \zeta_b + \zeta_o + \zeta_f$, it is meaningful and important to find the (P_p, P_v, t_1, t_2) that can make the peak-valley difference of the ζ minimum, that is to make the P_θ of ζ maximum.

Through the above analysis, we know λ is a function and $t_1, \Delta t, \beta$ are its independent variables, Δt is the duration of valley period, $\Delta t = t_2 - t_1$.

$$\lambda = f(t_1, \Delta t, \beta) \tag{11}$$

Therefore, in order to optimize the peak-valley difference, the following optimization model is built:

$$\max P_\theta(t_1, \Delta t, \beta)$$
$$\text{s.t.} \begin{cases} 0 \leq t_1 \leq 24 \\ 0 < \Delta t < 12 \\ 0 < \beta < \beta_{lim} \end{cases} \tag{12}$$

To take the maximum P_θ as the target, and use genetic algorithms to determine the start time t_1 and the duration of valley period Δt, as well as the peak-valley rate β.

The flow chart of genetic algorithm optimization and objective function is shown in Fig. 11.

6 Case study

In this example, the typical daily load curve of a city is selected as the basis. By assuming the specific parameters

of its wind farms and PV stations, and based on the historical data, we assume the city has one million EVs. EV constant charging power is $P_c = 2.5$ kW.

In this example, when driving demand is different, users' response to peak-valley rate β is shown in Fig. 12. When driving demand is common, the response curve is the average of the two curves. The maximum value is $\beta_{lim} = 1.5$.

At different time, users' driving demand for EVs is different in a day, so we divide one day into three periods according to EV users' schedule of the city: ①T_p: the period that users' driving demand for EVs is urgent (7:00–10:00 and 16:00–20:00); ②T_y: the period that users' driving demand for EVs is common (22:00–7:00); ③T_d: the period that users' driving demand for EVs is low (10:00–16:00 and 20:00–22:00).

According to the above parameters, the genetic algorithm toolbox of MATLAB is used to achieve the simulation, the optimization results are shown in Table 2.

From the above results, it can be seen that, for different peak-valley difference indices, an optimization result $(t_1, \Delta t, \beta)$ can be got corresponding to it, as shown in Fig. 13.

Take the optimization result when $\theta = 35\%$ as an example, $(t_1, \Delta t, \beta)$ is (0.926, 6.4787, 1.1822).

After using the genetic algorithm to optimize P_θ, the probability that the peak-valley difference is less than or equal to 35% is 79.69%, in the valley period of load, after guiding EV ordered charging, the load increases. It can be intuitively seen from Fig. 13b that, the optimized 'load belt'

(a) The optimization result when θ=40%

(b) The optimization result when θ=35%

(c) The optimization result when θ=30%

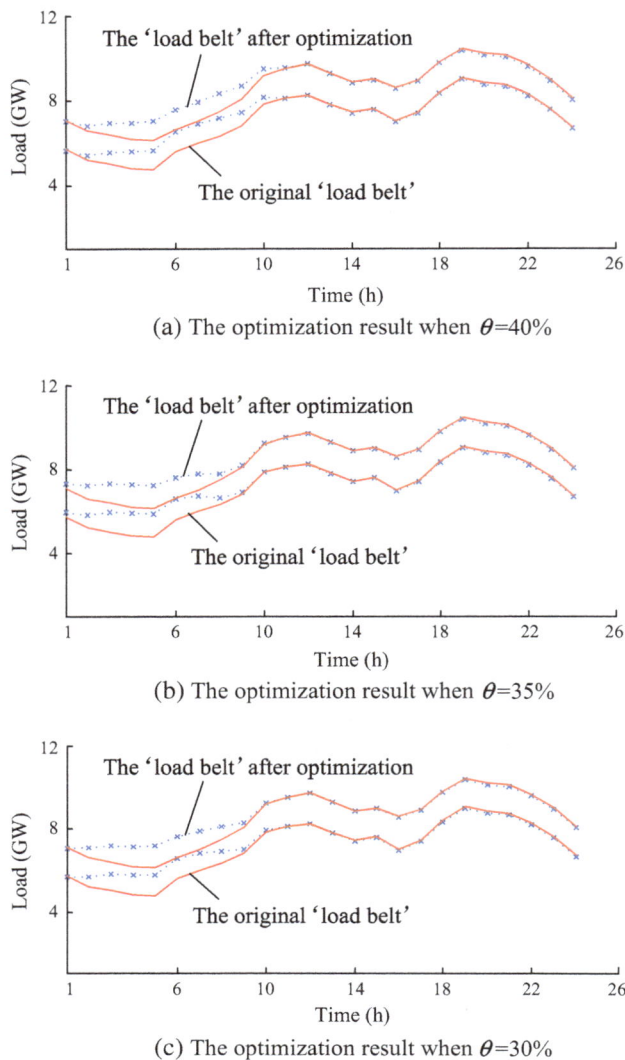

Fig. 13 Optimization 'load belts' of different peak-valley difference indices

tends to be smooth. It can be obtained from the optimization results that, when considering the regional wind and PV power outputs, the method of ordered charging in this paper can effectively optimize the peak-valley difference.

7 Conclusion

Through the probabilistic model of regional wind and PV power outputs and the model of peak-valley price in two-stage, this paper solves the problem of large scale EV ordered charging. Furthermore, it achieves the goal of stabilizing the power fluctuations caused by the regional wind and PV outputs and optimizing the peak-valley difference in the gird. As analyzed above, the effectiveness and rationality of the models and methods have been proved.

Acknowledgments This work is supported by National Natural Science Foundation of China (No. 51477116) and the Special Founding for "Thousands Plan" of State Grid Corporation of China (No. XT71-12-028).

References

[1] Zhang WL, Wu B, Li WF et al (2009) Discussion on development trend of battery electric vehicles in China and its energy supply mode. Power Syst Technol 33(4):1–5 (in Chinese)

[2] Zhuang X, Jiang KJ (2012) A study on the roadmap of electric vehicle development in China. Automot Eng 34(2):91–97 (in Chinese)

[3] Xu ZW, Hu ZC, Song YH et al (2012) Coordinated charging of plus-in electric vehicles in charging stations. Autom Electr Power Syst 36(11):38–43 (in Chinese)

[4] Ge SY, Huang L, Liu H (2012) Optimization of peak-valley TOU power price time-period in ordered charging mode of electric vehicle. Power Syst Protect Control 40(10):1–5 (in Chinese)

[5] Sun XM, Wang W, Su S et al (2013) Coordinated charging strategy for electric vehicles based on time-of-use price. Autom Electr Power Syst 37(1):191–195 (in Chinese)

[6] Yu DY, Song SG, Zhang B et al (2011) Synergistic dispatch of PEVs charging and wind power in Chinese regional power grids. Autom Electr Power Syst 35(14):24–28 (in Chinese)

[7] Wang CS, Zheng HF, Xie YH et al (2005) Probabilistic power flow containing distributed generation in distribution system. Autom Electr Power Syst 29(24):39–44 (in Chinese)

[8] Shi DY, Cai DF, Chen JF et al (2012) Probabilistic load flow calculation based on cumulant method considering correlation between input variables. Proc CSEE 32(28):104–113 (in Chinese)

[9] Zhang P, Lee ST (2004) Probabilistic load flow computation using the method of combined cumulants and Gram-Charlier expansion. IEEE Trans Power Syst 19(1):676–682

[10] Peterson SB, Whitacre JF, Apt J (2010) The economics of using plug-in hybrid electric vehicle battery packs for grid storage. J Power Sources 195(8):2377–2384

[11] Wang JH, Shahidehpour M, Li ZY (2008) Security constrained unit commitment with volatile wind power generation. IEEE Trans Power Syst 23(3):1319–1327

[12] Tian LT, Shi SL, Jia Z (2010) A statistical model for charging power demand of electric vehicles. Power Syst Technol 34(11):126–130 (in Chinese)

[13] Hu ZC, Song YH, Xu ZW et al (2012) Impacts and utilization of electric vehicles integration into power systems. Proc CSEE 32(4):1–10 (in Chinese)

[14] Zhao JH, Wen FS, Xue YS et al (2010) Power system stochastic economic dispatch considering uncertain outputs from plug-in electric vehicles and wind generators. Autom Electr Power Syst 34(20):22–29 (in Chinese)

[15] Saber AY, Venayagamoorthy GK (2011) Plug-in vehicles and renewable energy sources for cost and emission reductions. IEEE Trans Ind Electron 58(4):1229–1238

[16] Hübner M, Zhao L, Mirbach T, et al (2009) Impact of large-scale electric vehicle application on the power supply. In: Proceedings of the 2009 IEEE electrical power and energy conference (EPEC'09), Montreal, Canada, 22–23 Oct 2009, 6 pp

Hong LIU is the associate professor of electrical engineering in Tianjin University with research interests in urban power system planning and decision support theory on power system. In addition, he has been supported by Natural Scientific Funds of China and received the second prize of the Natural Scientific and Technological Progress.

Pingliang ZENG is the distinguished expert of "Thousands Plan" of Electric Power Research Institute of China. His current research interests include power system planning under uncertainty, system operations, power market, incentive arrangement, power system reliability, integration of electric vehicle and energy storage in low carbon electricity systems.

Jianyi GUO is the graduate of electrical engineering in Tianjin University, with research interests in EV ordered charging.

Huiyu WU is studying in University of Wisconsin Madison, with research interests in actuarial science and math.

Shaoyun GE is the professor of electrical engineering in Tianjin University with research interests in urban power system planning and the optimization of power system operation.

16

Energy management with dual droop plus frequency dividing coordinated control strategy for electric vehicle applications

Wuhua LI (✉), **Chi XU, Hongbin YU,**
Yunjie GU, Xiangning HE

Abstract In this paper, a dual droop-frequency diving coordinated control strategy is proposed for electric vehicle (EV) applications, where the hybrid energy storage system (HESS) with supercapacitors and batteries is integrated to prolong the life time of storage elements. The dynamic power allocation between the supercapacitor and batteries are obtained through the voltage cascaded control, upon which the high and low frequency power fluctuation are absorbed by the supercapacitors and batteries respectively to fully exploit the advantages of the supercapacitors and batteries. Moreover, the power capacity is scaled up by connecting storage blocks in parallel. A dual droop control scheme for parallel-connected energy storage system and its operation principle is introduced on the aspect of current sharing characteristic and state-of-charging (SOC) management. After detailed analysis and formula derivation, the corresponding loop parameters are designed. Through this control method, the current sharing performance is ensured and each block makes the self-adaptive adjustment according to their SOC. Consequently, the load power can be shared effectively, which helps to avoid the over-charge/over-discharge operation and contributes to the life cycle of the energy storage system. Each module is autonomous controlled without the necessity of communication, which is easy, economic and effective to realize. Finally, the simulation and experimental results are exhibited to verify the effectiveness of the proposed control scheme.

Keywords Dual droop control, Frequency diving coordinated control, Hybrid energy storage system, Electric vehicle

1 Introduction

The energy storage system (ESS) plays a key role in deciding the electric vehicle's performance and reliability [1–3]. Compared with the energy storage system solely structured by only batteries or supercapacitors, a hybrid energy storage system comprising both high power density and high energy density storage units, gains better performance, such as high power rating, high energy capacity, short response time, and extended the life cycle of the ESS [4–7]. Control, robustness, stability, efficiency, and optimization of HESS remain an essential research area [8]. How to reinforce the complementary advantages of batteries and supercapacitors, the current sharing and SOC management of battery are key issues that need to be considered [9–12].

Many studies have been done upon the dynamic power sharing between batteries and supercapacitors [13–15]. The main control objective is to achieve the dynamic allocation of high frequency power demands and low frequency power demands to the batteries and supercapacitors respectively. In [14], a piecewise linear relationship between the current of the battery and supercapacitor decides the power distribution. When the load power varies tremendously, the current of supercapacitor increases dramatically. As a result, the fluctuating power is mainly compensated by the supercapacitor with fast response. However, in the case of small variation of the load power, all the power is balanced by the battery, which leads to frequent charge and discharge cycles. This may do great harm to the battery and shorten its service life. To separate the responses in frequency between these two energy

W. LI, C. XU, H. YU, Y. GU, X. HE, Zhejiang University, Hangzhou 310027, China
(✉) e-mail: woohualee@zju.edu.cn

storages, the power demand are divided into low frequency and high frequency components in [15]. And through the bus and supercapacitor voltage cascaded control, the converter current references are set such that the low frequency component is supplied by the battery and the high frequency component is supplied by the supercapacitor [16]. However, only one battery and one supercapacitor are concerned, which limits its applications in the high power occasion.

In [16–21], the total current demand is derived from the difference between the energy source and load power. Then the low-frequency component of the current, namely the total current reference for the battery interfacing power converters, is generated by passing this current reference through a low-pass filter (LPF), while the rest of the current is the reference for supercapacitors. However, the calculation of the total current demand requires the information of all energy source and load power. It relies on a wide communication bandwidth with the increasing number of interfaces, which decreases the system reliability. Meanwhile, the average current control is adopted for current sharing. Such centralized energy management system relies heavily upon the energy control center itself, resulting in inadequate system reliability and scalability, which is not suitable for systems with dynamic structure.

On the other hand, the batteries have the operation limitations in practice. Without a proper battery management, it may come into being serious problems, such as the over-charging, over-discharging, imbalance of voltage among the cells. Therefore, it is a necessity to develop advanced control strategies for the battery equalization. In [16, 18–22], SOC is taken into account in energy control center with centralized control, whose drawback is the less flexible and not appropriate in an expandable system. An attenuation coefficient, with consideration of factors like battery terminal voltage and SOC, is multiplied by the battery current reference in the control loop to guarantee the efficient and safe operation of batteries in [23]. However, the current reference restriction that no power outputs once the battery voltage exceeds the bounds is too rigorous.

Combined with advantages of the conventional droop control, which features high expandability and reliability, the dual droop frequency dividing coordinated control is derived for HESS in this paper with consideration of power sharing between batteries and supercapacitors and SOC of batteries. By employing this control scheme, the dynamic power sharing between two energy storage mediums is ensured with the DC bus and supercapacitor voltage cascaded control. While dual droop control, as an extension of the conventional droop control, realizes the current sharing and SOC management of different power modules without

requirement of communication, which increases the system reliability and flexibility.

2 Proposed control strategy for HESS

Figure 1 shows an electric vehicle propulsion system with HESS, which contains both high energy density and high power density storage elements. The battery packs, which provide power during the intermittent of the renewable energy sources, are connected to the DC bus by modular bi-directional DC-DC converters for the convenience of battery capacity expansion. While the supercapacitors are connected to the DC bus by multiple converters because of the high power rating.

The detailed control block of HESS is depicted in Fig. 2. C_{Bus} is the bus capacitor, and the power fluctuation caused by the motor drive is presented by a current source i_{bus}. For controlling the bi-directional DC/DC converters (BDCs), a cascaded control scheme using the inner current-control loop and outer voltage-control loop is realized, where i_{L_ref} is the current reference and the current transfer function is defined as $G_{BDC}(s)$. i_{Bat}, i_{SC} are the currents of the batteries and supercapacitors, respectively. i_{Bus_Bat}, i_{Bus_SC} are the corresponding current into the DC bus. $G_{vcSC}(s)$, H_{vBus}, V_{Bus_ref} are the compensators, feedback coefficient and voltage reference of the DC bus voltage control loop. The corresponding definitions of supercapacitor voltage control loop are expressed as $G_{vcBat}(s)$, H_{vSC}, V_{SC_ref}. And V_{Bus}, V_{SC}, V_{Bat} are the voltage of DC Bus, supercapacitor and battery. The conventional droop control is employed to realize the current sharing among the BDCs for supercapacitor, and the dual droop control is adopted for the current sharing and SOC management of the batteries. The frequency diving control method is realized through the DC bus and supercapacitor voltage cascaded control.

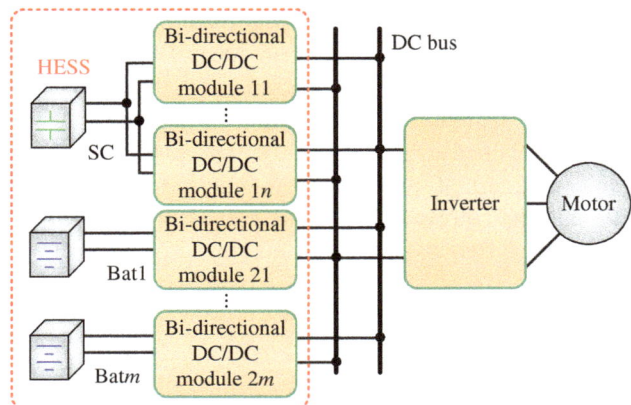

Fig. 1 Modular hybrid energy storage propulsion system

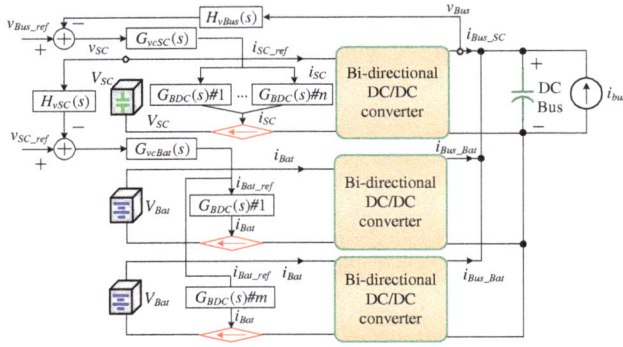

Fig. 2 Modular hybrid energy storage propulsion system

Table 1 Parameters of HESS

Components	Parameters
DC bus voltage feedback coefficient ($H_{vBus}(s)$)	8.533
Supercapacitor voltage feedback coefficient ($H_{vSC}(s)$)	100
DC Bus Capacitor (C_{Bus})	3000 µF
Supercapacitor capacity (C_{SC})	130 F
ESR of supercapacitor (R_{SC})	8m Ω
Supercapacitor voltage range (V_{SC})	30–60 V
Battery capacity	15 A h
Battery voltage range (V_{Bat})	40–56 V
DC bus voltage (V_{Bus})	400 V
Current transfer function gain ($G_{BDC}(s)$)	0.0444
Bandwidth of current control loop	400 Hz
Modular converter power (P_{out})	1500 W

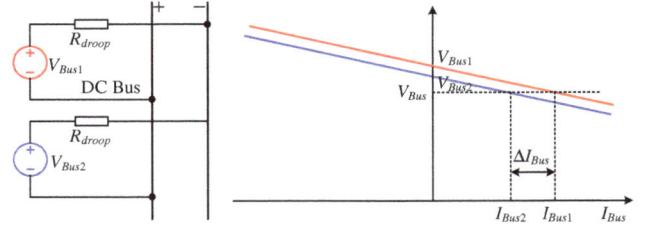

Fig. 3 Parallel operation of power converters with droop control and droop curves

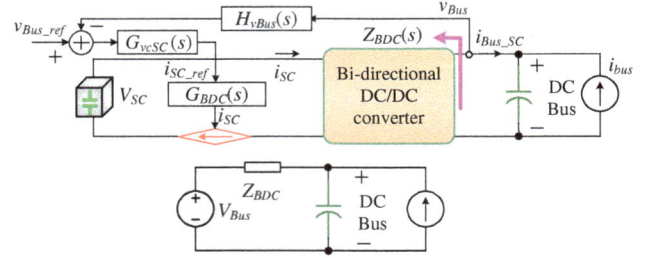

Fig. 4 Control scheme for supercapacitor converter

To analyze and verify the effectiveness of the control strategy, a HESS, with the combination of batteries and supercapacitor, is built with parameters listed in Table 1.

2.1 Control design of power modules for supercapacitor

A DC-DC converter with droop controller can be simplified as an electric circuit with an ideal DC voltage source in series with an output resistance, as in Fig. 3. One promising way to realize droop control is by setting the virtual output impedance R_{droop} of DC-DC converter. A schematic diagram of two parallel connected converter modules and the corresponding droop curves are depicted in Fig. 3. The difference of output currents ΔI_{Bus} can be derived as

$$\Delta I_{Bus} = \frac{V_{Bus1} - V_{Bus2}}{R_{droop}} \tag{1}$$

where V_{Bat1} and V_{Bat2} are the no-load voltage of each module.

As presented in [19], the input current reference i_{SC_ref} for the supercapacitor converter is used to regulate the output voltage (V_{Bus}) at the reference level. From Fig. 4, the virtual output impedance of the supercapacitor converter is obtained by

$$Z_{BDC}(s) = \frac{1}{H_{vBus}(s)G_{vcSC}(s)G_{BDC}(s)\frac{V_{SC}}{V_{Bus}}} \tag{2}$$

In order to realize the V-I droop curves as shown in Fig. 3, the compensator $G_{vcSC}(s)$ can be derived as

$$G_{vcSC}(s) = \frac{1}{H_{vBus}(s)G_{BDC}(s)\frac{V_{SC}}{V_{Bus}}}\frac{1}{R_{d\_SC}} \tag{3}$$

Hence the output impedance $Z_{BDC}(s)$ equals to R_{d_SC}.

Considering n converter modules in parallel, the open loop transfer function of DC bus voltage $G_{vBuspoc}(s)$ is n times of $G_{vBusoc}(s)$, which is given by

$$G_{vBuspoc}(s) = nG_{vBusoc}(s) = nG_{vcSC}(s)G_{BDC}(s)\frac{V_{SC}}{V_{Bus}}$$
$$\frac{1}{C_{Bus}s}H_{vBus}(s) = \frac{n}{R_{d\_SC}C_{Bus}s} \tag{4}$$

From (4), the bandwidth of DC bus voltage loop is decided by the number of branches in parallel n and the droop coefficient R_{d_SC}. The bandwidth of DC bus voltage loop is set at 5 Hz with consideration of the supercapacitor response time (in milliseconds). In the case of two converter modules in parallel, the droop coefficient R_{d_SC} is 21.2 Ω. If the voltage difference between V_{Bat1} and V_{Bat2} after calibration is 0.2% of the output voltage, namely 1.6

V, the difference of output current ΔI_{Bus} can be obtained by (1), which is 0.075 A and 4% of current in full load case. So the effect in distributing current equally is satisfied.

The open and close loop transfer function after compensation $G_{vBuspoc}(s)$, $G_{vBuspc}(s)$ are obtained by (5) and (6), while the current transfer function from battery to supercapacitor $G_{iSC\_iBatp}(s)$ is given by (7).

$$G_{vBuspoc}(s) = \frac{z_{c\_Bus}}{s} \tag{5}$$

$$G_{vBuspc}(s) = \frac{G_{vBuspoc}(s)/H_{vBus}(s)}{1+G_{vBuspoc}(s)} \approx \frac{z_{c\_Bus}}{s + z_{c\_Bus}} \frac{1}{H_{vBus}(s)} \tag{6}$$

$$\begin{aligned} G_{iSC\_iBatp}(s) &= -G_{vBuspc}(s)H_{vBus}(s)\frac{V_{Bat}}{V_{SC}} \\ &\approx \frac{z_{c\_Bus}}{s + z_{c\_Bus}} \frac{V_{Bat}}{V_{SC}} \end{aligned} \tag{7}$$

where z_{c_Bus} equals to $n/(R_{d\_SC}C_{Bus})$.

2.2 Control design of power modules for batteries

The equation of droop control for batteries is listed as (8), where V_{SC0} is the no-load voltage of each module, R_{d_Bat} is the drop coefficient and I_{Bat} is the battery current. If all the battery converters are controlled under the same droop curve, the batteries with lower SOC will go into over discharge while those with higher SOC will be over charged, which significantly decrease the system reliability. In order to balance SOC of each battery unit, a dual droop control scheme is proposed by introducing battery SOC into droop control scheme, which is shown in Fig. 5.

$$V_{SC} = V_{SC0} - R_{d\_Bat}I_{Bat} \tag{8}$$

V_{SC0_1}, V_{SC0_2} are the supercapacitor voltage reference for each battery converter, which are decided with their SOCs as in (9)

$$V_{SC0} = V_{SC0}^* + k_{v\_SOC}SOC \tag{9}$$

where V_{SC0}^* is the supercapacitor voltage reference under the circumstance of the minimum SOC and k_{v_SOC} is the secondary droop coefficient.

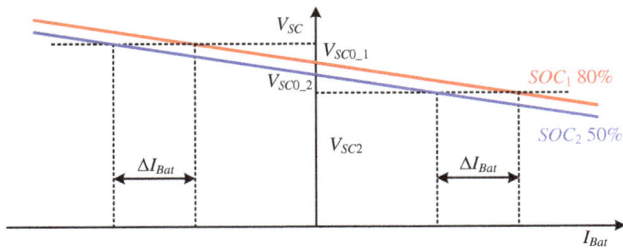

Fig. 5 Dual droop control scheme for battery converters

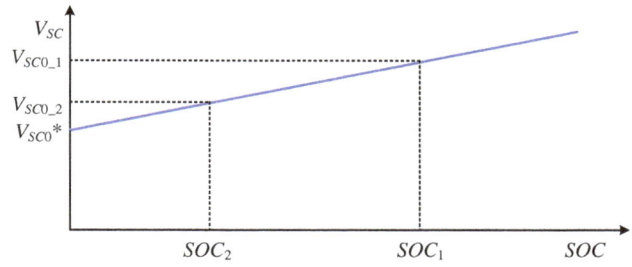

Fig. 6 Secondary droop between battery SOC and V_{SC0}^*

Equation (9) is drawn in Fig. 6, which achieves droop curves shifting up and down as presented in Fig. 5. Thus the battery with higher SOC takes in smaller current and gives out bigger current, which contributes to SOC balancing among the battery units.

The difference of battery currents ΔI_{Bat} is derived by

$$\Delta I_{Bat} = I_{Bat1} - I_{Bat2} = \frac{V_{SC0\_1} - V_{SC0\_2}}{R_{d\_Bat}} \tag{10}$$

$$\begin{aligned} V_{SC0\_1} - V_{SC0\_2} &= k_{v\_SOC}(SOC_1 - SOC_2) \\ &= k_{v\_SOC}\Delta SOC \end{aligned} \tag{11}$$

By substituting (11) into (12), ΔI_{Bat} is expressed as

$$\Delta I_{Bat} = \frac{k_{v\_SOC}\Delta SOC}{R_{d\_Bat}} \tag{12}$$

The battery SOC is related to its initial SOC, i.e. SOC_0 and the battery current as given by

$$SOC = SOC_0 - \frac{\int I_{Bat}dt}{C_{Bat}} \tag{13}$$

So the difference of SOCs can be calculated as following

$$\begin{aligned} \Delta SOC &= SOC_1 - SOC_2 = SOC_{0_1} - SOC_{0_2} \\ &- \frac{\int (I_{Bat1} - I_{Bat2})dt}{C_{Bat}} = \Delta SOC_0 - \frac{\int \Delta I_{Bat}dt}{C_{Bat}} \end{aligned} \tag{14}$$

Carrying out Laplace transform on (12) and (14), (15) and (16) can be derived

$$\Delta I_{Bat}(s) = \frac{k_{v\_SOC}\Delta SOC(s)}{R_{d\_Bat}} \tag{15}$$

$$\Delta SOC(s) = \Delta SOC_0 - \frac{\Delta I_{Bat}(s)}{C_{Bat}s} \tag{16}$$

By substituting (15) into (16), the difference of SOCs and battery currents are expressed as

$$\Delta SOC(s) = \frac{R_{d\_Bat}C_{Bat}s}{R_{d\_Bat}C_{Bat}s + k_{v\_SOC}}\Delta SOC_0 \tag{17}$$

$$\Delta I_{Bat}(s) = \frac{k_{v\_SOC}C_{Bat}s}{R_{d\_Bat}C_{Bat}s + k_{v\_SOC}}\Delta SOC_0 \tag{18}$$

According to (17), the SOC balance is a first order system so the difference of SOCs gradually deceases to zero with some decay time constant T_{SOC} as given in (19). Consequently the difference of battery currents turns to zero, realizing the SOC balancing as demonstrated in (18).

$$T_{SOC} = \frac{R_{d\_Bat}C_{Bat}}{k_{v\_SOC}} \tag{19}$$

3 Design considerations of proposed control strategy

3.1 Primary droop control parameter design

The block diagram of the supercapacitor voltage control loop is depicted in Fig. 7, where the supercapacitor is modeled as a capacitor C_{SC} and a resistor R_{SC} in series. And the impedance of the supercapacitor $Z_{SC}(s) = (R_{SC}C_{SC}s+1)/(C_{SC}s)$. The relationship between the supercapacitor voltage and the battery current can be derived by

$$v_{SC}(s) = \frac{v_{SC\_ref}}{H_{vSC}(s)} - \frac{1}{G_{vcBat}(s)G_{BDC}(s)H_{vSC}(s)}i_{Bat}(s) \tag{20}$$

In order to lower the gain around zero introduced by the supercapacitor impedance so as to suppress the batteries' response to the power fluctuation around the zero frequency, the compensator $G_{vcBat}(s)$ is designed as (20) where $z_{SC} = 1/(R_{SC}C_{SC})$ and the droop equation is derived by (22).

$$G_{vcBat}(s) = \frac{1}{H_{vSC}(s)G_{BDC}(s)}\frac{1}{R_{d\_Bat}}\frac{z_{SC}}{s+z_{SC}} \tag{21}$$

$$v_{SC}(s) = \frac{v_{SC\_ref}}{H_{vSC}(s)} - R_{d\_Bat}\frac{s+z_{SC}}{z_{SC}}i_{Bat}(s) \tag{22}$$

For the DC component, namely $s = 0$, the droop equation is simplified as following

$$v_{SC}(0) = \frac{v_{SC\_ref}}{H_{vSC}(0)} - R_{d\_Bat}i_{Bat}(0) \tag{23}$$

The bandwidth of the supercapacitor voltage loop is set at 0.05 Hz considering the batteries response time (in seconds). In the case of two converter modules in parallel, the droop coefficient R_{d_SC} is 0.05 Ω.

3.2 Secondary droop control parameter design

The secondary droop control between the supercapacitor voltage reference and SOC is achieved by adding a

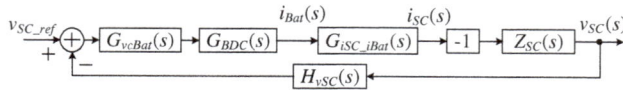

Fig. 7 Block diagram of supercapacitor voltage control loop

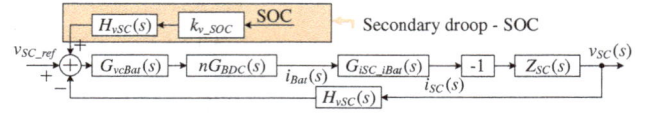

Fig. 8 Block diagram of dual droop control for battery units

feedforward loop into the supercapacitor voltage control loop, as presented in Fig. 8. The secondary droop equation in steady state is given by

$$V_{SC} = \frac{v_{SC\_ref}}{H_{vSC}} + k_{v\_SOC}SOC \tag{24}$$

In order not to influence the stability of the supercapacitor voltage control loop, the SOC balancing time constant T_{SOC} is set as 300 s, far larger than the corresponding time of the cross-over frequency in the supercapacitor voltage control loop (20 s). From (19), the secondary droop coefficient k_{v_SOC} can be obtained.

3.3 System response to power fluctuation

The relationship between the control parameters and frequency response range is analyzed to theoretically verify the proposed control strategy.

The transfer block diagram of response of the supercapacitor current and battery current to power fluctuation is plotted in Fig. 9.

The transfer function from the supercapacitor current to the battery current is obtained by

$$\begin{aligned} G_{i_{Bat}\_iSCP}(s) &= Z_{SC}(s)H_{vSC}(s)nG_{vcBat}(s)G_{BDC}(s) \\ &= \frac{z_{c\_SC}}{s}\frac{V_{SC}}{V_{Bat}} \end{aligned} \tag{25}$$

where z_{c_SC} is the crossover frequency of supercapacitor voltage control loop.

From Fig. 9, the output impedance of HESS can be calculated by

$$\begin{aligned} Z_{op}(s) &= \frac{v_{Bus}(s)}{i_{bus}(s)} = \frac{1}{1 + G_{vBuspoc}(s)(1 + G_{i_{Bat}\_iSCP}(s)\frac{V_{Bat}}{V_{SC}})} \\ \frac{1}{C_{Bus}s} &= \frac{s}{(s + z_{c\_SC})(s + z_{c\_Bus})C_{Bus}} \end{aligned} \tag{26}$$

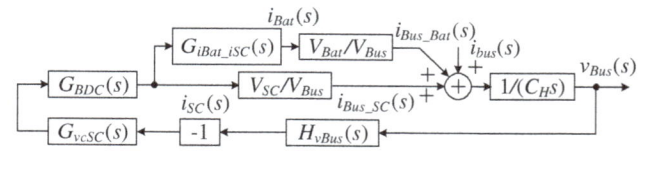

Fig. 9 Transfer block diagram of response of supercapacitor current and battery current to power fluctuation

Therefore, the transfer function of the DC bus capacitor current, supercapacitor current and batteries current to power fluctuation $G_{iCBus\_ibusp}(s)$, $G_{iBus\_SC\_ibusp}(s)$ and $G_{iBus\_Bat\_ibusp}(s)$ can be derived by

$$G_{i_{CBus}\_i_{bus}p}(s) = \frac{Z_{op}(s)}{\frac{1}{C_{Bus}s}} = \frac{s^2}{(s + z_{c\_SC})(s + z_{c\_Bus})} \quad (27)$$

$$G_{i_{Bus}\_SC\_i_{bus}p}(s) = Z_{op}(s)G_{vBuspoc}(s)C_{Bus}s$$
$$= \frac{z_{c\_Bus}s}{(s + z_{c\_SC})(s + z_{c\_Bus})} \quad (28)$$

$$G_{i_{Bus}\_Bat\_i_{bus}p}(s) = G_{i_{Bus}\_SC\_i_{bus}p}(s)\frac{V_{Bus}}{V_{Bat}}G_{i_{Bat}\_i_{SCP}}(s)\frac{V_{Bat}}{V_{Bus}}$$
$$= \frac{z_{c\_SC}z_{c\_Bus}}{(s + z_{c\_SC})(s + z_{c\_Bus})} \quad (29)$$

Figure 10 gives the frequency response range of the DC bus capacitors, supercapacitors and batteries to the power fluctuation. It can be seen that the batteries absorb the low frequency power fluctuation while supercapacitors provide the high frequency part and the rest power is supplied by DC bus capacitors. Meanwhile, the diving frequencies are decided by the crossover frequencies of the DC bus voltage control loop and supercapacitor voltage control loop. So the HESS can be fully used through proper controller design.

4 Experimental and simulation verification

In this section, a HESS, comprising two supercapacitor converters and two battery converters with parameters as listed in Table 1, is built to demonstrate the advanced performance of the proposed control strategy.

The current sharing effect and droop character of two supercapacitor converters are displayed in Fig. 11. The no-load voltage difference before and after calibration are

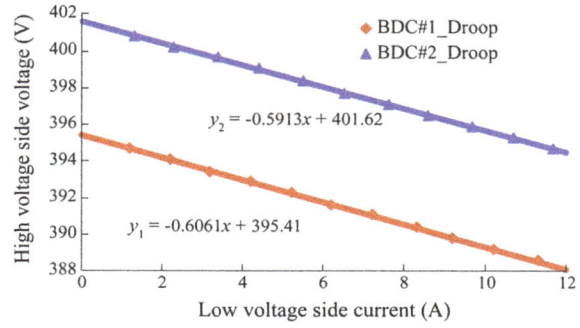

(a) before calibration

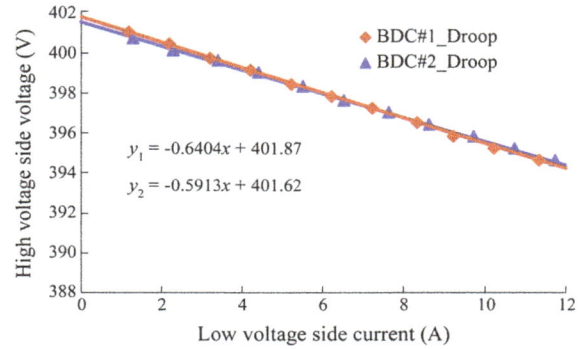

(b) after calibration

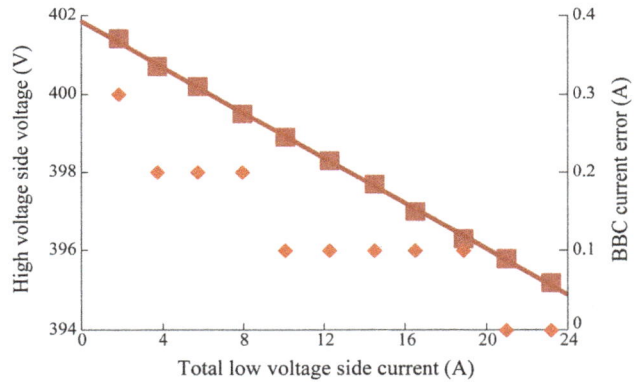

(c) droop curve and current difference

Fig. 11 Current sharing effect

shown in Fig. 11a, b respectively. And the droop curve and current difference are demonstrated in Fig. 11c, which presents a good current sharing result.

The process of SOC balancing and current equalization of batteries is shown in Fig. 12. The initial SOC difference is set to 5% and the measurement of SOC difference is calculated by the Ah-integration as presented in (14). With dual control scheme, the currents of battery units tends to equality and the SOC difference decreases gradually from the initial value 5% to 1.8% in 300 s, which is corresponding to the time constant T_{SOC}.

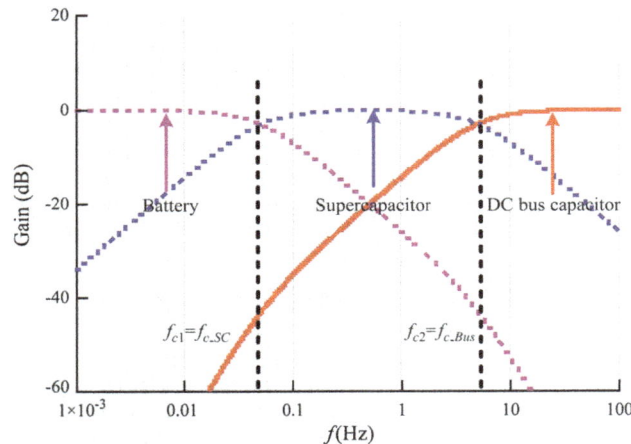

Fig. 10 Response of energy storage to DC bus power fluctuation

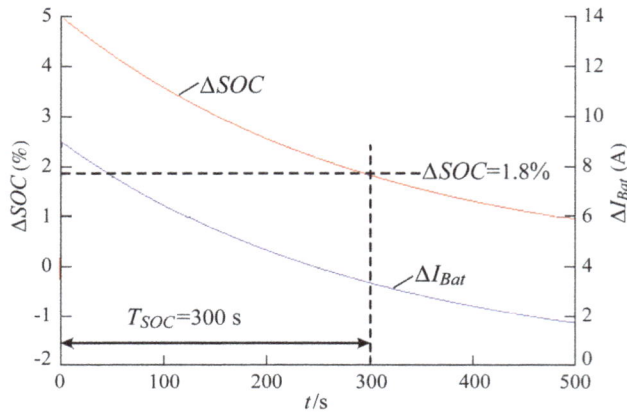

Fig. 12 Results of battery SOC equilibrium and current sharing

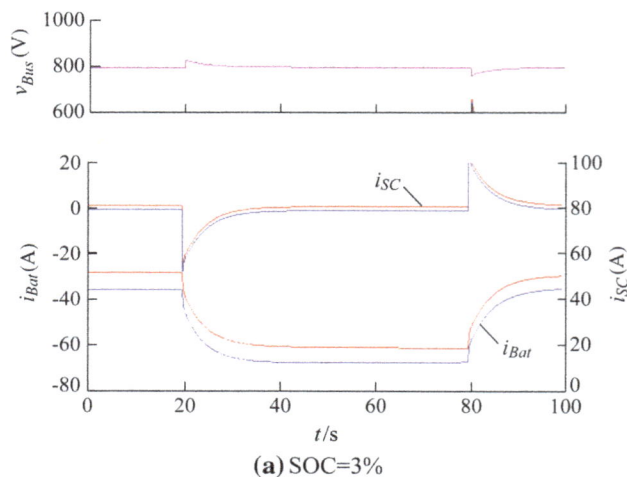

(a) SOC=3%

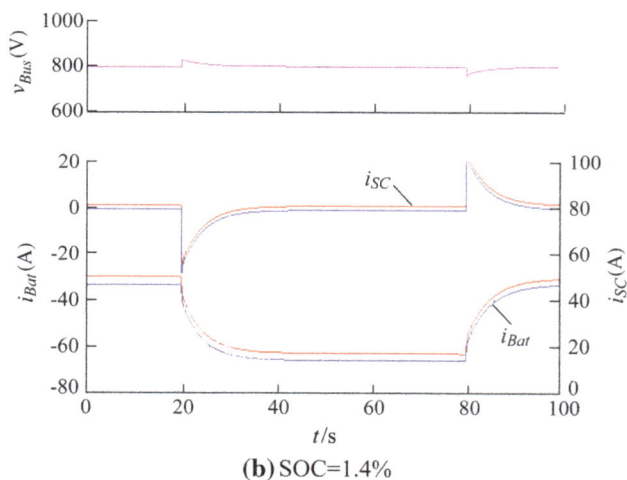

(b) SOC=1.4%

Fig. 13 Dynamic response of supercapacitor and battery current to power fluctuation

Figure 13 shows the dynamic power allocation between supercapacitor and batteries when a 1 kW step load is applied to the system. The supercapacitor responses to the

high frequency power fluctuation while the batteries response to the rest power. The output voltage is well regulated at DC bus voltage reference without steady state error since the supercapacitor does not response to the dc component. Meanwhile, it can be seen that the current difference in the case of $\Delta SOC = 3\%$ is larger than that in the case of $\Delta SOC = 1.4\%$. So the current difference is positively correlated to ΔSOC, which helps to SOC equalization.

5 Conclusion

This paper proposes a dual droop-frequency diving co-ordinated control strategy for HESS in electric vehicle application. The dynamic power allocation is realized through the voltage cascaded control, where the DC bus voltage control loop and supercapacitor voltage control loop set the frequency range for both supercapacitor and battery, make the most use of hybrid energy storage system. On the other aspect, the dual droop control scheme provides not only current sharing between battery units but also SOC equalization without the necessity of communication, which helps to lengthen the battery cycle life and increase the system reliability and flexibility. The experimental results have proven it to be a viable solution for energy management of HESS.

References

[1] Cao J, Emadi A (2012) A new battery/ultracapacitor hybrid energy storage system for electric, hybrid, and plug-in hybrid electric vehicles. IEEE Trans Power Electron 27(1):122–132

[2] Napoli AD, Crescimbini F, Capponi FG et al (2002) Control strategy for multiple input DC-DC power converters devoted to hybrid vehicle propulsion systems. In: Proceedings of the 2002 IEEE international symposium on industrial electronics, L'Aquila, 8–11 July 2002, pp 1036–1041

[3] Zhang SQ, Mishra Y, Ledwich G et al (2013) The operating schedule for battery energy storage companies in electricity market. J Mod Power Syst Clean Energy 1(3):275–284

[4] Venet LP, Guermazi A, Troudi A (2013) Battery/supercapacitors combination in uninterruptible power supply (UPS). IEEE Trans Power Electron 28(4):1509–1522

[5] Zandi M, Payman A, Martin JP et al (2011) Energy management of a fuel cell/supercapacitor/battery power source for electric vehicular applications. IEEE Trans Veh Technol 60(2):433–443

[6] Mendis N, Muttaqi KM, Perera S (2014) Management of battery-supercapacitor hybrid energy storage and synchronous

condenser for isolated operation of PMSG based variable-speed wind turbine generating systems. IEEE Trans Smart Grid 5(2):944–953

[7] Wang L, Li H (2010) Maximum fuel economy-oriented power management design for a fuel cell vehicle using battery and ultracapacitor. IEEE Trans Ind Appl 46(3):1011–1020

[8] Thounthong P, Pierfederici S, Martin JP et al (2010) Modeling and control of fuel cell/supercapacitor hybrid source based on differential flatness control. IEEE Trans Veh Technol 59(6):2700–2710

[9] Camara MB, Dakyo B, Gualous H (2012) Polynomial control method of DC/DC converters for DC-Bus voltage and currents management—battery and supercapacitors. IEEE Trans Power Electron 27(3):1455–1467

[10] Lukic SM, Wirasingha SG, Rodriguez F et al (2006) Power management of an ultra-capacitor/battery hybrid energy storage system in an HEV. In: Proceedings of the 2006 IEEE vehicle power and propulsion conference, Windsor, 6–8 Sep 2006, pp 1–6

[11] Jiang W, Fahimi B (2010) Active current sharing and source management in fuel cell-battery hybrid power system. IEEE Trans Ind Electron 57(2):752–761

[12] Wang GS, Ciobotaru M, Agelidis VG (2013) Power management of hybrid energy storage system for a MW photovoltaic system. In: Proceedings of the 2013 IEEE industrial electronics society Conference, Vienna, 10–13 Nov 2013, pp 6777–6782

[13] Mendis N, Muttaqi KM, Perera S (2014) Management of low- and high-frequency power components in demand-generation fluctuations of a DFIG-based wind-dominated RAPS system using hybrid energy storage. IEEE Trans Ind Appl 50(3):2558–2568

[14] Zhang Y, Jiang ZH (2009) Dynamic power sharing strategy for active hybrid energy storage systems. In: Proceedings of the 2009 IEEE vehicle power and propulsion conference, Dearborn, Michigan, 7–11 Sep 2009, pp 558–563

[15] Garcia FS, Ferreira AA, Pomilio JA (2009) Control strategy for battery-ultracapacitor hybrid energy storage system. In: Proceedings of the 2009 IEEE applied power electronics conference and exposition, Washington, DC, 15–19 Feb, pp 826–832

[16] Zhou HH, Bhattacharya T, Tran D et al (2011) Composite energy storage system involving battery and ultracapacitor with dynamic energy management in microgrid applications. IEEE Trans Power Electron 26(3):923–930

[17] Abbey C, Strunz K, Joós G (2009) A knowledge-based approach for control of two-level energy storage for wind energy systems. IEEE Trans Energy Convers 24(4):539–547

[18] Tani A, Camara MB, Dakyo B (2012) Energy management based on frequency approach for hybrid electric vehicle applications: fuel cell/Lithium-battery and ultracapacitors. IEEE Trans Veh Technol 61(8):3375–3386

[19] Schaltz E, Khaligh A, Rasmussen PO (2009) Influence of battery/ultracapacitor energy-storage sizing on battery lifetime in a fuel cell hybrid electric vehicle. IEEE Trans Veh Technol 58(8):3882–3891

[20] Wang GS, Ciobotaru M, Agelidis VG (2014) Power smoothing of large solar pv plant using hybrid energy storage. IEEE Trans Sustain Energy 5(3):834–842

[21] Yoo H, Sul SK, Park Y et al (2008) System integration and power-flow management for a series hybrid electric vehicle using supercapacitors and batterie. IEEE Trans Appl 44(1):108–114

[22] Dragicevic T, Guerrero JM, Vasquez JC et al (2014) Supervisory control of an adaptive-droop regulated DC microgrid with battery management capability. IEEE Trans Power Electron 29(2):695–706

[23] Chen Z, Ding M, Su JH (2011) Modeling and control for large capacity battery energy storage system. In: Proceeding of the 2011 IEEE electric utility deregulation and restructuring and power technologies, Weihai, 6–9 July 2011, pp 1429–1436

Wuhua LI received the B.Sc. and Ph.D. degree in Applied Power Electronics and Electrical Engineering from Zhejiang University, Hangzhou, China, in 2002 and 2008, respectively. From September 2004 to March 2005, he was a Research Intern, and from January 2007 to June 2008, a Research Assistant in GE Global Research Center, Shanghai, China. From July 2008 to April 2010, he joined the College of Electrical Engineering, Zhejiang University as a Post doctor. In May 2010, he became a faculty member at Zhejiang University as a Lecturer. In December 2010, he was promoted as an Associate Professor. Since December 2013, he has been a Full Professor at Zhejiang University. From July 2010 to September 2011, he was a Ryerson University Postdoctoral Fellow with the Department of Electrical and Computer Engineering, Ryerson University, Toronto, ON, Canada. His research interests include high power devices, advanced power converters and system operation optimization for renewable energy based power systems. Dr. Li has published more than 100 peer-reviewed technical papers and holds over 30 issued/pending patents. Due to his excellent teaching and research contributions, Dr. Li received the 2011 TOP TEN Excellent Young Staff Award and the 2012 Distinguished Young Scholar from Zhejiang University, the 2012 Outstanding Young Researcher Award from Zhejiang Province, the 2012 Delta Young Scholar from Delta Environmental & Educational Foundation and the 2012 National Outstanding Young Scholar. He received three Scientific and Technological Achievements Awards from Zhejiang Provincial Government and the State Educational Ministry of China in 2009 and 2011, respectively.

Chi XU was born in Jiangxi, China in 1990. He received the B.S. degrees from the College of Electrical and Electronic Engineering, Huazhong University of Science and Technology, Wuhan, China, in 2012. He is currently working toward the M.S. degree in the College of Electrical Engineering, Zhejiang University. His research interests include advanced power conversion and control for renewable energy systems.

Hongbin YU was born in Hunan, China, in 1987. He received the B.Sc. and M.Sc. degrees from the College of Electrical Engineering, Zhejiang University, Hangzhou, China, in 2010 and 2013. His current research interests include bidirectional dc–dc converter and energy storage system.

Yunjie GU was born in Hebei, China, in 1987. He received the B.Sc. degree from the Department of Electrical Engineering, Zhejiang University, Hangzhou, China, in 2010, where he is currently working toward the Ph.D. degree in electrical engineering. From May 2011 to January 2012, he was a Research Intern at GE Global Research Center, Shanghai, China. From July to September 2013, he was a Student Visitor at Newcastle University, Newcastle upon Tyne, U.K. His research interests include control and networking of power conversion systems.

Xiangning HE received the B.Sc. and M.Sc. degrees from the Nanjing University of Aeronautical and Astronautical, Nanjing, China, in 1982 and 1985, respectively, and the Ph.D. degree from Zhejiang University, Hangzhou, China, in 1989. From 1985 to 1986, he was an Assistant Engineer at the 608 Institute of Aeronautical Industrial General Company, Zhuzhou, China. From 1989 to 1991, he was a Lecturer at Zhejiang University. In 1991, he obtained a Fellowship from the Royal Society of U.K., and conducted research in the Department of Computing and Electrical Engineering, Heriot-

Watt University, Edinburgh, U.K., as a Postdoctoral Research Fellow for two years. In 1994, he joined Zhejiang University as an Associate Professor. Since 1996, he has been a Full Professor in the College of Electrical Engineering, Zhejiang University. He was the Director of the Power Electronics Research Institute and the Head of the Department of Applied Electronics, and he is currently the Vice Dean of the College of Electrical Engineering, Zhejiang University. His research interests are power electronics and their industrial applications. He is the author or coauthor of more than 280 papers and one book "Theory and Applications of Multi-level Converters." He holds 22 patents. Dr. He received the 1989 Excellent Ph.D. Graduate Award, the 1995 Elite Prize Excellence Award, the 1996 Outstanding Young Staff Member Award and 2006 Excellent Staff Award from Zhejiang University for his teaching and research contributions. He received seven Scientific and Technological Achievements Awards from Zhejiang Provincial Government and the State Educational Ministry of China in 1998, 2002, 2009 and 2011 respectively, and six Excellent Paper Awards. He is a Fellow of the IEEE and has been appointed as the IEEE Distinguished Lecturer by the IEEE Power Electronics Society, in 2011. He is also a Fellow of the Institution of Engineering and Technology (formerly IEE), U.K.

17

Centralized control of parallel connected power conditioning system in electric vehicle charge–discharge and storage integration station

Jiuqing CAI, Changsong CHEN (✉),
Peng LIU, Shanxu DUAN

Abstract This study presents a centralized control scheme that coordinates parallel operations of power conditioning system (PCS) for the grid interactions of electric vehicles (EVs) in EV charge–discharge and storage integration station. Key issues for the control and operation of PCS under various operation modes are discussed, including vehicle to grid (V2G) mode, stand-alone mode and seamless transfer mode. The intelligent multi-mode charge–discharge method is utilized for the V2G mode, and the parallel control method based on communication network is adopted for the stand-alone mode. In addition, a novel seamless transfer strategy is proposed, which is able to implement PCS transition between V2G mode and stand-alone mode. The detailed process of the seamless transfer between the two modes is illustrated. Experimental results are presented to show the performance and feasibility of this strategy.

Keywords Centralized control, Power conditioning system (PCS), Charge–discharge and storage integration station, Vehicle to grid (V2G), Seamless transfer

1 Introduction

The needs to reduce pollutant gas emissions and the increasing energy consumption have led to an increase of the electric vehicles (EVs) and renewable energy generation [1–4]. Large-scale utilization of EVs has the potential to reduce greenhouse gases emission, save fuel cost for EV drivers, enhance power system security, and increase penetration of renewable energy [5–7]. The development of the microgrid concept endows distribution networks with increased reliability and resilience and offers an adequate management and control solution for massive deployment of renewable energy generation and EVs [8–10]. EVs are considered as both a new type of load and flexible generation resources with vehicle-to-grid (V2G) technology.

The charge–discharge and storage integration station, consisting of bi-directional converters and hierarchical control structure, is able to realize bidirectional power flow between EVs and power grid. Many research works on designing the topologies and controllers of bi-directional power electronic converters for EV application, which are able to function as battery charger and to transfer electrical energy between battery pack and the grid [11–13]. In addition, the centralized control system for parallel operation of the converters during grid-connected and stand-alone operations has been conducted [14–16]. The issue of seamless transition between the V2G mode and stand-alone mode is discussed widely. A phase locked loop (PLL)-based seamless transfer control method between grid-connected and islanding modes is applied in a three-phase grid-connected inverter [17]. The performance of the transfer process is highly dependent on the characterization of PLL. In [18], a control strategy based on the frequency and magnitude droop control is used for the distributed generation (DG) to achieve a seamless transfer between grid-tied mode and islanding mode. However, both the magnitude and the frequency of the output voltage are varied due to the droop operation. In [19], a transfer strategy based on indirect current control is proposed for the three-phase inverter in the DG. However, it is difficult

J. CAI, C. CHEN, P. LIU, S. DUAN, State Key Laboratory of Advanced Electromagnetic Engineering and Technology, School of Electrical and Electronic Engineering, Huazhong University of Science and Technology, Wuhan, China
(✉) e-mail: ccsfm@163.com

to realize the current limiting control because the inductance current is not controlled in both operation modes. In [20], a transfer methodology is presented for the three-phase grid-tied inverter without static transfer switch. In this case, because the instantaneous grid current is not introduced in the control algorithm, the harmonic performance of the grid current may not be good. Compared with the existing strategies for the grid-connected inverters [17–20], the contribution of this paper on the mode transfer strategy can be concluded into two points. First, an improved transfer control strategy based on indirect current control is proposed according to the model of three-phase (power conditioning system) PCS in the synchronous reference frame. In the seamless transfer mode, double closed-loop control technology is applied, the outer loops track instructions mutually while the inner loop remains the same. Second, particular issues of bidirectional converter are considered, including the charging state and discharging state during seamless transfer between V2G mode and stand-alone mode.

In this paper, a flexible and efficient centralized control scheme is developed for the parallel connected PCS in EV charge–discharge and storage integration station. The system configuration and theoretical analysis of three operation modes principles are described. Moreover, a novel seamless transfer method of PCS between V2G and stand-alone modes is presented. Finally, the control scheme has been verified on a 1 MW parallel connected PCS prototype.

2 System configuration

Figure 1 shows the infrastructure of EV charge–discharge and storage integration station with parallel

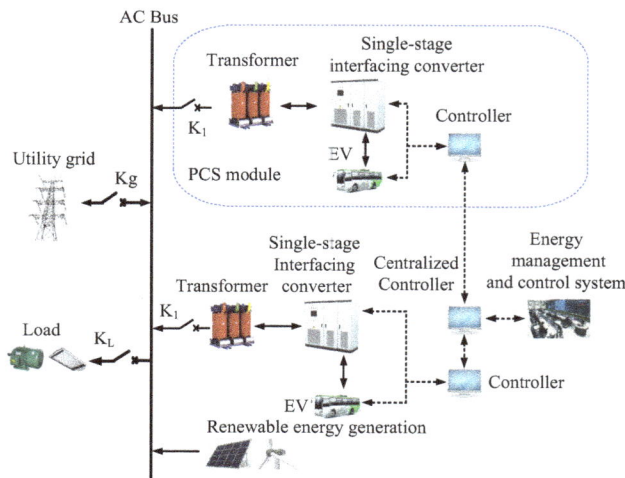

Fig. 1 Infrastructure of EV charge–discharge and storage integration station

connected PCS adopting centralized control architecture. In the V2G mode, integration station is connected to the utility grid. PCS can achieve several major functions: battery charger, active power regulation and reactive power compensation. In the stand-alone mode, integration station is separated from the utility grid. PCS functions as an uninterruptible power supply (UPS) to maintain the output voltage of the integration station. The PCS should transfer between the two modes in order to provide electrical power to the critical load during utility grid interruptions.

The topology of parallel connected PCS for EV charge–discharge and storage integration station is shown in Fig. 2, where V_{bat} is the battery voltage, L_{dc} and C_{dc} are the dc-link filter inductor and capacitor, L_1, L_2 and L are the ac filter inductors, and C is the ac filter capacitor. The dc power acquired by the battery packs is converted to three-phase ac power through two parallel connected IGBT full-bridge with a LCL filter. Compared with the L or LC filter, LCL filter is more suitable in high-power low-switching-frequency grid connected inverter applications due to its better performance on inhibiting grid current harmonics. Power transformer is selected to implement electrical isolation and voltage matching between battery packs and utility grid.

A three-level architecture is designed where complex control tasks are decomposed into simpler and manageable ones. The architecture consists of three levels, namely, level 1 converter control, level 2 centralized control and level 3 energy management system (EMS). The EMS is the highest level which ensures power balance within the EV charge–discharge and storage integration station. The centralized system as the middle level is responsible for coordinating the parallel operation of PCSs. The centralized control center uploads parallel system operation status and obtains the dispatch order from EMS through serial communication interface. The converter controller is the lowest level which handles the primary control of PCS. The converter controller sends the voltage–current instruction and receives the synchronization signal and operation mode instructions from centralized control center by optical fiber cable and controller area network (CAN) respectively.

3 Operational modes and control strategy

3.1 V2G mode

As shown in Fig. 3 the double-loop control uses the grid current loop to generate the reference for the inverter current loop under dq synchronous rotating coordinate. During the discharging process of Li-ion battery packs, PCS modules work as grid-connected inverters, the

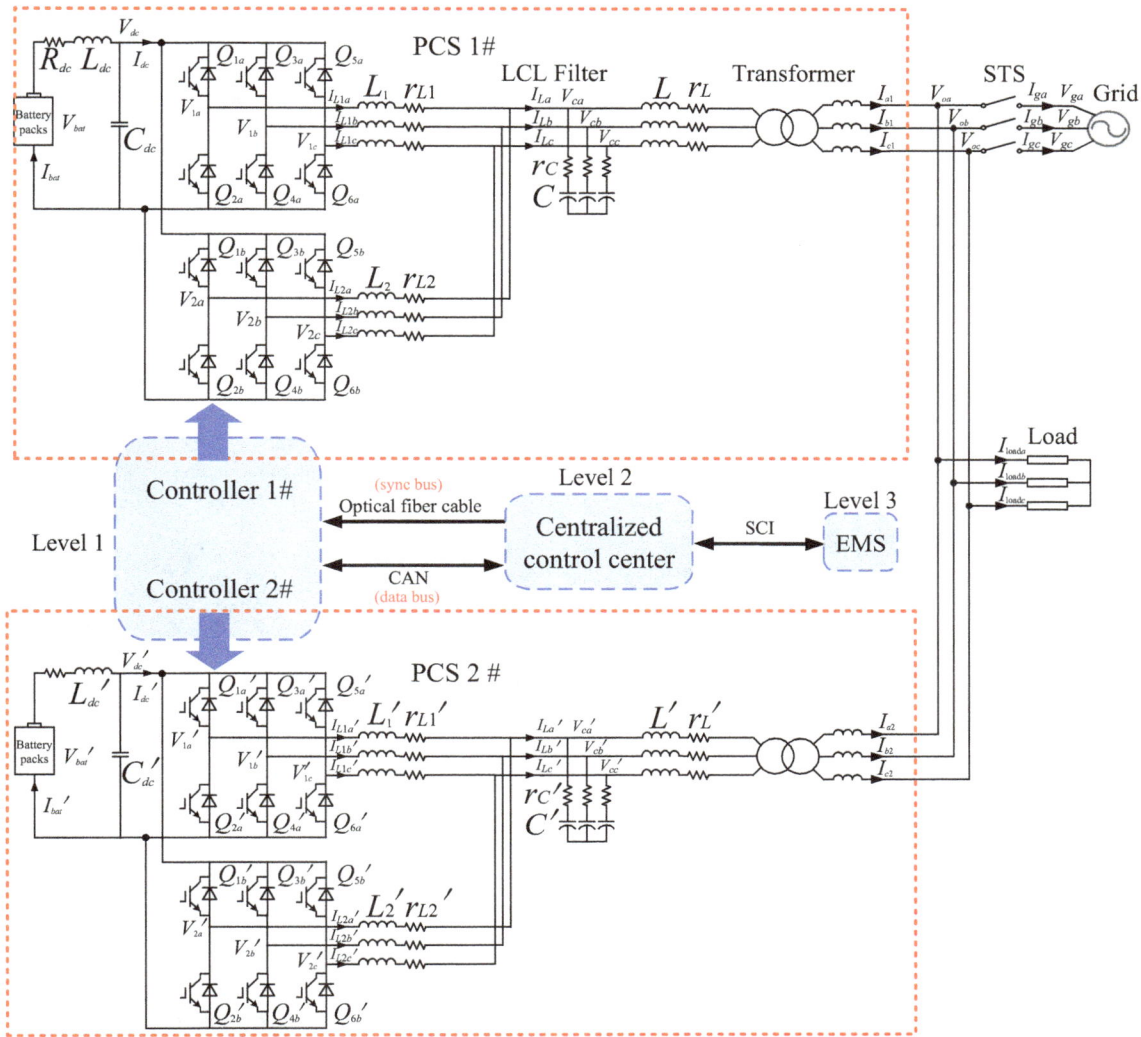

Fig. 2 Topology of parallel connected PCS

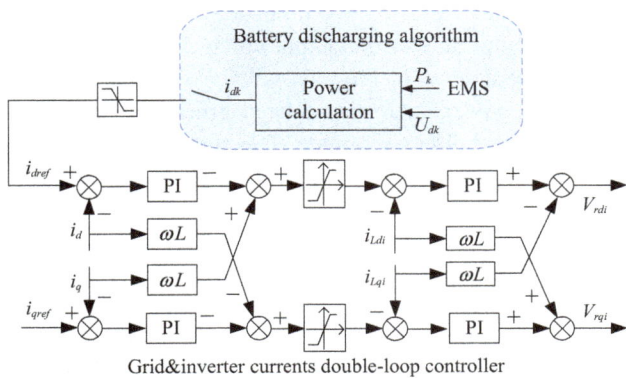

Fig. 3 Controller block of discharging in V2G mode

discharging power is determined by the EMS. The bi-directional converter will control the active–reactive power and insure the output current harmonics to be low.

During the charging process of Li-ion battery packs, PCS modules work as grid-connected rectifiers. The bidirectional converter will control the power factor to unity and insure the input current harmonics to be low. In addition to the double-loop controller above, external controllers are designed to regulate the battery current and battery voltage respectively as shown in Fig. 4. Considering the battery state and application requirement under various conditions, the battery charging algorithm should be flexibly selected.

Constant current–constant voltage (CC–CV) charging method is adopted as conventional charging method. Under the arrangement of CC–CV charging algorithm, a constant current are applied to charge the battery till the transition time from CC to CV determined by terminal voltage of the battery. Then constant voltage is held after reaching the terminal voltage and the charging current will reduce automatically. Finally, the battery packs are fully charged.

This charge strategy can effectively increase the battery life cycle and avoid overcharge.

However, a faster and more efficient charging algorithm is required. The pulse charge with constant voltage (CV-PC) charging method is adopted as advanced charging method. The basic idea of the CV-PC is to adjust the duty cycle of the pulse within a certain range and obverse the response of the charging current. This charge strategy can really retard the polarization and reduce the battery-charging time.

3.2 Stand-alone mode

The block diagram of double-loop control in the stand-alone mode is described in Fig. 5.

The capacitor voltage loop generates the reference for the inverter current loop under dq synchronous rotating coordinate. PCS module works as voltage source converter, the output voltage should keep strictly sinusoidal. When PCS modules are connected in parallel, circulating currents will inevitably occur due to the asynchronous switching process and module parameter difference. In order to effectively solve the impact of circulating current and to achieve superior accuracy of current sharing, the power-sharing controller is designed besides the double-loop controller.

Due to the voltage source nature, each PCS module has to be strictly consistent in output voltage amplitude, frequency and phase to suppress the circulating current. Synchronization is essential to achieve reliable parallel operation, which can be solved by a synchronization bus through the optical fiber cable. Rapid transmission rate of optical fiber can ensure the minimal synchronous error. The introduced power sharing strategy depends on the active-reactive power and output impedance of PCS module. The data exchange between centralized controller and converter

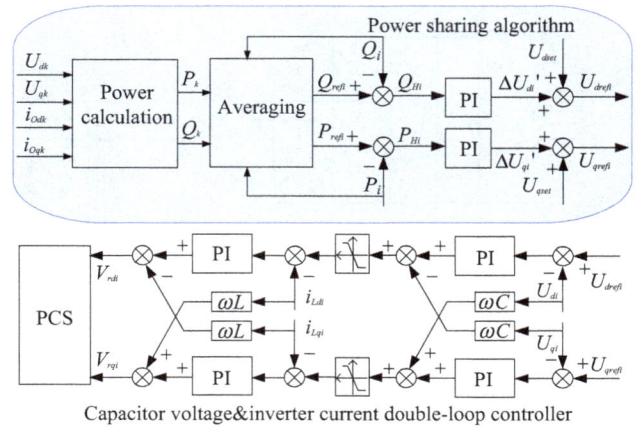

Fig. 5 Control strategy of stand-alone mode

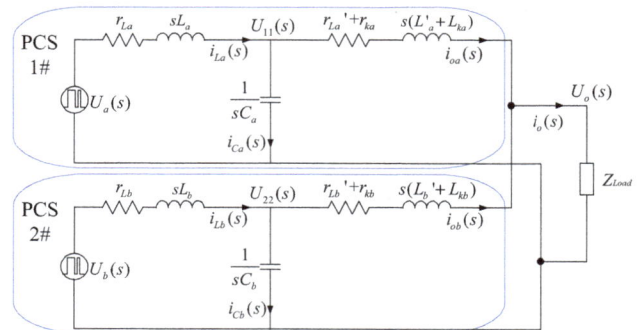

Fig. 6 Equivalent circuit of paralleled PCS in stand-alone mode

controllers are achieved by CAN, including active-reactive power, battery status and control instructions. The module output impedance is highly inductive because of the transformer leakage inductance and inductance on the grid side of LCL filter. Therefore, the active power flow and reactive power flow are mainly influenced by the phase angle and the amplitude of the output voltage respectively. The decoupling control of d/q-axis voltage reference and active-reactive power error are realized effectively under *dq* synchronous rotating coordinate. The equivalent circuit of parallel connected PCS in the stand-alone mode is shown in Fig. 6.

3.3 Seamless transfer mode

Figure 7 shows the proposed control block diagram for V2G and stand-alone operations of PCS module. When PCS transfers between V2G mode and stand-alone mode, the outer loop simultaneously changes from the grid current control mode for V2G operation to the capacitor voltage control mode for stand-alone operation. The voltage–current double-loop control in the stand-alone mode is a conventional strategy, widely used in three-phase voltage

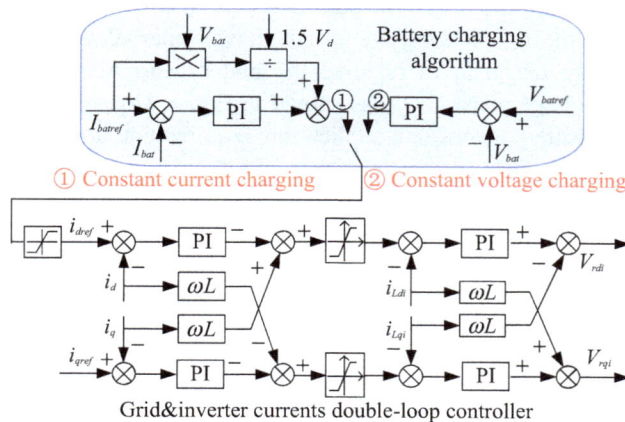

Fig. 4 Controller block of charging in V2G mode

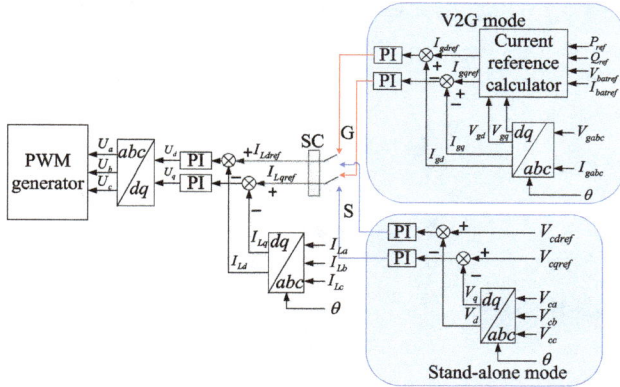

Fig. 7 Novel seamless transfer control strategy

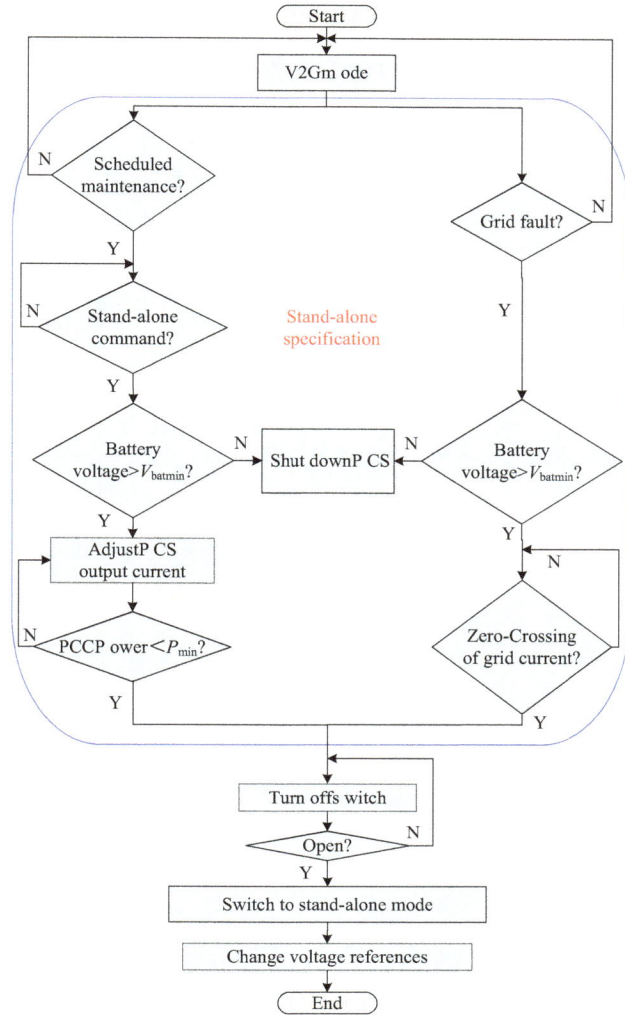

Fig. 8 Transfer sequence from V2G to stand-alone mode

source inverter. The parameter design and stability analysis of current double-loop control in the V2G mode is discussed in [21, 22]. During the process of transfer, the outer loops track current instruction mutually while the inner loop for inverter current control remains the same. In this case, it can be regarded as change in control instructions rather than change in controllers, the stability in both the stand-alone mode and the V2G mode guarantees the stable transition period between stand-alone mode and V2G mode.

A detailed sequence for the seamless transfer from V2G to stand-alone operation is summarized as follows. The process of the seamless transfer from V2G to stand-alone mode is illustrated in Fig. 8.

1) Judge a fault or scheduled maintenance on the grid.
2) Preparing for mode transfer adhere to stand-alone specification.
3) The static transfer switch (STS) is tripped in order to disconnect the PCS from the utility grid.
4) The control switch S_c is connected to 'S' from 'G', PCS changes its control from current control mode to voltage control mode.
5) Gradually change voltage references V_{cdref} and V_{cqref} to the desired values. The initial value of the V_{cdref} and V_{cqref} is determined by grid voltage V_{gd} and V_{gq}, which is calculated from the measured three-phase grid voltages at the transfer point.

In the meanwhile, a detailed sequence for the seamless transfer from stand-alone to V2G operation is summarized as follows. The process of the seamless transfer from stand-alone to V2G mode is illustrated in Fig. 9.

1) Detect that the grid voltage is within the normal operating range.
2) Preparing for mode transfer adhere to V2G specification.

3) When the frequency, phase and magnitude of the PCS output voltage match the grid voltage, the STS is closed. The PCS is connected to the utility grid afterwards.
4) The control switch S_c is connected to 'G' from 'S', PCS changes its control from voltage control mode to current control mode.
5) Gradually change the current reference I_{gdref} and I_{gqref} to the desired values.

4 Experimental results

An experimental device of parallel-connected PCS has been built to verify the proposed control method with the parameters and prototype shown in Table 1 and Fig. 10 respectively.

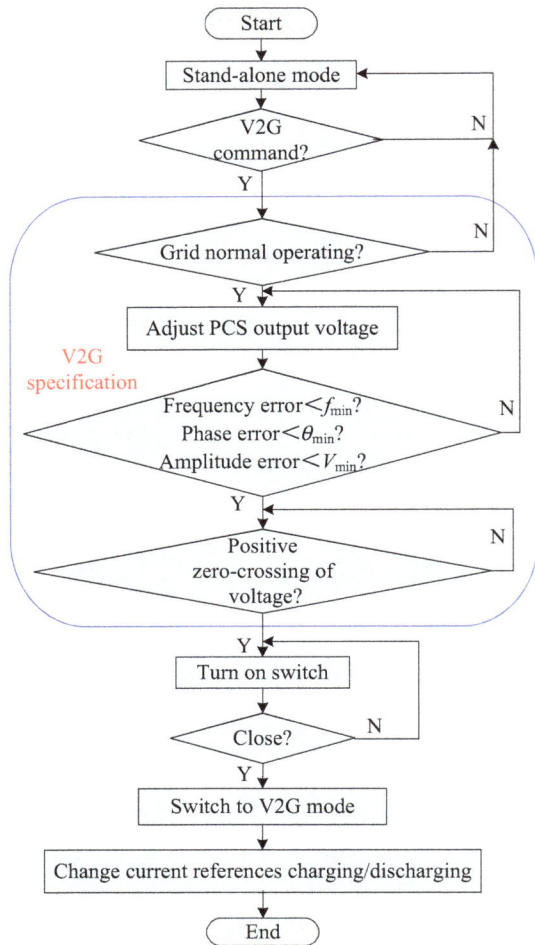

Fig. 9 Transfer sequence from stand-alone to V2G mode

Table 1 Experiment specification of PCS module

Parameter	Symbol	Value
Rated power	S	500 kVA
Grid voltage	Vg	380 V/50 Hz
Switching frequency	Fs	3 kHz
△-Y Transformer ratio	N	315:380
DC filter inductance	L_{dc}	0.17 mH
DC bus capacitor	C_{dc}	22 mF
Inverter-side inductance	L_1/L_2	0.36 mH
Grid-side inductance	L	0.05 mH
AC filter capacitor	C	200 μF

Figure 11 shows the experimental waveforms of both charging and discharging process in the V2G mode. The THD of grid-side current fulfills the grid standard, and DC bus voltage ripple is limited.

Fig. 10 Experimental device of parallel-connected PCSs

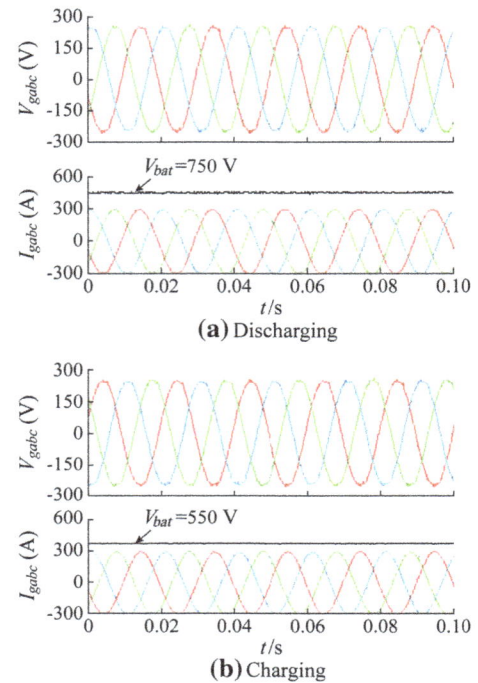

(a) Discharging

(b) Charging

Fig. 11 Experimental waveforms of charging and discharging process in the V2G mode

Figure 12 shows the experimental waveforms of load current sharing in the stand-alone mode. It is clearly that the steady-state and dynamic performance of parallel connected PCS is excellent during load variation.

Figure 13 shows the experimental waveforms of the seamless transfer process from V2G mode to stand-alone mode. Figure 14 shows the experimental waveforms of the seamless transfer process from stand-alone mode to V2G mode when grid fault occurs. The proposed transfer strategy is capable of providing the critical loads with a stable and seamless sinusoidal voltage during the whole transition period.

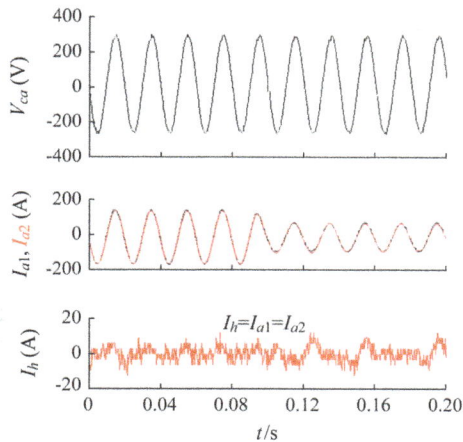

Fig. 12 Experimental waveforms of load sharing in the stand-alone mode

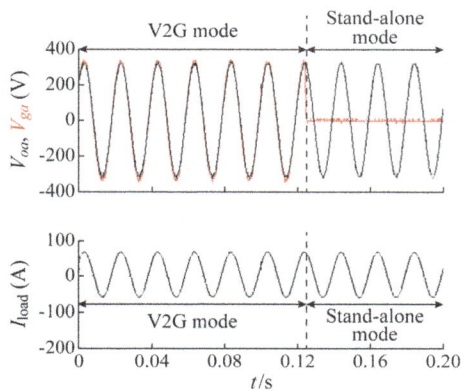

Fig. 13 Experimental waveforms of the proposed control showing a mode transfer from V2G mode to stand-alone mode

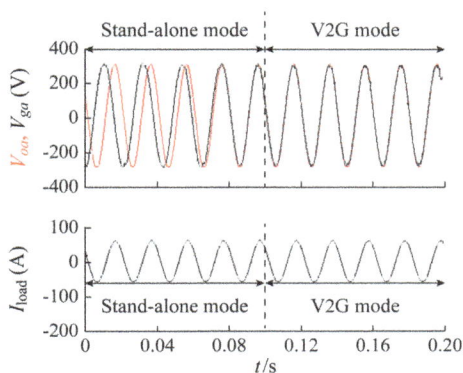

Fig. 14 Experimental waveforms of the proposed control showing a mode transfer from stand-alone mode to V2G mode

5 Conclusion

In this paper, a centralized control strategy for parallel connected PCS in EV Charge–discharge and storage

integration station is designed. The PCS infrastructure and operational principles are illustrated, controllers are analyzed in both V2G mode and stand-alone mode. A novel seamless transfer strategy and detailed sequence have been proposed and utilized to achieve better performance between V2G mode and stand-alone mode. The effectiveness of the proposed control strategies have been validated with comprehensive experimental results under various operations.

Acknowledgment This work was supported in part by the National Natural science Foundation of China under Grant 51361130150 and Grant 51477067, in part by the Fundamental Research Funds for the Central Universities under Grant 2014QN219.

References

[1] Luo ZW, Song YH, Hu ZC et al (2011) Forecasting charging load of plug-in electric vehicles in China. In: Proceedings of the 2011 IEEE power and energy society general meeting, San Diego, CA, USA, 24–29 Jul 2011, 8 pp

[2] Madawala UK, Thrimawithana DJ (2011) A bidirectional inductive power interface for electric vehicles in V2G systems. IEEE Trans Ind Electron 58(10):4789–4796

[3] Khayyam H, Ranjbarzadeh H, Marano V (2012) Intelligent control of vehicle to grid power. J Power Sources 201:1–9

[4] Chen CS, Duan SX, Cai T et al (2011) Optimal allocation and economic analysis of energy storage system in microgrids. IEEE Trans Power Electron 26(10):2762–2773

[5] Ortega-Vazquez MA, Bouffard F, Silva V (2012) Electric vehicle aggregator/system operator coordination for charging scheduling and services procurement. IEEE Trans Power Syst 28(2):1806–1815

[6] Zheng Y, Dong ZY, Xu Y et al (2014) Electric vehicle battery charging/swap stations in distribution systems: comparison study and optimal planning. IEEE Trans Power Syst 29(1): 221–229

[7] Boulanger AG, Chu AC, Maxx S et al (2011) Vehicle electrification: status and issues. Proc IEEE 99(6):1116–1138

[8] Gouveia C, Leal Moreira C, Peças Lopes JA et al (2013) Microgrid service restoration: the role of plugged-in electric vehicles. IEEE Ind Electron Mag 7(4):26–41

[9] Lopez MA, Martin S, Aguado JA et al. (2011) Optimal microgrid operation with electric vehicles. In: Proceedings of the 2nd IEEE PES international conference and exhibition on innovative smart grid technologies (ISGT Europe'11), Manchester, UK, 5–7 Dec 2011, 8 pp

[10] Chen CS, Duan SX (2014) Optimal integration of plug-in hybrid electric vehicles in microgrids. IEEE Trans Ind Inform 10(3):1917–1926

[11] Lee YJ, Emadi A (2014) Integrated bi-directional AC/DC and DC/DC converter for plug-in hybrid electric vehicle conversion.

In: Proceedings of the 2007 IEEE vehicle power and propulsion conference (VPPC'07), Arlington, TX, USA, 9–12 Sept 2007, pp 215–222

[12] Onar O, Kobayashi J, Khaligh A (2011) A multi-level grid interactive bi-directional AC/DC-DC/AC converter and a hybrid battery/ultra-capacitor energy storage system with integrated magnetics for plug-in hybrid electric vehicles. In: Proceedings of the 26th IEEE annual applied power electronics conference and exposition (APEC'11), Fort Worth, TX, USA, 6–11 Mar 2011, pp 829–835

[13] Zhou XH, Lukic S, Bhattacharya S et al (2009) Design and control of grid-connected converter in bi-directional battery charger for plug-in hybrid electric vehicle application. In: Proceedings of the 2009 IEEE vehicle power and propulsion conference (VPPC'09), Dearborn, MI, USA, 7–10 Sept 2009, pp 1716–1721

[14] Tan KT, Peng XY, So PL et al (2012) Centralized control for parallel operation of distributed generation inverters in microgrids. IEEE Trans Smart Grid 3(4):1977–1987

[15] Guerrero JM, Hang LJ, Uceda J (2008) Control of distributed uninterruptible power supply systems. IEEE Trans Ind Electron 55(8):2845–2859

[16] Liu BY, Duan SX, Cai T (2011) Photovoltaic DC-building-module-based BIPV system: concept and design considerations. IEEE Trans Power Electron 26(5):1418–1429

[17] Tran TV, Chun TW, Lee HH et al (2014) PLL-based seamless transfer control between grid-connected and islanding modes in grid-connected inverters. IEEE Trans Power Electron 29(10):5218–5228

[18] Guo ZQ, Sha DS, Liao XZ (2014) Voltage magnitude and frequency control of three-phase voltage source inverter for seamless transfer. IET Power Electron 7(1):200–208

[19] Liu Z, Liu JJ (2014) Indirect current control based seamless transfer of three-phase inverter in distributed generation. IEEE Trans Power Electron 29(7):3368–3383

[20] Ochs DS, Mirafzal B, Sotoodeh P (2014) A method of seamless transitions between grid-tied and stand-alone modes of operation for utility-interactive three-phase inverters. IEEE Trans Ind Appl 50(3):1934–1941

[21] Azani H, Massoud A, Benbrahim L et al (2014) An LCL filter-based grid-interfaced three-phase voltage source inverter: performance evaluation and stability analysis. In: Proceedings of the 7th IET international conference on power electronics, machines and drives (PEMD'14), Manchester, UK, 8–10 Apr 2014, 6 pp

[22] Hu XF, Wei Z, Chen YH et al (2012) A control strategy for grid-connected inverters with LCL filters. Proc CSEE 32(27):142–148 (in Chinese)

Jiuqing CAI received the B.S. degree in electrical engineering and automation from Huazhong University of Science and Technology, Wuhan, China in 2011. He is currently working for Ph.D. degree in School of Electrical and Electronics Engineering, Huazhong University of Science and Technology. His research interests include power conditioning system, UPS and renewable energy applications.

Changsong CHEN received the Ph.D. degrees in electrical engineering from Huazhong University of Science and Technology, Wuhan, China, in 2011. He was a Postdoctoral Research Fellow with the Department of Control Science and Engineering, Huazhong University of Science and Technology, from 2011 to 2013. He is currently a faculty member in the School of Electrical and Electronics Engineering, Huazhong University of Science and Technology. His current research interests include renewable energy applications, microgrid and power electronics applied to electric vehicles.

Peng LIU received the B.S. degree in electrical engineering from Huazhong University of Science and Technology, Wuhan, China in 2013.He is currently working for Ph.D. degree in School of Electrical and Electronics Engineering, Huazhong University of Science and Technology. His research interests include power conditioning system, UPS and modular DC–DC applications.

Shanxu DUAN received the B.Eng., M. Eng., and Ph.D. degrees in electrical engineering from Huazhong University of Science and Technology, Wuhan, China, in 1991, 1994, and 1999, respectively. Since 1991, he has been a Faculty Member in the College of Electrical and Electronics Engineering, Huazhong University of Science and Technology, where he is currently a Professor. His research interests include stabilization, nonlinear control with application to power electronic circuits and systems, fully digitalized control techniques for power electronics apparatus and systems, and optimal control theory and corresponding application techniques for high-frequency pulsewidth-modulation power converters. Dr. Duan is a Senior Member of the Chinese Society of Electrical Engineering and a Council Member of the Chinese Power Electronics Society. He was selected as one of the New Century Excellent Talents by the Ministry of Education of China in 2007. He was also the recipient of the honor of "Delta Scholar" in 2009.

18

A novel genetic algorithm based on all spanning trees of undirected graph for distribution network reconfiguration

Jian ZHANG (✉), Xiaodong YUAN,
Yubo YUAN

Abstract Network reconfiguration is of theoretical and practical significance to guarantee safe and economical operation of distribution system. In this paper, based on all spanning trees of undirected graph, a novel genetic algorithm for electric distribution network reconfiguration is proposed. Above all, all spanning trees of simplified graph of distribution network are found. Tie branches are obtained with spanning tree subtracted from simplified graph. There is one and only one switch open on each tie branch. Decimal identity number of open switch on each tie branch is taken as the optimization variable. Therefore, the length of chromosome is very short. Each spanning tree corresponds to one subpopulation. Gene operations of each subpopulation are implemented with parallel computing method. Individuals of offspring after gene operation automatically meet with radial and connected constraints for distribution network operation. Disadvantages of conventional genetic algorithm for network reconfiguration that a large amount of unfeasible solutions are created after crossover and mutation, which result in very low searching efficiency, are completely overcome. High calculation speed and superior capability of the proposed method are validated by two test cases.

Keywords Network reconfiguration, Genetic algorithm, Paralleling computing, All spanning trees of undirected graph, Decimal coding, Distribution network

1 Introduction

In order to improve the reliability of power supply, urban distribution network is designed to be a looped net structure. For the sake of reducing short circuit current and facilitating protection relay settings, distribution network takes open loop operation mode. On distribution network, there are a large number of sectionalizing switches and a small amount of tie switches. By adjusting switches state, distribution network reconfiguration can reduce power loss, isolate fault, balance load, and improve voltage. At present, distribution automation demonstration projects is in full swing in many large and medium sized cities in China. Distribution automation system (DAS) can manually, interactively or automatically adjust switches state which lay a solid foundation for the application of distribution network reconfiguration.

Network reconfiguration is a large scale, nonlinear, mixed integer programming problem, mainly including branch-exchange method [1–3], optimal flow method [4, 5], genetic algorithm [6–15], heuristic method [16–18], mixed method [19], etc. Because genetic algorithm has the merits such as independency of initial values, good robustness, capabilities of finding global solutions, it attracts many scholars' attention.

In the early genetic algorithm for network reconfiguration, binary coding scheme is widely used, in which, each gene of chromosome corresponds to one switch. If gene is 0, then it indicates switch open. While if gene is 1, it indicates switch closed. This coding scheme is easy to implement and understand. However, a large amount of switches that cannot be opened in practice participate in coding. Therefore, the chromosome is too long, and in the process of crossover and mutation, too much unfeasible solutions are brought about, which result in extremely low search efficiency.

J. ZHANG, X. YUAN, Y. YUAN, Jiangsu Electric Power Company Research Institute, Nanjing 211103, China
(✉) e-mail: z_jj1219@sina.com

In [7], an improved genetic algorithm for network configuration is proposed, in which, only the switches located on a loop of distribution network participate in binary coding, unfeasible solutions after crossover and mutation are ameliorated. Nevertheless, the ameliorating process is time consuming. In [8], coding genes of switches located on the same loop are in one block, and coding genes of adjacent switches adjoin are proposed. But in this method, at each generation, crossover operation can only be implemented on one gene block, and mutation operation can only be done on one gene. Furthermore, in the process of mutation, unfeasible solutions are brought about. Consequently, the search efficiency of this method is comparatively low. In [9], a decimal coding scheme that the identity number of open switch located on loop is coded with an integer is proposed. In this method, if there is not any one switch located on more than one loops, then no unfeasible solutions are produced in the process of gene operation. Unfortunately, if there is switch located on more than one loops, many unfeasible solutions are produced.

In this paper, a novel genetic algorithm based on all spanning trees of undirected graph for distribution network reconfiguration is proposed. Above all, in the graph of distribution network, branches not located on any loop are removed, and adjacent branches whose crossing points have a degree of two is incorporated into one edge. Thus, an undirected simplified graph of distribution network is obtained. All spanning trees of this simplified graph are found with a program using backtracking and a method for detecting bridges based on depth-first search. Tie branches are obtained with spanning tree subtracted from the simplified graph. There is one and only one switch open on each tie branch. Once the switch on each tie branch is open, all other switch must be closed so as to meet the radial and connected constraints for distribution network operation. Which tie branches and which switch on each tie branch is chosen to be open are determined by genetic algorithm.

In this paper, a decimal coding scheme that the whole switch number of each tie branch constitutes the base vectors, and identity number of open switch on each tie branch taken as the optimization variables are proposed. With such a coding scheme, the length of chromosome is equal to the number of independent meshes on the graph of distribution network. Therefore, it is very short. In the genetic algorithm, one sub-population corresponds to one spanning tree. Selection, crossover, mutation, reinsertion operations are all taken in sub-population with paralleling computing method. Individuals of offspring after gene operation automatically meet radial and connected constraints for distribution network operation. None unfeasible solutions are produced and none ameliorating operation is needed. The disadvantages of conventional genetic algorithm for distribution network reconfiguration that a huge number of unfeasible solutions

which result in very low search efficiency brought in the process of crossover and mutation operation are completely overcome. When the specified number of evolution generation is reached, the chromosome whose objective function value is minimal in the whole population is output as the global optimal solution. Two simulation cases indicate the proposed algorithm greatly improve calculation speed.

2 Model of network reconfiguration

The objective is that power loss is minimal after network reconfiguration under necessary constraints for distribution network operation. The objective function is formulated as

$$\min f = \sum_{i=1}^{n} \frac{P_i^2 + Q_i^2}{U_i^2} r_i k_i \tag{1}$$

where P_i, Q_i are active and reactive power flowing through the terminal of branch i, respectively; U_i is the terminal node voltage of branch i; n is the total number of branches; r_i is the resistance of branch i; k_i is the switch state of branch i, which is a 0 or 1 discrete variable, with 0 indicating switch open, and 1 indicating switch closed; f is the active power loss of network, which can be obtained by power flow calculation.

Node voltage and branch power must meet the following constrains:

1) Node voltage constraint

$$U_{i\min} \leq U_i \leq U_{i\max}$$

where $U_{i\min}$, $U_{i\max}$ are the upper and lower voltage constraints for node i, respectively.

2) Branch power constraint

$$S_i \leq S_{i\max}$$

where S_i, $S_{i\max}$ are the calculated and allowed maximal power value flowing through branch i respectively.

When node voltage or branch power exceeds constraint, a penalty function will be taken into account in f. The more is the constraints being exceeded, the more is penalty function.

3 All spanning trees of simplified graph and relevant concepts

3.1 All spanning trees and tie branches of simplified graph

In order to decrease the number of spanning trees and reduce calculation complexity, adjacent branches located on a loop whose crossing point have a degree of 2 are incorporated into one edge, and branches not located on any loop can be removed from the graph of distribution network. Thus, the graph of distribution network can be simplified to a graph G. Spanning tree of G is a sub-graph of G, in which, any two nodes have one and only one simple path. Spanning tree of G has all nodes of G, but do not have all edges of G. Different spanning trees are composed of different edges. Tie branches are supplementary set of spanning tree. That is, tie branches equal to spanning tree subtracted from G. The number of edges in tie branches equals to the number of independent mesh in G. In this paper, we introduce the method in [20] to find all the spanning trees and corresponding tie branches of G.

Fig. 1 is IEEE typical three feeder test system, in which, dotted line are the branches on which tie switches located. In order to facilitate finding all the spanning tree, bus 1, 2 and 3 are connected together. With adjacent branches located on a loop whose crossing point have a degree of 2 incorporated into one edge. With branches not located on any loop removed, Fig. 1 is simplified to Fig. 2. It can be seen from Fig. 2, edges ①, ③, ④, ⑤ constitute a spanning tree while edges ②, ⑥, ⑦ are the corresponding tie branches. Edges ①, ②, ④, ⑤ constitute another spanning tree while edges ③, ⑥, ⑦ are the corresponding tie branches. All spanning trees and the corresponding tie branches of Fig. 2 are listed in 2nd and 3rd column of Table 1.

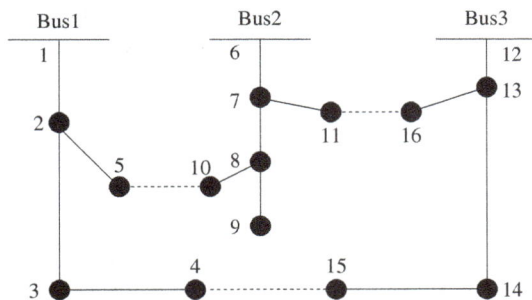

Fig. 1 IEEE typical three feeder test system

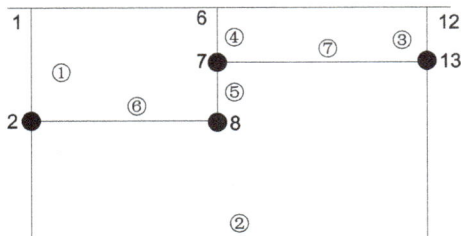

Fig. 2 Simplified system

3.2 Base vector of tie branches

Base vector is composed of total switch number on each tie branch. Taking tie branches ①, ②, ④ for example, as shown in the second row, third column of Table 1, there is 1 switch on edge ①, 5 switches on edge ②, and 1 switch on edge ④. Thus, the base vector of this tie branches is (151). All the base vectors of Fig. 2 are listed in 4th column of Table 1.

3.3 Candidate solutions

Any combination of open switches that meet the radial and connected constraints for distribution network operation is defined as a candidate solution for network reconfiguration. The optimal solution can only be one of the candidate solutions. One candidate solution can be created if and only if open one switch on each tie branch. As to each spanning tree, the number of candidate solutions equals to the product of each component of the corresponding base vector. For example, shown as the 2nd row, 5th column of Table 1, number of candidate solutions

Table 1 Spanning trees, tie branches and base vectors

No.	Spanning tree	Tie branches	Base vector	Candidate solutions
1	③ ⑤ ⑥ ⑦	① ② ④	(1 5 1)	5
2	② ③ ⑤ ⑦	① ④ ⑥	(1 1 3)	3
3	① ③ ⑤ ⑦	② ④ ⑥	(5 1 3)	15
4	② ③ ⑥ ⑦	① ④ ⑤	(1 1 1)	1
5	① ③ ⑥ ⑦	② ④ ⑤	(5 1 1)	5
6	② ③ ⑤ ⑥	① ④ ⑦	(1 1 3)	3
7	② ③ ④ ⑥	① ⑤ ⑦	(1 1 3)	3
8	② ③ ④ ⑤	① ⑥ ⑦	(1 3 3)	9
9	③ ④ ⑤ ⑥	① ② ⑦	(1 5 3)	15
10	① ③ ④ ⑤	② ⑥ ⑦	(5 3 3)	45
11	① ③ ④ ⑥	② ⑤ ⑦	(5 1 3)	15
12	① ③ ⑤ ⑥	② ④ ⑦	(5 1 3)	15
13	② ④ ⑥ ⑦	① ③ ⑤	(1 1 1)	1
14	② ④ ⑤ ⑦	① ③ ⑥	(1 1 3)	3
15	④ ⑤ ⑥ ⑦	① ② ③	(1 5 1)	5
16	① ④ ⑤ ⑦	② ③ ⑥	(5 1 3)	15
17	① ④ ⑥ ⑦	② ③ ⑤	(5 1 1)	5
18	② ④ ⑤ ⑥	① ③ ⑦	(1 1 3)	3
19	① ② ④ ⑤	③ ⑥ ⑦	(1 3 3)	9
20	① ② ④ ⑥	③ ⑤ ⑦	(1 1 3)	3
21	① ⑤ ⑥ ⑦	② ③ ④	(5 1 1)	5
22	① ② ⑤ ⑥	③ ④ ⑦	(1 1 3)	3
23	① ② ⑥ ⑦	③ ④ ⑤	(1 1 1)	1
24	① ② ⑤ ⑦	③ ④ ⑥	(1 1 3)	3
Total				190

corresponding to 1st spanning tree is $1 \times 5 \times 1 = 5$. That is, there are 5 combinations of open switches meeting the radial, connected constraints for distribution network operation. The total number of candidate solutions of network reconfiguration equals to the sum of candidate solutions corresponding to each spanning tree. Shown as the 26th row, 5th column of Table 1, there are 190 candidate solutions for the IEEE typical 3 feeder test system.

4 Coding scheme and gene operation

4.1 Coding scheme

In this paper, decimal coding scheme is introduced. Subpopulations are one-to-one relevant to spanning trees. The length of chromosome is equal to the number of tie branches, which is equal to the number of independent meshes of distribution network. In contrast to conventional binary coding scheme, the proposed decimal coding scheme can greatly shorten the length of chromosome. The value of each gene is a non negative integer, indicating the open switch identity number on corresponding tie branch. The value of each gene is smaller than corresponding component of corresponding base vector. When we talk about a chromosome, we must point out the subpopulation it belongs to. If two chromosomes have the same genes but belong to different subpopulations, they have different meanings. For example, if chromosome (000) belongs to 1st subpopulation corresponding to 1st spanning tree in Table 1, it implies the 0th switch of tie branch ①, ②, ④ are open. Otherwise, if chromosome (000) belongs to 2nd subpopulation corresponding to 2nd spanning tree in Table 1, it implies the 0th switch of tie branch ①, ④, ⑥ are open.

4.2 Creation of initial population

Number of individuals in each subpopulation can be different according to the candidate solutions for the corresponding spanning tree. For example, because there are 15 candidate solutions correspondent to the 3rd spanning tree in Table 1, individuals number of the 3rd subpopulation can be chose to be 2. While there are 45 candidate solutions correspondent to the 10th spanning tree in Table 1, the individuals number of the 10th subpopulation can be chose to be 6.

4.3 Fitness value of chromosome

The open switches in the distribution network can be obtained by decoding the chromosome in each subpopulation. Network power loss is calculated with parallel back forward sweep power flow calculation method introduced in [21]. Power losses, with their corresponding chromosomes are sorted in ascending order. Order numbers in descending order are uniformly spaced mapped to real numbers $0 \sim 2$. That is, the fitness value of chromosome with biggest power loss is mapped to 0, while the fitness value of chromosomes with smallest power loss is mapped to 2. The intervals of fitness value between any two adjacent chromosomes after sorted are equal.

4.4 Gene operation

Gene operation includes selection, crossover, mutation, and reinsertion, all of which are implemented in each subpopulation with parallel computing. After specified number of evolution generation, the individual whose objective value is minimal in the whole population (not in subpopulation) is output as the optimal solution.

1) Selection

In each subpopulation, because numbers of individuals are different, numbers of individuals selected to perform gene operation are different. Roulette method are introduced for selection.

2) Crossover

Crossover is to exchange the number at the same position of parents, with a specified probability. For example, shown as Table 2, if the crossover probability is specified to 0.7, the offspring of parents chromosomes (421), (212) may be (211), (422).

3) Mutation

Mutation is to replace the gene of parent with a non negative integer smaller than the corresponding component of base vector. For example, shown as Table 3, if the mutation rate is specified as 0.01, and the base vector is (533), the chromosomes (421), (212), after mutation may be (321), (222).

4) Reinsertion

As to each subpopulation, after crossover and mutation, specified numbers of individuals whose fitness values are smallest are removed from the parents. Then, the offspring are inserted into the parents. Thus, in each subpopulation, new individuals are created while multiple old elite individuals are retained and population sizes are kept constant.

4.5 Parallel computing

Coding, creation of initial population, calculation of fitness values and gene operations are all implemented in subpopulations. There is little coupled calculation between subpopulations. Therefore, it is very suitable for parallel computing. The calculation speed is highly improved, with parallel computing.

Table 2 Crossover operation

Parents			Offspring		
4	2	1	2	1	1
2	1	2	4	2	2

Table 3 Mutation operation

Parents			Offspring		
4	2	1	3	2	1
2	1	2	2	2	2

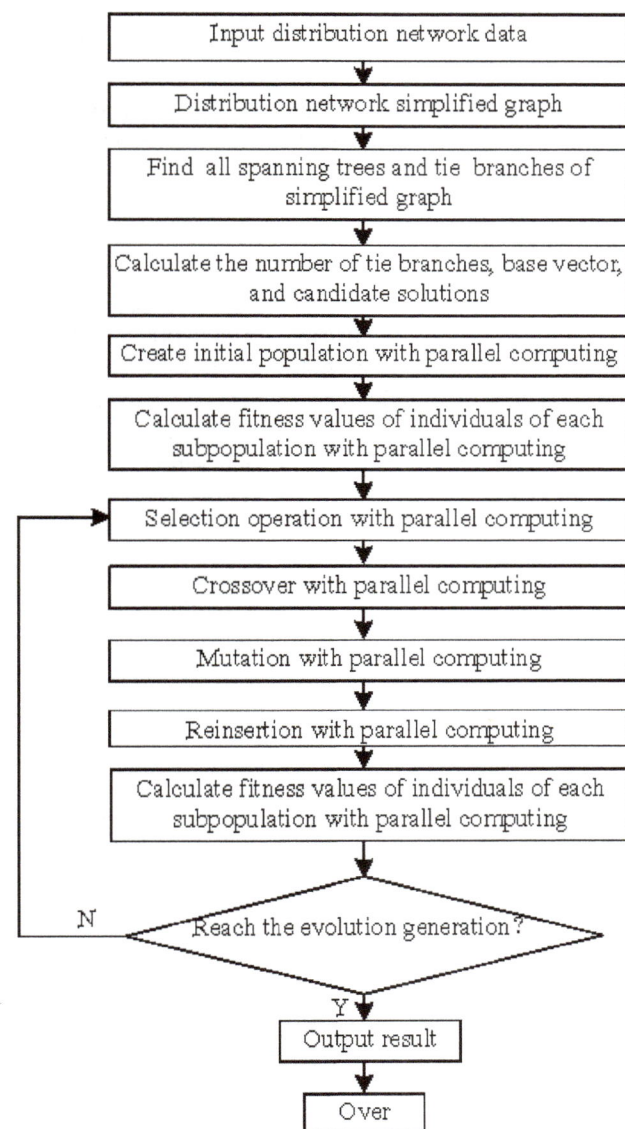

Input distribution network data

Distribution network simplified graph

Find all spanning trees and tie branches of simplified graph

Calculate the number of tie branches, base vector, and candidate solutions

Create initial population with parallel computing

Calculate fitness values of individuals of each subpopulation with parallel computing

Selection operation with parallel computing

Crossover with parallel computing

Mutation with parallel computing

Reinsertion with parallel computing

Calculate fitness values of individuals of each subpopulation with parallel computing

N — Reach the evolution generation?

Y

Output result

Over

Fig. 3 Flow chart of the algorithm

4.6 Flow chart of the proposed algorithm

The flow chart of the proposed algorithm is shown as Fig. 3. When the specified number of evolution generation is reached, the individual whose objective value is smallest in the whole population (not in the subpopulation) is output as the optimal solution of the network reconfiguration problem.

5 Simulation case

Two simulation cases are taken to test the capability of the proposed method under the condition that there is single power supply and multiple power supply. In order to facilitate recording data, the optimal solution is determined with the proposed method. Then the optimal solution is taken as the program termination condition. All the relevant programs are written by MATLAB language.

Case 1: shown as Fig. 1, there are 3 feeders, 16 nodes, 3 independent meshes in this distribution network. Fig. 2 is the simplified graph of Fig. 1. There are 24 spanning trees and total 190 candidate solutions for Fig. 2. The length of each chromosome is 3. Because the total amount of candidate solutions is very small, in each subpopulation, chromosomes are chosen to be one-to-one relevant to candidate solutions. That is, a method of exhaustion is introduced to directly find the optimal solution.

Case 2: shown as Fig. 4, there are 69 nodes, 73 branches, 5 independent meshes in this distribution network. Fig. 5 is the simplified graph of Fig. 4. There are 463 spanning trees and 377417 candidate solutions for Fig. 5. The length of each chromosome is 5. The number of subpopulations equals to that of spanning trees. In the 37 subpopulations which correspond to the most candidate solutions, there are 4 individuals for each population. While in the rest 426 subpopulations, there are 2 individuals for each subpopulation. Therefore, there are 1,000 individuals for the whole population. The crossover rate is 0.7, and the mutation rate is 0.01. The program is performed 50 times.

Reconfiguration results and relevant statistical data for case 1 and 2 are shown as Table 4 and Table 5.

As for the 16 nodes system, it takes 0.012 s to find the optimal solution. As for 69 nodes system, it takes 6.22 times of average evolution generation and 0.534 s to find the optimal solution, much better than the results of [8]. The reason why the reconfiguration results of the proposed method are different from that of [8] is that because there are no loads at nodes 45, 46, 47, open switches on the branches 44-45, 45-46, 46-47, 47-48 have the same effect.

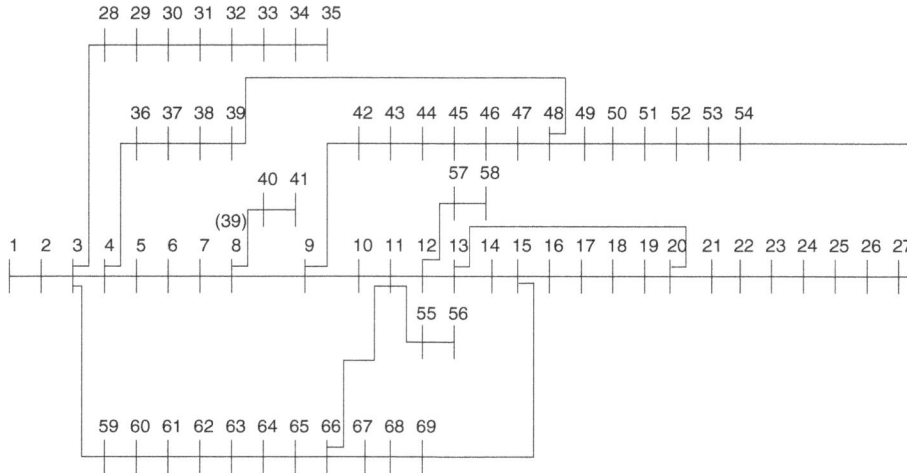

Fig. 4 American PG&E 69 nodes distribution network

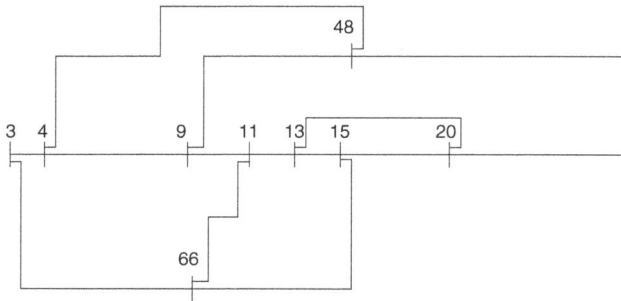

Fig. 5 Simplified graph of Fig. 4

It can be seen from the cases that the proposed method has very high search efficiency and calculation speed.

6 Conclusions

In this paper, a genetic algorithm based on all spanning trees of simplified graph of distribution network for network reconfiguration is proposed. The total candidate solutions and their distribution can be precisely calculated, which can supply information for determining the numbers of subpopulations and individuals in each subpopulation.

Table 4 Reconfiguration results of the two distribution network

Distribution system	Open switches before reconfiguration		Open switches after reconfiguration	Power loss before reconfiguration/kW	Power loss after reconfiguration/kW	Power loss descent rate/%
16 nodes	5-10, 11-16, 4-15		4-15, 8-10, 7-11	511.4	466.1	8.85
69 nodes	11-66, 15-69, 27-54, 13-20, 39-48	Results in [8]	11-66, 14-15, 50-51, 13-20, 47-48	229.5	102.10	55.5
		Results in this paper	11-66, 14-15, 50-51, 13-20, 46-47	229.5	101.01	56.0

Table 5 Evolution statistics

Distribution system		Minimal generation to get global optimum results/generation	Maximal generation to get global optimum results/generation	Average generation to get global optimum results/generation	Average calculation time/s
16 nodes	1		1	1	0.012
69 nodes	Results in [8]	21	32	22.40	25.423
	Results in this paper	1	16	6.22	0.534

The disadvantages of conventional genetic algorithm for distribution network reconfiguration that a large amount of unfeasible solutions are created in the process of crossover and mutation is completely overcome. Therefore, search efficiency of the proposed algorithm is highly improved. The identity number of open switch on each tie branch is taken as the optimization variable and decimal coding scheme is introduced. As a consequence, the length of chromosome is greatly shortened and storage space is minimized while calculation speed is improved. There is little calculation coupled between subpopulations. So it is very suitable for parallel computing. Two cases indicate that the proposed algorithm has the merits of high calculation speed and superior capabilities.

References

[1] He YQ, Peng JC, Wen M et al (2010) Minus feasible analysis unit and fast algorithm for distribution network reconfiguration. P CSEE 30(31):50–56 (in Chinese)

[2] Civanlar S, Grainger JJ, Yin H et al (1988) Distribution feeder reconfiguration for loss reduction. IEEE Trans Power Deliv 3(3):1217–1223

[3] Bi PX, Liu J, Zhang WY (2001) A refined branch-exchange algorithm for distribution networks reconfiguration. P CSEE 21(8):98–103 (in Chinese)

[4] Lei JS, Deng YM, Zhang BM (2001) Hybrid flow pattern and its application in network reconfiguration. P CSEE 21(1):57–62 (in Chinese)

[5] Liu W, Han ZX (2004) Distribution network reconfiguration based on optimal flow pattern algorithm and genetic algorithm. Power Syst Technol 28(19):29–33 (in Chinese)

[6] Liu J, Bi PX, Dong HP (2002) Simplified analysis and optimization of complicated distribution network. China Electric Power Press, Beijing, China (in Chinese)

[7] Bi PX, Liu J, Zhang WY (2001) Study on algorithms of distribution network reconfiguration. Autot Electr Power Syst 25(14):54–60 (in Chinese)

[8] Li XM, Huang YH, Yin XG (2004) Genetic algorithm based on improvement strategy for power distribution network reconfiguration. P CSEE 24(2):49–54 (in Chinese)

[9] Bi PX, Liu J, Liu CX et al (2002) Refined genetic algorithm for power distribution network reconfiguration. Autom Electr Power Syst 26(2):57–61 (in Chinese)

[10] Ma XF, Zhang LZ (2004) Distribution network reconfiguration based on genetic algorithm using decimal encoding. Trans China Electrotech Soc 19(10):65–69 (in Chinese)

[11] Nara K, Shiose A, Kitagawa M et al (1992) Implementation of genetic algorithm for distribution system loss minimum reconfiguration. IEEE Trans Power Syst 7(3):1044–1051

[12] Xu K (2008) Study on network reconfiguration based on improved adaptive genetic algorithms. Master Thesis, Guangxi University, Nanning, China (in Chinese)

[13] Wang YS, Chen GM, Zhang JS et al (2006) Optimal switching device placement based on niche genetic algorithm in distribution networks. Trans China Electrotech Soc 21(5):82–86 (in Chinese)

[14] Liu L, Cheng XY (2000) Reconfiguration of distribution network based on fuzzy genetic algorithms. P CSEE 20(2):66–69 (in Chinese)

[15] Yu YX, Duan G (2000) Shortest path algorithm and genetic algorithm based distribution system reconfiguration. P CSEE 20(9):44–49 (in Chinese)

[16] Ge SY, Liu ZF, Yu YX (2004) An improved Tabu search for reconfiguration of distribution systems. Power Syst Technol 28(23):22–26 (in Chinese)

[17] Wang SX, Wang CS (2000) A novel network reconfiguration algorithm implicitly including parallel searching for large-scale unbalanced distribution systems. Autom Electr Power Syst 24(19):34–38 (in Chinese)

[18] Taleski R, Rajicid D (1997) Distribution network reconfiguration for energy loss reduction. IEEE Trans Power Syst 12(1):398–406

[19] Chen GJ, Li JG, Tang GQ (2002) Tabu search approach to distribution network reconfiguration for loss reduction. P CSEE 22(10):28–33 (in Chinese)

[20] Gabow HN, Myers EW (1978) Finding all spanning trees of directed and undirected graphs. SIAM J Comput 7(3):280–287

[21] Yan W, Liu F, Wang GJ et al (2003) Layer-by-layer back/forward sweep method for radial distribution load flow. P CSEE 23(8):76–80 (in Chinese)

Jian ZHANG received the Bachelors, Masters and Doctors degrees in electric engineering from the department of electric engineering, Wuhan University, in 2005, 2007 and 2011 respectively. At present, He works at Jiangsu Electric Power Company Research Institute. His research interests comprise load modeling, power system transient analysis and control, renewable distributed generation and distribution network technology.

Xiaodong YUAN received the Bachelors and Masters degree in electric engineering from the department of electric engineering, Southeast University, in 2001 and 2004 respectively. At present, He works at Jiangsu Electric Power Company Research Institute. His research interests comprise power quality, renewable distributed generation and distribution network technology.

Yubo YUAN received the Bachelors, Masters and Doctors degree in electric engineering from the department of electric engineering, Hohai University, Hohai University and Southeast University, in 1998, 2001 and 2004 respectively. At present, He works at Jiangsu Electric Power Company Research Institute. His research interests comprise protection relay, renewable distributed generation and distribution network technology.

19

Reflections on stability technology for reducing risk of system collapse due to cascading outages

Yongjie FANG (✉)

Abstract This paper aims at putting forward viewpoints regarding the use of stability technology to prevent and control cascading outages by examining recent blackout events. Based on the inquiry reports of the 2011 Southwest America blackout and the 2012 India power blackouts, event evolution features are first summarized from a stability perspective. Then a comparative analysis is conducted so as to propose suggestions of effective measures, either preventive or emergency, which could have avoided the blackouts. It is shown that applications of several mature technologies can create opportunities of preventing or interrupting the cascading development. These include offline dynamic simulation, online stability analysis and preventive control, real-time situational awareness and automatic emergency control. Further R & D directions are given to address the challenges of modern power systems as well. They cover system fault identification criterion of protection and control devices, verification of adaptability of control effect to system operating conditions, real-time operational management of emergency control measures and improvement of simulation accuracy.

Keywords Cascading outages, Blackout, System collapse, Stability technology, Protection and control

1 Introduction

Major blackouts are rare, and no two blackout scenarios are the same. Severe blackouts in modern power systems, often with catastrophic consequences, were usually caused by a cascading development of emergency situation with system collapse as the final result. A cascade occurs when there is a sequential tripping of numerous transmission lines and generators in a widening geographic area [1], often with one event or state leading to another in a cause and effect manner [2].

Any occurrence of real blackout events always draws attention to the issue of power system stability and at the same time provides invaluable source information for assessment of stability technologies. This paper takes the 2011 Southwest America blackout and the 2012 India power blackouts as examples for analysis. On the afternoon of September 8, 2011, an 11-minute system disturbance occurred in the Pacific Southwest and affected parts of Arizona, Southern California, and Baja California, Mexico, leading to cascading outages and leaving approximately 2.7 million customers without power [3]. All of the San Diego area lost power, and it took 12 h to restore 100 % of its load. The India power blackouts on the 30th and 31st of July in 2012 followed the same pattern of fast cascading evolution, resulting in collapse of Northern Regional Grid with a load loss of 36 GW and collapse of Northern, Eastern and North-Eastern regional grids with a load loss of 48 GW respectively [4].

Although no single factor was responsible for these blackouts and weak grid structures as well as lack of coordination of system management and operation were key causes of failing to prevent cascading outages, this paper focuses on aspects which are closely related to stability technology for prevention and control of cascading outages. Although each event has a distinct feature, all these events share some common characteristics. A comparative study can therefore reveal control measures that are effective for all events as well as control measures that are effective only for each individual event.

The post-event analysis of all the above-mentioned blackout events showed that the system concerned was not

Y. FANG, NARI Group Corporation, Nanjing 211106, China
(✉) e-mail: fangyongjie@sgepri.sgcc.com.cn

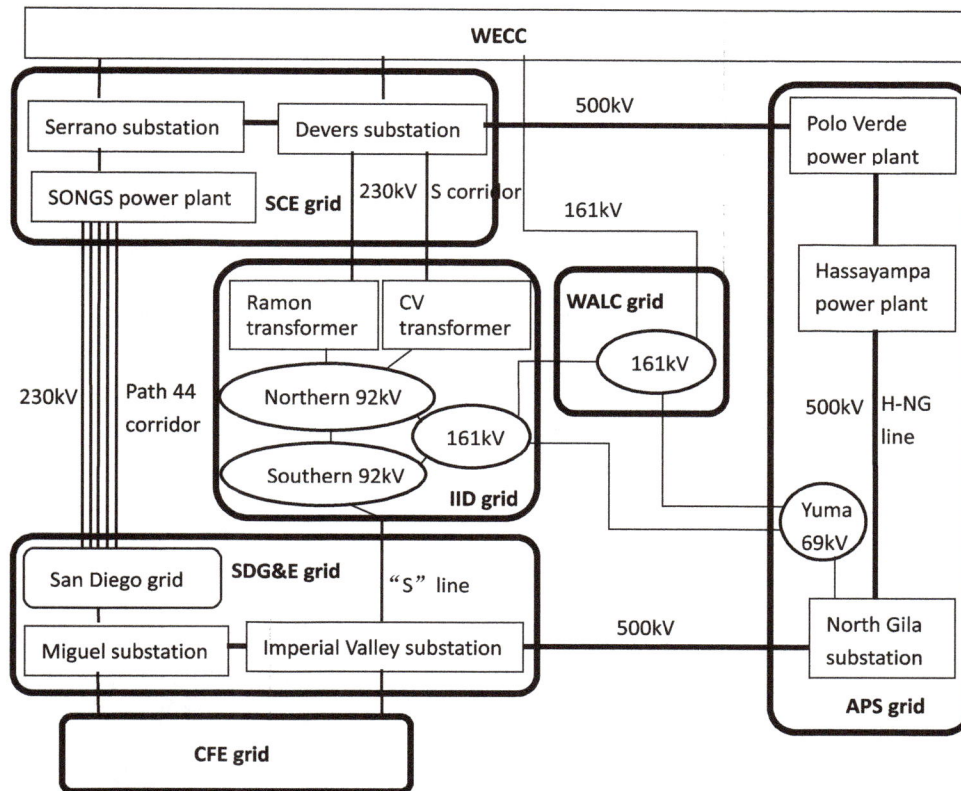

Fig. 1 Schematic structure of Southwest power grid in USA

being operated in a secure N − 1 state. However, this did not prevent the inquiry reports from giving detailed in-depth cause analysis and recommendations [3, 4]. Based on the inquiry reports, this paper concentrates on presenting both advanced but mature stability technology applications and fundamental measures necessary to be strengthened, aiming to provide reference for both industrial practice and academic research.

2 Event evolution features

A brief summary of the sequence of events for each exemplar case is given below to clarify the mechanisms of cascading evolution and analyze potentially possible ways of prevention and control of system collapse.

2.1 The 2011 Southwest America blackout [3]

A schematic structure of Southwest power grid is shown in Fig. 1, which consists of Imperial Irrigation District (IID) grid, San Diego Gas & Electric (SDG&E) grid, Arizona Public Service Co.(APS)grid, Southern California Edison (SCE) grid, Comisión Federal de Electricidad (CFE) grid and Western Area Power Administration—Lower Colorado

(WALC)grid. There are three parallel transmission corridors through which power flows into the area where the blackout occurred. The first transmission corridor consists of a single 500 kV Hassayampa–North Gila line (H–NG), a major transmission corridor that transports power in an east–west direction. The second corridor is Path 44 which includes the five 230 kV lines in the northernmost part of the SDG&E system that connect SDG&E with the San Onofre Nuclear Generating Station (SONGS). The third transmission corridor, shown as the "S Corridor" on Fig. 1, consists of lower voltage (230, 161 and 92 kV) facilities and feeds power to WALC and north IID.

The loss of the H–NG 500 kV line (taken as 0 s) initiated the event. Power flows instantaneously redistributed throughout the system, increasing flows through IID 92 kV and 161 kV systems to the north of the southwest power link and creating sizeable voltage deviations and equipment overloads. Significant overloading occurred on three of IID's 230/92 kV transformers located at the Coachella Valley (CV) and Ramon substations, as well as on Path 44. The flow redistributions, voltage deviations, and resulting overloads had a ripple effect, as transformers, transmission lines, and generating units tripped offline, initiating automatic load shedding throughout the region in a relatively short time span.

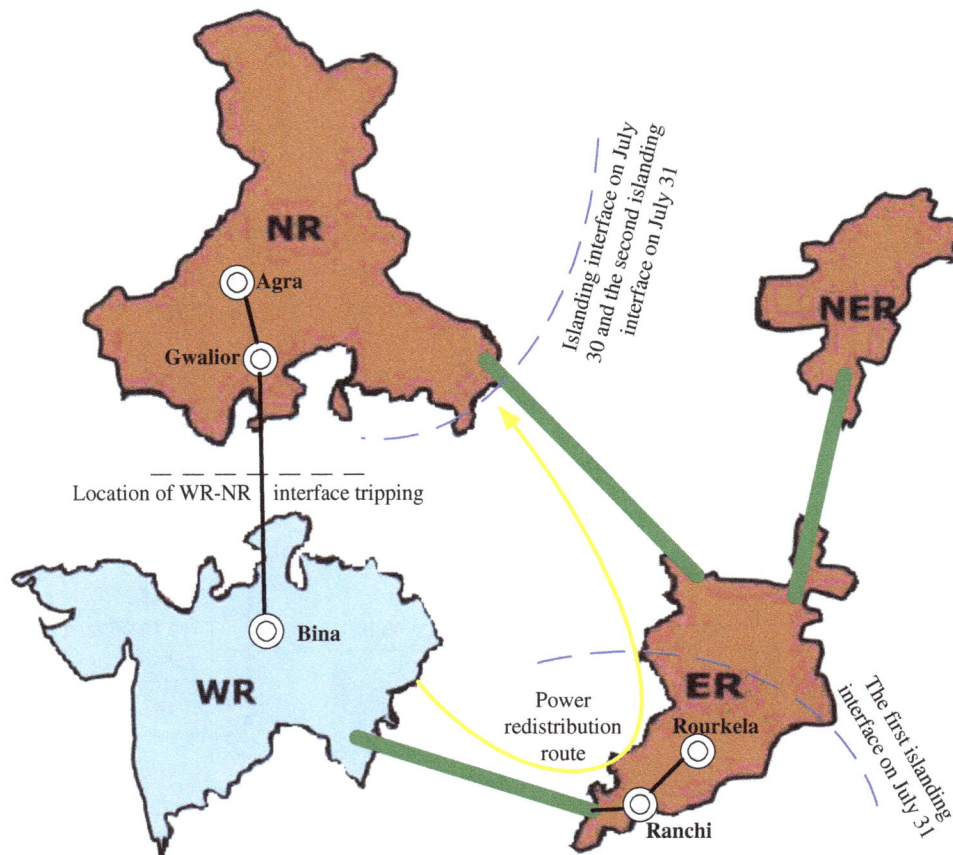

Fig. 2 Sketch of India power blackout event evolution

For instance, in 37.5 and 38.2 s, IID's CV transformer banks No. 2 and No. 1 tripped by overload protection relays; in less than 5 min, IID's Ramon 230/92 kV transformer tripped leading to local voltage collapse and followed by automatic under-voltage load shedding, multiple generator tripping and line tripping in IID's northern 92 kV system; in about 8 min, WALC's Gila 161/69 kV transformers tripped due to time-overcurrent protection; in about 9 min, the Yucca 161/69 kV transformers 1 and 2 and Pilot Knob-Yucca 161 kV transmission line tripped by overload protection followed by tripping of the YCA combined cycle plant on the Yuma 69 kV system, hastening the collapse of the Yuma load pocket; in 10 min, the El Centro-Pilot Knob 161 kV line tripped by Zone 3 protection isolating the southern IID 92 kV system onto a single transmission line from SDG&E-the S Line, forcing all of the remaining load in IID to draw through the SDG&E system and pushing the aggregate current on Path 44–8,400 amps, well above the trip point of 8,000 amps, and 3 s later, the S Line RAS at Imperial Valley Substation initiated the tripping of two combined cycle generators at Central La Rosita in Mexico driving Path 44 flow to about 9,500 amps, and another 4 s later, the S Line RAS tripped

the S Line itself creating an IID island leading to load tripping mostly in its southern 92 kV system.

Just seconds before the blackout, Path 44 carried all flows into the San Diego area as well as parts of Arizona and Mexico. In less than 11 min after H–NG tripped, the excessive loading on Path 44 initiated an intertie separation scheme at SONGS, which separated SDG&E from Path 44, led to the loss of the SONGS nuclear units, and eventually resulted in the complete blackout of San Diego, CFE and Yuma grids.

2.2 The 2012 India power blackouts [4]

A sketch of India power blackout event evolution is shown in Fig. 2. The India power grid consists of five regional grids, namely Northern (NR), Western (WR), Southern (SR), Eastern (ER) and North Eastern (NER) grids. SR grid is connected to ER and WR grids through asynchronous links, and the remaining four regional grids operate in synchronism.

The characteristics of pre-disturbance system conditions on 30th and 31st July 2012 were similar. The system was weakened by multiple outages of transmission lines in the WR–NR interface and effectively the 400 kV Bina-Gwalior-Agra line was the only main AC circuit available between WR–NR interface with high loading due to power overdrawn by some of the NR utilities. Triggering events of the two blackout events were the same, i.e., tripping of 400 kV Bina-Gwalior line by its zone-3 protection of distance relay due to load encroachment (taken as 0 s).

The WR–NR power transmission interface tripping immediately led to such an emergency state: there was a sudden large power imbalance between the sending WR grid and the receiving NR grid and the NR loads were met through WR–ER–NR route which was not purely AC transmission lines but a complex long-distance network consisting of generators, lines, transformers, distribution feeders, customers and so on. As a result, fast relative motion between WR generators and NR generators led to loss of synchronism and out-of-step oscillation. The system was then split by line tripping near the oscillation center due to operation of line distance relays.

In the July 30 blackout, the oscillation center was in the NR–ER interface and in 4 s the corresponding tie lines tripped, isolating the NR system from the WR–ER–NER system with a power imbalance of 5,800 MW. The NR grid system collapsed due to under frequency and further power swing within the region. The WR-ER system survived without any further splitting within the WR-ER grid due to tripping of a few generators in this region on high frequency.

In the July 31 blackout, the oscillation center was in the ER, near ER–WR interface, and in 6 s, a small part of ER (Ranchi and Rourkela), along with WR, got isolated from the rest of the NR–ER–NER grid with a power imbalance of 3,000 MW. Power deficiency in the NR–ER–NER grid led to multiple tripping of lines and generators attributed to internal power swings, under frequency and overvoltage at different places. In 1 min, the power swing in the NR–ER interface resulted in further separation of the NR from the ER–NER system. Subsequently, all the three grids (NR, ER and NER) collapsed. The WR system, however, survived due to over-frequency generator tripping.

In both blackout events, the SR system, which was getting power from ER and WR, survived after automatic under-frequency load shedding (UFLS) and HVDC power ramping.

2.3 Comparison of the blackouts from a stability perspective

The above-mentioned cases have some similarities. Systems in all events were operated in an unsecured N-1 state prior to the initial disturbance. The cascading processes were triggered by the tripping of a key transmission

Table 1 Comparative analysis of characteristics and control

Exemplar case	2011 Southwest America Blackout	2012 India power blackouts
Common characteristics	Unsafe N − 1 state prior to the initial disturbance. Triggered by the tripping of a key transmission interface and escalated as a result of power flow rerouting and inadvertent operation of protection or control devices	
Effective control	Preventive control	
Salient characteristics	Lasting for 11 min before collapse with overloading as the main stability problem	Lasting for 1 min before collapse with complex stability phenomena
Particularly effective control	Manual emergency control	Automatic emergency control

interface and developed as a result of power flow rerouting and inadvertent operation of protection or control devices. Therefore, preventive control measures would have effectively avoided the cascading outages. (Table 1)

The most prominent difference is the speed with which the cascade evolved. In the 2011 Southwest America blackout, it was a relatively slow process with equipment overloading as the main stability problem. There did exist opportunities for interrupting the cascading development with the assistance of proper technology. In the 2012 India power blackouts, however, triggering event and subsequent events happened so fast and the interrelated stability phenomena were so complex that the only effective measures seemed to be either preventive control in normal operating state or emergency control immediately after the first event occurrence.

3 Self-evident stability technology applications

It is quite obvious from analysis based on the event inquiry reports [3, 4] that applications of some mature technologies should be very effective for prevention and control of cascading outages, as summarized and emphasized below.

3.1 Offline dynamic simulation

A deep and thorough system study is the most fundamental measure for secure system operation, and offline dynamic simulation studies should be carried out in order to grasp the system characteristics under different grid conditions and anticipated scenarios.

While the inquiry report of the India blackouts calls for the formation of a task force to study the grid security issues, that of the Southwest America blackout lists in detail quite a few key findings and recommendations in this respect. For instance, in various time-scale system studies including long-term, seasonal, day-ahead and online analysis, many relevant power companies did not reflect external grids and lower voltage level grids in a sufficiently detailed and accurate way in their computer simulation models, resulting in a lack of understanding of dynamic mutual influence among different regions and voltage levels. This in turn led to problems such as not sharing study results and real-time information with neighboring grids, inappropriate relay settings of low-voltage facilities and so on. In addition, Western Electricity Coordinating Council (WECC) did not perform the required review and assessment of all NERC-defined Special Protection Systems (SPSs) in their areas. This serious negligence of mutual influence between stability control systems and system dynamics made it impossible to grasp system dynamics appeared during the cascading process or to determine proper control measures.

Therefore, whether it is offline study, online analysis or real-time monitoring, all facilities that impact bulk power system reliability should be covered, which should be selected not by geographic location, voltage level or functional classification but by deep understanding of system characteristics.

3.2 Online stability analysis and preventive control

Since the 2003 North America blackout [1], there have been extensive activities worldwide in research, development and application of online stability analysis techniques [5–8], a result of the necessity of assessing system security and making control decisions based on actual operating conditions.

Unfortunately, the India control centers did not have any analytical tool to periodically conduct an assessment of the system security condition and the state estimator results were not quite reliable [4]. Deployment of dynamic security assessment (DSA) has been listed as one of the recommendations.

In the 2011 Southwest America blackout case, real-time tools of the affected grids were inadequate and nonoperational when needed to provide early-warning information and preventive control decisions. For instance, if IID's DSA system had identified in advance the impact of H–NG line tripping and given preventive control suggestions, the operators could have taken action before the line tripping to adjust the system condition and thus the overload tripping of CV transformers could have been avoided so that the cascading evolution could have been interrupted at its initial stage. In addition, if IID's DSA system had alarmed the operators that on that day, total power flow of the two

CV transformers was at such a high level that only preventive control measures could have avoided overload tripping of the second transformer upon the tripping of the first, the operators could have dispatched additional generation to alleviate the transformer loading to avoid the loss of both transformers. Moreover, if IID had in its DSA system modeled its neighboring network using APS's complete topological model with real-time measurement data instead of a highly simplified equivalent model of pseudo-generators plus tie lines, the DSA could have assessed the impact of a sudden loss of H–NG on its power system before it actually happened and IID could have taken proper preventive control actions before the H–NG line tripped.

3.3 Real-time situational awareness

The importance of network visualization ability was recognized and emphasized by the inquiry committee of the 2012 India power blackouts, as strengthening the communication network, ensuring reliability of data at load dispatch centers and deploying WAMS were put forward as recommendations [4].

In the 2011 Southwest America blackout, changes in power flow and voltage magnitude as a result of H–NG line tripping were noticed by affected power companies but many control centers were unable to know in real time the causes and impacts of these changes because of not having adequate situational awareness of their neighboring systems. The result was inability to take action or even misjudgment of the event evolution trend.

In the post-event analysis of the September 8th blackout, phase measurement unit (PMU) data proved extremely valuable in constructing the sequence of events and validating simulation results. However, PMUs did not play a part in observing system behaviors in real time. For instance, tripping of the H–NG line led to the increase of phase angle difference between voltages of the two terminals. Post-event simulation showed that this angle difference exceeded APS's synchro-check relay setting of 60 degrees and re-dispatch of a significant amount of generation was required to reduce this difference to the acceptable value. Without situational awareness of the real-time value of this phase angle difference, WECC and California Independent System Operator(CAISO)were informed by APS that the line could be quickly reconnected. If PMUs were used to monitor the voltage angle difference between two ends of the tripped line, the time required to reclose the H–NG line would not have been wrongly estimated and WECC probably would have taken action to reduce loading on Path 44 to prevent its automatic tripping.

During the fast cascade stage, inadequate situational awareness led to inability to take timely emergency control measures. Neither the operators' monitoring nor the automatic alarming of Path 44 power flow exceeding the limit was correlated to SONGS separation scheme, which was a lack of real-time early warning against the effect of this limit violation. Otherwise, the automatic operation of the SONGS separation scheme could have been prevented by earlier emergency control actions such as manual load shedding.

3.4 Automatic emergency control

It is interesting to note that the inquiry report of the 2012 India power blackouts [4] is of the opinion that after loss of about 5,000–6,000 MW to Northern Region, had the UFLS relays operated, the grid could have been saved but the report of the 2011 Southwest America blackout [3], much more detailed and thorough, shows that the post-event simulation analysis could not explain why UFLS performance could not have prevented the SDG&E system from frequency collapse.

In fact, all the blackout events showed that equipment along the power flow rerouting path during the dynamic process could be tripped by protection relays earlier than UFLS actions. Therefore the cascading development such as successive islanding could not possibly be interrupted by solely depending on the distributed control measures such as UFLS. Moreover, the emergency control problem should be formulated as that of the blackout defense mechanism in case of sudden loss of a key power transmission interface which although was a single transmission line in all the above cases. Therefore, the primary control objective should be to prevent cross-grid rerouting of large power flow and the follow-up uncontrolled equipment tripping and system islanding, and automatic generator tripping and load shedding triggered by the event of a sudden loss of transmission corridor is the most effective emergency control measure to prevent similar cascading outages. The economic rationality of deploying this type of emergency control can be analyzed from the point of view of risk reduction.

Even at the later stage of cascade, this could be also effective for preventing collapse of the islanded system and reducing loss of load. For example, active load shedding triggered by Path 44 tripping signal would have been such a measure.

4 In-depth reflections on further R&D directions

From the point of view of system operation, those listed in the previous section are mature technologies to enhance protection against cascading outages. However, in face of the increasing scale, complexity and flexibility of modern power systems, several fundamental measures including

further R&D efforts should be taken to ensure the effectiveness of technological measures.

4.1 System fault identification criterion of protection and control devices

It was common that in blackout events, many Zone 3 distance protection relays perceived by mistake the current and voltage conditions due to load encroachment or system oscillations as line fault. Also, it is still not clear whether ineffectiveness of UFLS schemes had something to do with the device operation criterion. All these deserve further studies.

In addition, in modern power systems, the AC system fault characteristics can be severely impacted by adjacent HVDC links or renewable energy source injection together with its complex control systems. Traditional fault identification criterion is based on AC measurements and should be reviewed so as to prevent protection and control devices from unintended operation or failing to operate as a result of being unable to adapt to new system characteristics.

4.2 Analysis of mutual influence between control logic and system response

Even if protection and control devices can properly identify system fault scenarios and operate correctly according to predesigned logic of device-level or local grid level, the concept of integrity of system stability should be always borne in mind and thus the device operation logic should be taken as an integral part for grasping global characteristics of the system as a whole, in order to perform coordination among different control actions and between control action and system dynamics.

It is clear from the blackout cases that offline simulation studies failed to predict such complex system behavior and so much uncontrolled equipment tripping. As more and more strong interconnection in modern power systems leads to ever-increasing cross-regional influence of protection and control operation, mutual interaction via system responses and coordination among distributed and independently operated protection relays or regional emergency control systems must be stressed. This is to prevent system real-time responses provoked by local control actions from going beyond the scope of offline studies and pre-designed control logic, thus avoiding the passive situation of uncontrolled equipment tripping.

4.3 Verification of adaptability of control effect to system operating conditions

It can also be seen from the blackout cases that there could hardly be any records of correct active control actions during the cascading evolution. Even if the fault identification criterion and control logic are appropriate judged by the locally measured system conditions, the flexible system state changes can still make the pre-specified control actions have deficient, excessive or even unintended or undesirable effects. One analogy is the totally different location of the oscillation center of the India system dynamics on 30 July from that on 31 July 2012.

Therefore, adaptability of control effects of the protection and control actions to the actual system operating condition should be verified in an online environment and once there is any problem, the system operating state should be adjusted or the automatic control decisions should be refreshed according to the online analysis results [5, 6]. In this way, control measures can be both proactive and orderly in case of system emergency.

The typical time needed to complete one round of calculation is 5–15 min in the current on-line DSA system and the contingency list normally contains only $N - 1/N - 2$ scenarios. Therefore, adaptability of control effect can be assessed only when the power system reaches a relatively stable condition after the already occurred disturbance and before the happening of another pre-specified disturbance. It needs further consideration how to include in the contingency list scenarios of cascading nature which best reflect the most up-to-date system state.

4.4 Real-time operational management of emergency control measures

The 2012 India blackouts revealed that practically there was no load relief from defense mechanisms like UFLS, which was a natural result of violation of the various system security related standards and power companies' attention solely on overdrawing from the grid.

Therefore, measures to ensure stability control's effectiveness in system emergency should not limited to safeguarding reliability of control devices on the grid side through operational maintenance, and real-time monitoring and management should be extended to control execution terminals on the generation and load side such as monitoring available quantity of generator tripping or load shedding in order to guarantee sufficient control effects once necessary. To this end an effective management mechanism is necessary and functional authority of dispatch control centers for adjusting inappropriate control measures should be ensured.

The centralized management function of stability control system can be further elevated to more advanced function such as real-time correlated monitoring. For instance, correlated monitoring of power quantities (such as Path 44 loading) and operation logic of control systems

(such as SONGS separation scheme) can be implemented to give timely warning about effects of control system operation as a result of grid operation state changes.

4.5 Improvement of simulation accuracy

Simulation is the fundamental tool in various stages of grid planning, system operation analysis, protection and control decision-making, online DSA, post-event analysis and so on. Each occurrence of large disturbance in actual power systems provides an opportunity for checking accuracy of simulation via actual system responses.

Analysis in the inquiry report of the 2012 India blackouts is rather qualitative because of insufficient field data records and lack of detailed post-event simulation study results.

The inquiry report of the 2011 Southwest America blackout gives examples of insufficiency of system simulation models in reproducing the actual event. For instance, it was unable to predict either the tripping of the SONGS generation units or the collapse of SDG&E and CFE systems by using WECC dynamic simulation models for near-term and long-term planning purposes. In the post-event simulations the SDG&E system should have been prevented from frequency collapse by the UFLS operational performance, which was quite the opposite of reality. Only by further addition of recorded details of the actual event to the system model, including UFLS activation programs and automatic switching logics of capacitors, did the simulation results of the islanded region become better aligned to the actual event following operation of the intertie separation scheme at SONGS. In addition, the impedance value of IID's CV transformers was 0.1 per unit in WECC's planning model but 0.05 per unit in its DSA model. As a result of this difference, there was an error of approximately 16 % of the CV transformer loading in the online DSA calculation result compared to the result obtained using the planning model. This demonstrates the importance of data and modeling accuracy as CV transformers were such key facilities during the event evolution process.

There were large deviations of grid voltage and frequency from nominal values in all the blackout cases, particularly at the final stage of cascading development. Whether simulation algorithms can appropriately deal with abnormal excursions of voltage and frequency should also be reviewed.

5 Conclusions

The 2011 Southwest America blackout and the 2012 India blackouts are used as examples to form the discussion in this paper. Through analysis of system behavior and event evolution causes of these two events, the study shows that opportunities of interrupting the cascading failure can be found in various stages, whether they belong to preventive or emergency control technologies.

In order to use stability technologies to prevent and control cascading outages, it is important to construct a power system security defense infrastructure based on advanced conceptual design. However, to ensure the effectiveness of technological measures during cascading evolution process, several fundamental measures need to be strengthened first. These include a complete, thorough and timely understanding of system dynamic characteristics, system fault identification criterion of protection and control devices, analysis of mutual influence between control logic and system response, verification of adaptability of control effect to system operating conditions and improvement of simulation analysis.

Centralized generator tripping and load shedding automatically triggered by the event of sudden loss of a transmission interface is the most effective emergency control measure for prevention of cross-grid rerouting of large power flows, uncontrolled equipment tripping and successive grid islanding. To ensure the dependability and reliability of control effects, operational management of stability control measures should be strengthened and authorization for control centers to adjust these measures should be guaranteed.

Acknowledgments This work is supported by State Grid Corporation of China (No. SGCC-MPLG003-2012).

References

[1] US-Canada Power System Outage Task Force (2004) Final report on the August 14, 2003 blackout in the United States and Canada: Causes and recommendations

[2] Voropai NI, Efimov DN (2008) Analysis of blackout development mechanisms in electric power systems. In: Proceedings of the 2008 IEEE power and energy society general meeting— conversion and delivery of electrical energy in the 21st century, Pittsburgh, 20–24 July 2008, 7pp

[3] Federal Energy Regulatory Commission and the North American Electric Reliability Corporation (2012) Arizona-Southern California outages on September 8 2011: Causes and recommendations

[4] Enquiry Committee (2012) Report of the Enquiry Committee on grid disturbance in Northern Region on 30th July 2012 and in Northern, Eastern & North-Eastern region on 31st July 2012

[5] Xue YS (2003) The way from a simple contingency to system-wide disaster– Lessons from the eastern interconnection blackout in 2003. Autom Electr Power Syst 27(18):1–5 (in:Chinese)

[6] Xue YS 2005) Defense schemes against power system blackouts in China with high load growth. Invited paper. In: Proceedings of the 15th power system computation conference (PSCC'05), Liege, 22–26 Aug 2005

[7] CIGRE Task Force WG C4.601 (2007) Review of on-line dynamic security assessment tools and techniques. CIGRE Technical brochure no. 325, Paris

[8] CIGRE Task Force WG C4.601 (2007) Wide area monitoring and control for transmission capability enhancement. CIGRE Technical brochure no. 330, Paris

Yongjie FANG obtained his BSc and MSc from North China University of Electric Power in 1984 and 1987 respectively, and his PhD from Imperial College (London) in 1996, all in Electrical Engineering. His professional experience, mainly at NARI Group Corporation, SGCC, China, spans the R & D, engineering and management of power system stability analysis and control. Currently, he is the Deputy Manager of Power System Stability Control Branch Company of NARI Technology Development Co. Ltd. He is a member of CIGRE, IEEE and IET.

20

Binary glowworm swarm optimization for unit commitment

Mingwei LI (✉), Xu WANG, Yu GONG,
Yangyang LIU, Chuanwen JIANG

Abstract This paper proposes a new algorithm—binary glowworm swarm optimization (BGSO) to solve the unit commitment (UC) problem. After a certain quantity of initial feasible solutions is obtained by using the priority list and the decommitment of redundant unit, BGSO is applied to optimize the on/off state of the unit, and the Lambda-iteration method is adopted to solve the economic dispatch problem. In the iterative process, the solutions that do not satisfy all the constraints are adjusted by the correction method. Furthermore, different adjustment techniques such as conversion from cold start to hot start, decommitment of redundant unit, are adopted to avoid falling into local optimal solution and to keep the diversity of the feasible solutions. The proposed BGSO is tested on the power system in the range of 10–140 generating units for a 24-h scheduling period and compared to quantum-inspired evolutionary algorithm (QEA), improved binary particle swarm optimization (IBPSO) and mixed integer programming (MIP). Simulated results distinctly show that BGSO is very competent in solving the UC problem in comparison to the previously reported algorithms.

Keywords Binary glowworm swarm optimization, Correction method, Priority list, Unit commitment

M. LI, X. WANG, Y. GONG, Y. LIU, C. JIANG, Department of Electrical Engineering, Shanghai Jiao Tong University, Shanghai 200240, China
(✉) e-mail: lmw0546@126.com

1 Introduction

Unit commitment is an important optimization problem in the power system. Its objective is to determine the on/off status of each unit and the economic dispatch of power demand in a scheduling period in order to minimize the total system production cost under generating units' constraints and power system's constraints. Since the unit commitment (UC) problem has the characteristic of high-dimension, discreteness and non-linearity, it takes lots of time to get the exact best solution of this problem by the enumeration method, and the computation time increases dramatically with the size of unit [1].

Because of its significant economic benefits, researchers around the world have done a lot of research and proposed many methods. Reference [2] used the dynamic programming method to solve the UC problem. In order to save the computation time, the units were classified and all the units formed different kinds of groups. As a result, the combinations of the units, as well as the computation time decreased. The extended priority list (EPL) method was introduced in [3]. The EPL method consisted of two steps. At first, disregarding the operational constraints, we got the original solutions by priority list (PL) algorithm very quickly, secondly, some heuristic processes were used to ensure that all the solutions satisfy the operational constraints. Reference [4] concentrated on the implementation aspects of Lagrangian relaxation (LR) method applied to realistic and practical UC problem, which aided in confirming the viability of this technique especially for large scale thermal UC programs. On this basis, [5] presented the enhanced adaptive Lagrangian relaxation (ELR) method with novel method to decide the on/off status of the units, new way of initializing the Lagrangian multipliers, unit classification, and adaptive adjustment of Lagrangian multiplier. As a result, the production cost was less expensive than the Lagrangian relaxation method. Furthermore, the CPU time is much smaller. Although these conventional optimization algorithms have the advantage of high speed and accuracy in solving small-scale UC problem, with the increase in the size of the generating units, the quality of the solution decreases and some of these algorithms are easy to fall into the "curse of dimensionality".

Reference [6] presented a genetic algorithm (GA) to solve the UC problem. Since the selection scheme, mutation operation and the corresponding correction method were used, the GA provided flexibility in modeling both time-dependent and coupling constraints. Simulated annealing (SA) method had the probabilistic jumping property which existed in the whole searching process and varied with time. When it was applied to the UC problem, it helped to keep the diversity of the feasible solutions and improved the probability of getting the best solution [7, 8]. Particle swarm optimization (PSO) method, first proposed by Eberhart and Kennedy, was easy to code and did not have many parameters to adjust in comparison with the GA mentioned above. Therefore, it not only obtained the better solution, but also considered more constraints such as the realistic nonlinear time-dependent startup cost, limits of the ramp rate and the prohibited zone [9]. Instead of the random mutation in the PSO method, the improved particle swarm optimization algorithm (IPSO) takes into account more information of the particles, thus the particles had more probability of moving to the better solution. Penalty factor was applied to the solutions that violate any of the constraints [10]. On this basis, [11] presented a new improved binary PSO (IBPSO) method, which was used to deal with the on/off status of the units. Meanwhile, the Lambda-iteration method was adopted to dispatch the load economically. In the iterative process, some heuristic strategies were used to repair the solutions that violate the system constraints or operational constraints. Reference [12] proposed the evolutionary programming (EP) method. All the feasible solutions changed randomly and competed with each other; then the better solutions were selected and got into the next iteration. The power output of the units in the whole scheduling period was represented by a string of symbols. The quantum-inspired evolutionary algorithm (QEA) proposed in [13] was improved by integrating the quantum theory. To be specific, the on/off status of the units was represented by the quantum bits and used the rotation gates to keep the diversity of the feasible solutions and move to the better solutions. The best solution of the QEA method was less expensive than that of the previous methods and the execution time increased linearly with the size of the generating units. To a certain extent, these intelligent algorithms solve the problem brought by the augment of the scale of the generating units. However, they have the disadvantage of falling into local optimal solution prematurely. With the

population of the software CPLEX, the MIP method to solve the UC problem became very effective. However, the accuracy of convergence had great impact on the computation time and the quality of the solution [14].

Glowworm swarm optimization (GSO) is a new swarm intelligence optimization algorithm [15]. The optimization process is as follows. At first, all the glowworms are randomly generated in the search space. Each of them carries luciferin, which represents the brightness of the light send out by the glowworms, then they look for the glowworms that have higher brightness within their own range of view, and move towards one of them using the roulette approach. After the move, the luciferin of the glowworms is updated. In case of having too many glowworms within their view range, every one of them adjust their view range after the move. Many researchers have applied the GSO algorithm to solve practical problems. The 0–1 knapsack problems were effectively solved by using GSO algorithm [16]. The GSO algorithm was also used to find the optimal solution for the continuous optimization problem. The results above showed that GSO performed much better than many other algorithms, especially for different kinds of various global optimization problems [17].

This paper proposes a binary glowworm swarm optimization (BGSO) algorithm to solve the UC problem. Each glowworm in the BGSO algorithm is a T·N matrix that represents all the units' on/off status in the whole scheduling period. Since the on/off status of the units are binary variables, we propose the Hamming distance to represent the distance between the glowworms creatively, instead of the Euclidean distance adopted in the original GSO. Furthermore, we thought of a new way to update the on/off status of the units in the form of probability. Meanwhile, the Lambda-iteration method is adopted to solve the economic dispatch problem. The Lambda-iteration method and the BGSO algorithm are run at the same time for the purpose of finding the solution that has the least total production cost. Furthermore, the correction method and several adjustment techniques are proposed to ensure that the solutions are diverse in the iterative process and satisfy all the constraints.

This paper is organized as follows. The mathematical formulation of the UC problem including the objective function and the constraints is illustrated in Section 2. Section 3 describes the procedure and principle of GSO. Section 4 proposes the BGSO applied to the UC problem. Furthermore, the correction method to guarantee that the solutions are feasible and several techniques to keep the diversity of the solutions and contribute to better solutions are also illustrated. The BGSO method is tested with the number of generating units in the range of 10–140 and the results are compared with the other algorithms in Section 5. The conclusion is given in Section 6.

2 Formulation of UC problem

2.1 Objective function

The objective of the UC problem is to minimize the total production cost consisting of the generation cost and the start-up cost of the generating units under the circumstance where the operational constraints and the constraints of the generating units are satisfied in the scheduling period. The objective function is expressed as

$$F = \sum_{t=1}^{T} \sum_{i=1}^{N} [C_i(p_i^t) \cdot u_i^t + S_i u_i^t (1 - u_i^{t-1})] \quad (1)$$

where F is the total production cost; T the number of hours in the scheduling period; N the number of generating units; and u_i^t on/off status of the unit i at hour t, 1 represents the on status of the unit i at hour t, 0 represents the off status of the unit i at hour t. $C_i(p_i^t)$ is the generation cost function of unit i. It is normally a quadratic polynomial represented by

$$C_i(p_i^t) = a_i(p_i^t)^2 + b_i(p_i^t) + c_i \quad (2)$$

where p_i^t generation output of unit i at hour t; and a_i, b_i, c_i are parameters of unit i.

S_i is the start-up cost of unit i which is related to the duration time of the off state of unit i. It can be expressed by

$$S_i = \begin{cases} H_{SCi} & M_{DTi} < X_{OFFi}^t \leq M_{DTi} + C_{SHi} \\ C_{SCi} & M_{DTi} + C_{SHi} < X_{OFFi}^t \end{cases} \quad (3)$$

where H_{SCi} is hot start-up cost of unit i; C_{SCi} the cold start-up cost of unit i; X_{OFFi}^t the duration time during which unit i keeps off status at hour t; C_{SHi} cold start time of unit i; and M_{DTi} the minimum down time of unit i.

2.2 Constraints

The constraints of the UC problem are listed as follows:

1) System power balance constraint

$$\sum_{i=1}^{N} u_i^t p_i^t = D^t \quad (4)$$

2) Spinning reserve constraint

$$\sum_{i=1}^{N} u_i^t p_i^{max} \geq D^t + R^t \quad (5)$$

3) Generation limit constraint

$$p_i^{min} \leq p_i^t \leq p_i^{max} \quad (6)$$

4) Minimum up time constraint

$$(u_i^{t-1} - u_i^t)(X_{ONi}^{t-1} - M_{UTi}) \geq 0 \quad (7)$$

5) Minimum down time constraint

$$(u_i^t - u_i^{t-1})(X_{OFFi}^{t-1} - M_{DTi}) \geq 0 \quad (8)$$

where D^t is power demand at hour t; R^t the spinning reserve at hour t; p_i^{max} the maximum power generation of unit i; p_i^{min} the minimum power generation of unit i; M_{UTi} the minimum up time of unit i; and X_{ONi}^t the duration time during which unit i keeps on status at hour t.

3 Glowworm swarm optimization

In the GSO algorithm, a group of glowworms are initialized randomly in the solution space of the objective function and each of them has the same value of luciferen. The brightness of the glowworm is proportional to the value of luciferen. Moreover, the fitness value of the glowworm is closely related to the luciferen. The larger the value of a glowworm's luciferen is, the more strongly it attracts the other glowworms within their own scope, which is called the local-decision range. In the iterative process, glowworm i moves towards one of the glowworms that both have better fitness value and are within the ith glowworm's local-decision range with a certain probability. Then the ith glowworm's local-decision range is adjusted for the purpose of controlling the quantity of the glowworms within it. The procedure of GSO algorithm is presented as:

1) Luciferin update phase

$$l_i(t) = (1 - \rho) \cdot l_i(t - 1) + \gamma \cdot J(x_i(t)) \quad (9)$$

where $x_i(t)$ is the location of glowworm i at iteration t; $l_i(t)$ the luciferin of glowworm i at iteration t; ρ the luciferin decay constant; γ the luciferin enhancement constant; and $J(x_i(t))$ the objective function of glowworm i.

2) Movement phase

Within glowworm i's local-decision range, it selects glowworm j from all the glowworms that have larger value of luciferin by the way of roulette probability.

Roulette probability formula

$$p_{ij}(t) = \frac{l_j(t) - l_i(t)}{\sum_{k \in N_i(t)} l_k(t) - l_i(t)} \quad (10)$$

Distance formula

$$d_{ij}(t) = ||x_i(t) - x_j(t)|| \quad (11)$$

Location update formula

$$x_i(t + 1) = x_i(t) + s\left(\frac{x_j(t) - x_i(t)}{d_{ji}(t)}\right) \quad (12)$$

where $N_i(t)$ is the numbers composed by all the glowworms that have larger value of luciferin within glowworm i's local-decision range; $d_{ij}(t)$ the distance between glowworm i and glowworm j; $\|.\|$ the standard Euclidean norm operator; and s the size of step.

3) Local-decision range update phase

$$r_d^i(t+1) = \min\{r_s, \max\{0, r_d^i(t) + \beta(n_t - |N_i(t)|)\}\} \tag{13}$$

where $r_d^i(t)$ is the local-decision range of glowworm i at iteration t; r_s the maximum local-decision range parameter used to control the rate of changing local-decision range; n_t the parameter used to control the number of glowworms within the local-decision range; and $|N_i(t)|$ the total number of glowworms that have the larger luciferin within the local-decision range.

4 BGSO for UC problem

4.1 Binary glowworm swarm optimization

The GSO algorithm is used to solve the problems that contain continuous variables. When it comes to the UC problem, the variables representing the on/off state of the units are binary, hence the BGSO is proposed to solve the UC problem. The modification of GSO is shown as follows.

1) Computation of distance

Instead of the Euclidean distance adopted in the GSO, Hamming distance is proposed to represent the distance between glowworm i and glowworm j. The Hamming distance between two glowworms is the number of locations where one has a "0" and the other a "1" [18]. It can be expressed by

$$hm\_d_{ij}(t) = hamming\_distance(x_i(t), x_j(t)) \tag{14}$$

2) Location update

The location of every glowworm is composed of m binary variables. In the location update process, the moving step is ignored and each dimension of glowworm i's location is updated in the form of the probability. It can be expressed by

$$x_{i,k}(t+1) = \begin{cases} x_{i,k}(t) & r(k) < p_1 \\ x_{j,k}(t) & p_1 \le r(k) \le p_2 \\ round(\xi) & p_2 \le r(k) \end{cases} \tag{15}$$

where $x_{i,k}(t)$ is the location of dimension k of glowworm i at iteration t; $r(k)$ the parameter generated randomly $r(k) \in [0, 1](1 \le k \le m)$; p_1, p_2 the parameters used to control the update probability; and ξ the random variable generated between 0 and 1.

4.2 Initialization of glowworms for UC problem

In this paper, the initialization process is not only to generate a quantity of initial feasible solutions that satisfy all the constraints, but also to keep the diversity of the solutions. At the end of the initialization process, the total production cost corresponding to each glowworm is calculated, which is the basis of the following iterative process.

4.2.1 Structure of glowworms

Each glowworm is a $T \cdot N$ matrix, the elements of which represent all the units' on/off status in the whole scheduling period. For example, u_i^t in row t and column i represents the on/off status of unit i at hour t.

$$U = \begin{bmatrix} u_1^1 & u_2^1 & \cdots & u_N^1 \\ u_1^2 & u_2^2 & \cdots & u_N^2 \\ \vdots & \vdots & \vdots & \vdots \\ u_1^T & u_2^T & \cdots & u_N^T \end{bmatrix}$$

4.2.2 Initialization of glowworms

In order to improve the quality of the initial solutions, the priority list and the decommitment of redundant unit are applied. In this paper, the priority list is based on the capacity of units. The unit that has the maximum capacity has the highest priority. If two units have the same capacity, the one that has lower average full-load cost has the higher priority. The procedure of the initialization of glowworms is as follows.

Step 1: Set $t = 1$.

Step 2: If $t = 1$, set the units whose initial status is a positive number and less than its minimum up time be on status. Else duplicate the on/off status of the units at hour $t - 1$.

Step 3: Check the maximum output of the committed units. If the committed units cannot satisfy the spinning reserve constraint (5) at hour t, commit the unit in the ascending order of the priority list until (5) is satisfied.

Step 4: Search for redundant unit that have the following properties in the descending order of the priority list at hour t.

1) This unit satisfies minimum up time constraint.

2) After this unit is decommitted, the spinning reserve constraint (5) is still satisfied.

If such a unit is found, it is decommitted with fifty percent probability, ensuring the diversity of the initial solutions.

Step 5: Update the status of the unitstatus when they meet the

$$
\begin{cases}
X^t_{ONi} = \begin{cases} X^{t-1}_{ONi} + 1 & if \ u^t_i = 1 \\ 0 & if \ u^t_i = 0 \end{cases} \\
X^t_{OFFi} = \begin{cases} 0 & if \ u^t_i = 1 \\ X^{t-1}_{OFFi} + 1 & if \ u^t_i = 0 \end{cases}
\end{cases} \tag{16}
$$

Step 6: If $t < T$, set $t = t + 1$ and return to step 2. Else go to step 7.

Step 7: Computation of total production cost
The total production cost consists of generation cost and start-up cost. These two parts are discussed separately as follows.

1) Generation cost
We have already got the on/off status of the units at every hour so we can apply the Lambda-iteration method to dispatch the load economically; then use (2) to compute the generation cost of each hour

2) Start-up cost
According to the changing time of the units' on/off status we can use (3) to compute the start-up cost very easily.

Step 8: Update the value of luciferin by using (9).

4.3 Iterative process for UC problem

After the initialization of the glowworms, the glowworms move in the iterative process. However, they may move out of the solution space, which means that the solutions violate at least one of the constraints so the correction method is adopted to keep the solutions feasible. Furthermore, several techniques are proposed in case of falling into the local optimal solution.

4.3.1 Formation of new glowworm by BGSO

In the iterative process, each of the glowworms moves towards another glowworm. The procedure is as follows.
Take glowworm i for example.

Step 1: Computation of the hamming distances between glowworm i and the other glowworms by using (14).

Step 2: Since the objective function is to minimize the production cost, glowworm i is attracted by the glowworms that have less luciferin within its own local-decision range.

Step 3: Use (10) to compute the possibility of moving to glowworm j.

Step 4: Select a glowworm with the roulette approach and use (15) to update the location of glowworm i.

4.3.2 Correction of newly formed glowworm

Instead of the penalty factor, the correction method is adopted to make the newly formed glowworm satisfy all the constraints.
Take the units' on/off status at time t for example.

Step 1: Set the units that violate (7) be on status and the units that violate (8) be off status at hour t.

Step 2: Set the units be off status when they meet the following two conditions.

1) The load at hour t is less than the load at hour $t - 1$.
2) The unit is off status at hour $t - 1$.

Step 3: Check whether the spinning reserve constraint (5) is satisfied.
If not, firstly commit the units that are on status at hour $t - 1$ ($t > 1$) and off status at hour t; then commit the units in the ascending order of the priority list until (5) is satisfied.

Step 4: Update the status of the unit as shown in (16).

4.3.3 Adjustment techniques of newly formed glowworm

In the iterative process these techniques are applied to the feasible solutions in order to provide the probability of better solution and guarantee the diversity of the solutions. The detailed discussion is shown below.

1) Decommitment of redundant unit

Step 1: Search for the redundant unit that have the following two properties in the descending order of the priority list at hour t.

a) This unit satisfies (7).
b) After this unit is decommitted, the spinning reserve constraint (5) is still satisfied.

Step 2: If such a unit is found, use (17) to determine this unit's on/off status at hour t.

$$
u^t_i = \begin{cases} 0 & rand < (j - 0.5)/M_{DTi} \\ 1 & rand \ge (j - 0.5)/M_{DTi} \end{cases} \tag{17}
$$

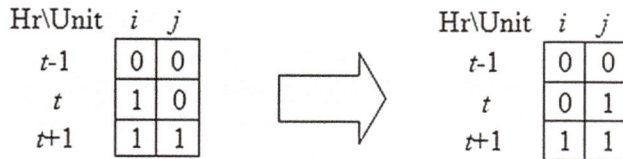

Fig. 1 Switch of the commitment order

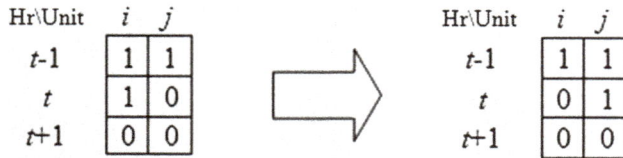

Fig. 2 Switch of the decommitment order

The way to decide the value of j.

  a) Set $k = 1, j = 1$.
  b) If $k + t = 25$, set $j = M_{DTi}$ and break.
  Else if $D^{t+k} > D^t$, set $j = k$ and break.
  c) If $k \leq M_{DTi}$, set $k = k + 1$ and return to step b).
  Else set $j = k$ and break.

Equation (17) takes into account the load information in the following hours so it is more likely to produce better solution; moreover, the usage of the form of probability and random number ensures the diversity of the feasible solutions.

2) Conversion from cold start to hot start

Normally, cold start cost is much larger than hot start cost. Therefore, for the unit i that is committed at hour t when it has been kept off status for exactly $M_{DTi} + C_{SHi} + 1$ hours, if it is committed one hour ahead, the total production cost may decrease. The detailed procedure is as follows

Step 1: Calculate the generation cost at hour $t - 1$ represented by C_{g1}.

Step 2: Commit unit i at hour $t - 1$ and recalculate the generation cost at hour $t - 1$ represented by C_{g2}.

Step 3: Compare $C_{g2} - C_{g1}$ and the difference between cold start cost and hot start cost of unit i. If the latter is larger, it is reasonable to commit unit i one hour ahead, which means unit i's conversion from cold start to hot start.

3) Switch of two units' commitment order in the adjacent two hours

Take the unit i committed at hour t and the unit j committed at hour $t + 1$ shown in Fig. 1 for example. If the commitment of unit j instead of unit i at hour t does not violate (5) and (8), then compute the total production cost at hour t and $t + 1$ in these two different situations and choose the better one with seventy percent probability.

4) Switch of two units' decommitment order in the adjacent two hours

Take the unit j decommitted at hour t and the unit i decommitted at hour $t + 1$ shown in Fig. 2 for example. If the decommitment of unit i instead of unit j at hour t does not violate (5) and (7), then compute the generation cost at hour t in these two different situations and choose the better one with seventy percent probability.

5) Replacement of the committed unit based on the minimum up time

If unit i is committed at hour t and decommitted at hour $t + M_{UTi}$, which indicates that unit i may be redundant from hour $t + 1$ to $t + M_{UTi}$, replace unit i with the units that have less minimum up time and satisfy (8) at hour t with fifty percent probability.

4.4 Implementation of BGSO for solving UC problem

The procedure of the proposed BGSO for solving UC problem is presented as follows.

Table 1 10-unit system data

Unit	P_{max}/MW	P_{min}/MW	a/($/MW^{2}h)	b/($/MWh)	c/($/h)	Min up/h	Min dn/h	Hot start cost/$	Cold start cost/$
1	455	150	0.00048	16.19	1000	8	8	4500	9000
2	455	150	0.00031	17.26	970	8	8	5000	10000
3	130	20	0.002	16.6	700	5	5	550	1100
4	130	20	0.00211	16.5	680	5	5	560	1120
5	162	25	0.00398	19.7	450	6	5	900	1800
6	80	20	0.00712	22.26	370	3	3	170	340
7	85	25	0.00079	27.74	480	3	3	260	520
8	55	10	0.00413	25.92	660	1	1	30	60
9	55	10	0.00222	27.27	665	1	1	30	60
10	55	10	0.00173	27.79	670	1	1	30	60

Step 1: Set counter = 1.
Step 2: Initialize the glowworms as in Section 4.2.2.
Step 3: Use (9) to update every glowworm's luciferin value.
Step 4: Every glowworm move towards one of the other glowworms that have less production cost.

Step 5: Modify the glowworms as in Section 4.3.2 to make them satisfy all the constraints.
Step 6: Adjust the glowworms as in Section 4.3.3.
Step 7: Calculate every glowworm's total production cost.
Step 8: Use (13) to update their local-decision range.
Step 9: If counter < maximum iterations, counter = counter + 1 and go to step 3. Else, go to step 10.
Step 10: Display the optimal solution.

5 Numerical results

The BGSO algorithm is tested on the power system with 10, 20, 40, 60, 80, 100, 120 and 140 generating units in the 24-h scheduling period. The 10-unit data is shown in Table 1 and the power demand in the scheduling period is shown in Table 2. The 20, 40, 60, 80, 100, 120 and 140-unit data are obtained by duplicating the 10-unit data, whereas the power demand is proportional to the number of units. In Addition, the spinning reserve is set to be 10% of power demand.

Table 2 Load demand

Hour	Demand/MW	Hour	Demand/MW
1	700	13	1400
2	750	14	1300
3	850	15	1200
4	950	16	1050
5	1000	17	1000
6	1100	18	1100
7	1150	19	1200
8	1200	20	1400
9	1300	21	1300
10	1400	22	1100
11	1450	23	900
12	1500	24	800

Table 3 Unit output of the 20-unit system's best solution

Hour	Generating unit																			
	1	2	3	4	5	6	7	8	9	10	11	12	13	14	15	16	17	18	19	20
1	455	455	245	245	0	0	0	0	0	0	0	0	0	0	0	0	0	0	0	0
2	455	455	295	295	0	0	0	0	0	0	0	0	0	0	0	0	0	0	0	0
3	455	455	382.5	382.5	0	0	0	0	25	0	0	0	0	0	0	0	0	0	0	0
4	455	455	455	455	0	0	0	0	40	40	0	0	0	0	0	0	0	0	0	0
5	455	455	455	455	0	0	130	0	25	25	0	0	0	0	0	0	0	0	0	0
6	455	455	425	425	130	0	130	130	25	25	0	0	0	0	0	0	0	0	0	0
7	455	455	455	455	130	0	130	130	45	45	0	0	0	0	0	0	0	0	0	0
8	455	455	455	455	130	130	130	130	30	30	0	0	0	0	0	0	0	0	0	0
9	455	455	455	455	130	130	130	130	97.5	97.5	20	20	25	0	0	0	0	0	0	0
10	455	455	455	455	130	130	130	130	162	162	33	33	25	25	10	10	0	0	0	0
11	455	455	455	455	130	130	130	130	162	162	73	73	25	25	10	10	10	10	0	0
12	455	455	455	455	130	130	130	130	162	162	80	80	25	25	43	43	10	10	10	10
13	455	455	455	455	130	130	130	130	162	162	33	33	25	25	10	10	0	0	0	0
14	455	455	455	455	130	130	130	130	97.5	97.5	20	20	25	0	0	0	0	0	0	0
15	455	455	455	455	130	130	130	130	30	30	0	0	0	0	0	0	0	0	0	0
16	455	455	310	310	130	130	130	130	25	25	0	0	0	0	0	0	0	0	0	0
17	455	455	260	260	130	130	130	130	25	25	0	0	0	0	0	0	0	0	0	0
18	455	455	360	360	130	130	130	130	25	25	0	0	0	0	0	0	0	0	0	0
19	455	455	455	455	130	130	130	130	30	30	0	0	0	0	0	0	0	0	0	0
20	455	455	455	455	130	130	130	130	162	162	43	43	0	0	10	10	10	10	10	0
21	455	455	455	455	130	130	130	130	105	105	20	20	0	0	10	0	0	0	0	0
22	455	455	455	455	0	0	130	0	105	105	20	20	0	0	0	0	0	0	0	0
23	455	455	432.5	432.5	0	0	0	0	25	0	0	0	0	0	0	0	0	0	0	0
24	455	455	345	345	0	0	0	0	0	0	0	0	0	0	0	0	0	0	0	0

Table 4 Cost of the 20-unit system's best solution

Hour	Generation cost	Start-up cost	Spinning reserve	On/off status
1	27366.26	0	420	11110000000000000000
2	29109.00	0	320	11110000000000000000
3	33111.24	900	282	11110000100000000000
4	37195.34	900	244	11110000110000000000
5	39457.23	560	274	11110010110000000000
6	44157.72	2220	334	11111011110000000000
7	46008.84	0	234	11111011110000000000
8	48300.68	1100	264	11111111110000000000
9	53838.78	1200	309	11111111111110000000
10	60115.10	640	304	11111111111111110000
11	63832.12	120	314	11111111111111111100
12	67780.33	120	324	11111111111111111111
13	60115.11	0	304	11111111111111110000
14	53838.78	0	309	11111111111110000000
15	48300.68	0	264	11111111110000000000
16	43027.32	0	564	11111111110000000000
17	41283.65	0	664	11111111110000000000
18	44774.09	0	464	11111111110000000000
19	48300.68	0	264	11111111110000000000
20	61047.05	640	299	11111111111100111110
21	53891.99	0	279	11111111111100100000
22	44328.11	0	234	11110010111100000000
23	34862.51	0	182	11110000100000000000
24	30854.84	0	220	11110000000000000000

Table 5 Comparison of simulation results for different systems

Unit	Algorithm	Cost			Mean time
		Best	Worst	Mean	
10	MIP [13]	564647			2
	QEA [12]	563938	564672	563969	19
	IBPSO [10]	563977	565312	564155	27
	BGSO	563938	564226	563952	3
20	MIP [13]	1123908			5
	QEA [12]	1123607	1125715	1124689	28
	IBPSO [10]	1125216	1125730	1125448	55
	BGSO	1123297	1124081	1123771	12
40	MIP [13]	2243020			11
	QEA [12]	2245557	2248296	2246728	43
	IBPSO [10]	2248581	2249302	2248875	110
	BGSO	2242882	2244573	2243582	31
60	MIP [13]	3361614			29
	QEA [12]	3366676	3372007	3368220	54
	IBPSO [10]	3367865	3368779	3368278	172
	BGSO	3361683	3364103	3363115	52
80	MIP [13]	4483194			38
	QEA [12]	4488470	4492839	4490126	66
	IBPSO [10]	4491083	4492686	4491681	235
	BGSO	4482003	4486739	4484513	76
100	MIP [13]	5601857			47
	QEA [12]	5609550	5613220	5611797	80
	IBPSO [10]	5610293	5612265	5611181	295
	BGSO	5601281	5608327	5604186	104
120	BGSO	6722634	6732546	6726644	128
140	BGSO	7891543	7905542	7898763	154

Parameters are set as follows: the number of glowworms is 50; the luciferin decay constant $\rho = 0.4$; the luciferin enhancement constant $\gamma = 0.6$; the local-decision range is twice the number of the units; $\beta = 0.08$; $n_t = 5$; $p_1 = 0.1$; $p_2 = 0.9$. The program is written in MATLAB R2011a and executed on a 2.5 GHz CPU with 4-GB RAM personal computer. In order to have a comprehensive understanding of the BGSO method, 50 trials are done on every test system.

Since the best solution of the 10-unit system of BGSO is the same as that of QEA, the units' power output of the best solution can be seen in [13].The best solution of the 20-unit system is shown in Table 3 and 4, which have never been illustrated in detail before. We can see that thegeneration cost in the scheduling period is 1114879 and the start-up cost is 8400 so the total production cost is 1123297.

The best, worst and mean values of the total production cost, together with the mean computation time by MIP, QEA, IBPSO and BGSO for different test systems are shown in Table 5. We can see that the best solution of the BGSO algorithm is better in most of the test systems and

the best solution of BGSO algorithm is very close to that of the MIP method in the 60-unit test system. From Fig. 3, we can see that the proposed method is faster than the IBPSO method in all the test systems and QEA algorithm in 10, 20, 40 and 60-unit test systems. Although the calculation time of BGSO is longer than that of the MIP method, the calculation time of BGSO increases almost linear with the number of the units, which means that it has the capacity of solving large-scale UC problems.

6 Conclusion

A BGSO has been proposed for solving the UC problem. The distance between the glowworms is represented by the Hamming distance instead of the Euclidean distance and the update of the glowworm's location is expressed by the way of probability. The priority list and decommitment of redundant unit make a big contribution to the high quality

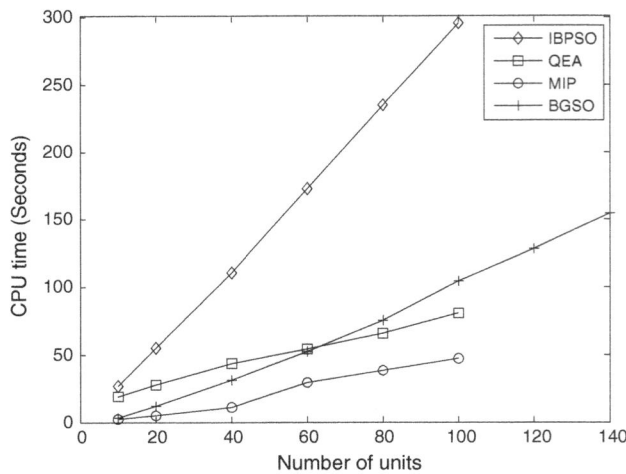

Fig. 3 Compare of different algorithms' computation time

and diversity of the initial solutions. Furthermore, in the iterative process, the correction method and several adjustment techniques help to search for the better feasible solutions. The simulated results show that the total production cost of BGSO is less expensive than those of the other methods in the range of 10–100 units except for the MIP method in the 60-unit system. In addition, the CPU time of BGSO increases almost linear with the size of the units, which is favorable for the large-scale power systems.

References

[1] Wood AJ, Wollenberg BF (2012) Power generation, operation, and control. Wiley, New York

[2] Snyder WL, Powell HD, Rayburn JC (1987) Dynamic programming approach to unit commitment. IEEE Trans Power Syst 2(2):339–348

[3] Senjyu T, Shimabukuro K, Uezato K et al (2003) A fast technique for unit commitment problem by extended priority list. IEEE Trans Power Syst 18(2):882–888

[4] Virmani S, Adrian EC, Imhof K et al (1989) Implementation of a Lagrangian relaxation based unit commitment problem. IEEE Trans Power Syst 4(4):1373–1380

[5] Ongsakul W, Petcharaks N (2004) Unit commitment by enhanced adaptive Lagrangian relaxation. IEEE Trans Power Syst 19(1):620–628

[6] Kazarlis SA, Bakirtzis AG, Petridis V (1996) A genetic algorithm solution to the unit commitment problem. IEEE Trans Power Syst 11(1):83–92

[7] Wong KP, Wong YW (1995) Thermal generator scheduling using hybrid genetic/simulated-annealing approach. IEE Proc-Gener Transm Distrib 142(4):372–380

[8] Wong KP, Wong YW (1996) Combined genetic algorithm/simulated annealing/fuzzy set approach to short-term generation

scheduling with take-or-pay fuel contract. IEEE Trans Power Syst 11(1):128–136

[9] Gaing ZL (2003) Particle swarm optimization to solving the economic dispatch considering the generator constraints. IEEE Trans Power Syst 18(3):1187–1195

[10] Jeong YW, Park JB, Jang SH et al (2010) A new quantum-inspired binary PSO: Application to unit commitment problems for power systems. IEEE Trans Power Syst 25(3):1486–1495

[11] Yuan XH, Nie H, Su AJ et al (2009) An improved binary particle swarm optimization for unit commitment problem. Expert Syst Appl 36(4):8049–8055

[12] Juste KA, Kita H, Tanaka E et al (1999) An evolutionary programming solution to the unit commitment problem. IEEE Trans Power Syst 14(4):1452–1459

[13] Lau TW, Chung CY, Wong KP et al (2009) Quantum-inspired evolutionary algorithm approach for unit commitment. IEEE Trans Power Syst 24(3):1503–1512

[14] Wang N, Zhang LZ, Xie GH (2010) An improved mixed integer quadratic programming algorithm for unit commitment. Autom Electr Power Syst 34(15):28–32 (in Chinese)

[15] Krishnanand KN, Ghose D (2009) Glowworm swarm optimization for simultaneous capture of multiple local optima of multimodal functions. Swarm Intell 3(2):87–124

[16] Gong QQ, Zhou YQ, Yang Y (2011) Artificial glowworm swarm optimization algorithm for solving 0-1 knapsack problem. Adv Mater Res 143(144):166–171

[17] Wu B, Qian CH, Ni WH et al (2012) The improvement of glowworm swarm optimization for continuous optimization problems. Expert Syst Appl 39(7):6335–6342

[18] Steane AM (1996) Error correcting codes in quantum theory. Phys Rev Lett 77(5):793–797

Mingwei LI received the B.S. degree in electrical engineering from Shanghai Jiao Tong University, Shanghai, China in 2012. In 2013, he was a visiting student with the Department of Electrical and Computer Engineering, Institution of Georgia Technology, Atlanta, USA. He is currently pursuing his M.S. degree in Shanghai Jiao Tong University, Shanghai, China. His research interests include power system optimization, renewable energy and electricity market.

Xu WANG received the B.S. degree in electrical engineering from Southeast University, Nanjing, China in 2010. He is currently pursuing his Ph.D. degree in Shanghai Jiao Tong University, Shanghai, China. His research interests include electricity market and power system risk management.

Yu GONG received the B.S. degree in electrical engineering from Shanghai Jiao Tong University, Shanghai, China in 2012. He is currently pursuing his M.S. degree in Shanghai Jiao Tong University, Shanghai, China. His research interests include electricity and low carbon electricity.

Yangyang LIU received the B.S. and M.S. degree in electrical engineering from Shanghai Jiao Tong University, Shanghai, China in 2009 and 2011. He is currently pursuing his Ph.D. degree in Shanghai Jiao Tong University, Shanghai, China. His research interests include electricity market and power system optimization.

Chuanwen JIANG received the M.S. and Ph.D. degrees from Huazhong University of Science and Technology, Hubei, China, in 1996 and 2000, respectively. He is currently a professor with Shanghai Jiao Tong University, Shanghai, China. His research interests include power system optimization, load forecast in power system, and electricity market.

21

Control and protection strategy for MMC MTDC system under converter-side AC fault during converter blocking failure

Puyu WANG, Xiao-Ping ZHANG (✉),
Paul F. COVENTRY, Zhou LI

Abstract This paper investigates a control and protection strategy for a four-terminal modular multilevel converter (MMC) based high-voltage direct current (HVDC) system under a converter-side AC fault. Based on the system operating condition, a control and protection strategy against the fault with normal blocking of the converter is proposed. In practical, applications encountering such a fault, the MMC at the fault side may experience different conditions of blocking failure. The blocking failures may occur on: ① the whole converter; ② one converter arm; ③ one sub-module (SM)/several SMs of one converter arm; ④ other conditions. The phenomenon of the multi-terminal HVDC (MTDC) system following the fault is analyzed under the first three conditions with real-time simulations using the real-time digital simulator (RTDS). Based on the impact of different conditions on the MTDC system, the necessity of utilizing special control and protection is discussed. A special control and protection strategy is proposed for emergency conditions, and its effectiveness is verified by real-time simulation results.

Keywords Converter blocking failure, Converter-side AC fault, Control and protection, Modular multilevel converter (MMC), Multi-terminal high-voltage direct current (MTDC), Real-time digital simulation

1 Introduction

There has been a rather long history of research on the control and protection of AC faults in electric power transmission systems [1, 2]. Regarding the high-voltage direct current (HVDC) including multi-terminal HVDC (MTDC) technologies, research effort has been made on the protection against DC faults [3–6]. As for AC faults, control strategies of voltage source converter (VSC) for improving the AC fault ride-through capability of VSC-HVDC systems were proposed in [7, 8]. An approach of limiting the AC unbalanced fault on the VSC MTDC grids was proposed in [9]. Control and protection strategies were investigated in [10] for a hybrid MTDC system against AC faults. In most situations, AC faults can be isolated by AC circuit breakers. However, when an AC fault occurs at the nearby AC side of the converter, which is beyond the protection area of the nearby AC circuit breakers, the fault isolation is achieved by both blocking of the converter to prevent current flowing from the DC side and tripping the AC side circuit breakers. In the above situation, previous work has assumed that the blocking of the converter is successful. However, this assumption is not necessarily valid in terms of practical operation of converters, especially in the more advanced modular multilevel converter (MMC) [11, 12] in the current system. The MMC at the fault side has different conditions of blocking failure: ① for the communication outage of control signals where the whole converter can be out of control and cannot be blocked, although the occurrence of such a condition is considered to be rare; ② a more common condition is that one of the six converter arms is failed to be blocked; ③ the most common condition is that one sub-module (SM)/several SMs on one converter arm is/are failed to be blocked; ④ other conditions. Hence, it is worth identifying the potential risk to the MTDC system, analyzing the dynamic performance to reveal the fact that special control and

P. WANG, X.-P. ZHANG, University of Birmingham, Birmingham B15 2TT, UK
(✉) e-mail: x.p.zhang@bham.ac.uk
P. F. COVENTRY, National Grid, Warwick CV34 6DA, UK
Z. LI, Southeast University, Nanjing 210096, China

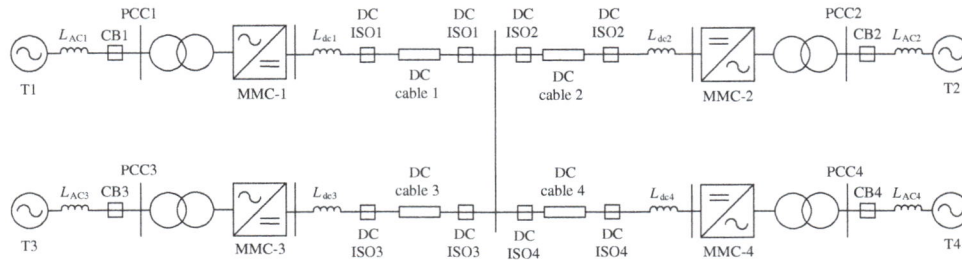

Fig. 1 Configuration of four-terminal MMC HVDC system

protection is necessary to be conducted when the MTDC system is significantly affected under emergency conditions, while under other non-emergency conditions, the special control and protection may not be essential.

This paper investigates the control and protection strategy against a converter-side AC fault during the blocking failure of the converter of a four-terminal MMC HVDC system. The MMC technology was firstly introduced in 2001 [13] and its advantageous over the traditional VSC technology [11, 12], such as low switching losses and small harmonic proportions, have been widely recognized. The number of HVDC projects deployed the MMC technology [14, 15] has gradually been increasing. The occurrence of the aforementioned blocking failure conditions may exist and deserve to be investigated.

The rest of this paper is outlined as follows. Section 2 introduces the MTDC system configuration and control strategy. Section 3 presents the control and protection strategy against a converter-side AC fault with normal blocking operation. In Section 4, the dynamic performance of the MTDC system during different conditions of MMC blocking failure following the fault is analyzed and the impact of each condition on the MTDC system is revealed by real-time simulations. An associated special control and protection strategy is proposed for certain emergency conditions where the MTDC system is significantly affected. Conclusions are drawn in Section 5.

2 MTDC system

A single-line diagram of a four-terminal MMC HVDC system is shown in Fig. 1. Tn ($n = 1, 2, 3, 4$) denotes the HVDC terminal. On the AC side, each AC source is modeled as an ideal voltage source with a line-to-neutral voltage of 138 kV. CBn ($n = 1, 2, 3, 4$) represents the AC circuit breaker. On the DC side, DC ISOn denotes the DC isolation switch. The length of each DC cable is 100 km. The nominal DC voltage is ±50 kV. Four converters (MMC-n) are seven-level, half-bridge MMC converters. Each MMC consists of six converter arms where each arm consists of six SMs and one arm inductor.

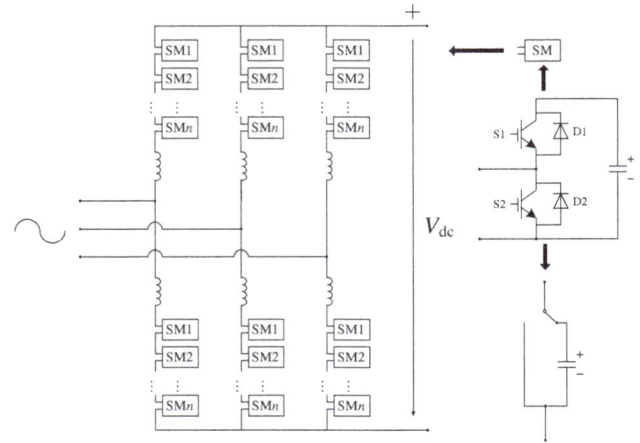

Fig. 2 Schematic diagram of MMC system

Table 1 Parameters of MTDC system

Quantity	Value
MMC rating	150 MVA
Nominal AC voltage (L-N)	138 kV
AC system inductance L_{AC}	150 mH
Nominal AC frequency	50 Hz
Transformer voltage ratio (L-N)	138 kV/30 kV (Y/Δ)
Transformer rating	150 MVA
Transformer leakage inductance	5%
Nominal DC voltage	±50 kV
Arm inductance	3 mH
SM capacitance	2,500 μF

Figure 2 shows a schematic diagram of the MMC system. To achieve the capacitor voltage balancing in each converter arm, the charging and discharging operations depend on the arm current direction and the capacitor voltage of each SM [12, 13] where the SM with lower capacitor voltage is charged first, while the SM with higher capacitor voltage is discharged first. Detailed parameters of the MTDC system are shown in Table 1.

The control of the MMC converter station is achieved in a dq reference frame and the well-known dq decoupled control strategy is applied [16–18]. The d-axis control regulates either the active power or the DC voltage, while the q-axis control regulates either the reactive power or the AC voltage magnitude. In the MTDC system, MMC-1, MMC-3 and MMC-4 use the active power control to regulate the active power at the converter AC terminals. MMC-2 applies DC voltage control to maintain the voltage of the MTDC grid. In order to reduce the power losses caused by the reactive current, the reactive power control is applied by all the four converter stations where the reactive power reference is set to 0 Mvar.

3 Control and protection strategy against converter-side AC fault with normal blocking operation

When an AC fault occurs at the converter AC side of one terminal of the MTDC system, say the most severe three-phase short-circuit fault, the voltage at the faulted point will drop, and the fault current will flow into the faulted point from both the grid AC side and the MMC AC side.

Figure 3 shows the diagram of the fault occurring at the converter AC side. The MMC AC side current is from the DC grid flowing through the MMC. In the MTDC system without appropriate control and protection, the occurrence of the AC fault at one terminal will affect the interconnected terminals. In order to isolate the fault and protect power electronic devices, the insulated gate bipolar transistors (IGBTs) of the MMC at the fault side are rapidly blocked when the fault occurs. The unilateral conductive characteristics of the diode will prevent the DC current flowing from the DC grid to the MMC AC side. In order to isolate the fault from the grid AC side, the AC circuit breaker at the fault terminal is tripped. The blocking of the MMC and tripping of the AC circuit breaker to isolate the fault is shown in Fig. 4.

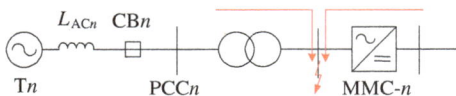

Fig. 3 Three-phase short-circuit fault at MMC AC side

Fig. 4 Fault isolation by blocking of MMC and tripping of AC circuit breaker

If the fault is temporary, the fault terminal can be resumed to the normal operating condition once the fault is cleared. If it is a permanent fault, the MTDC system will be re-configured as a three-terminal HVDC system. In the following case studies, the AC fault is applied as a permanent fault at T1.

The control and protection strategy against a converter side AC fault comprises a series of control and protection operations, i.e., control and protection sequence. The control and protection sequence of the four-terminal MMC HVDC system under a permanent three-phase short-circuit fault at the MMC AC side of T1 is shown in Fig. 5. The fault is applied at 2 s. The IGBTs can normally be blocked within $1 \sim 2$ ms, while the AC circuit breakers can normally be tripped within half to one cycle, i.e., $10 \sim 20$ ms [19]. In the control and protection sequence discussed in this paper, MMC-1 is assumed to be blocked within 2 ms, while CB1 is tripped at 20 ms after the fault. Simulations are carried out using the real-time digital simulator (RTDS) to present the system dynamic performance and verify the proposed control and protection strategy.

Initially, T1 and T3 are importing powers to the DC grid, while T2 and T4 are exporting powers from the DC grid. The system dynamic performance is shown in Fig. 6 applying the control and protection sequence of Fig. 5. Figure 6a shows the active power of each terminal measured at the point of common coupling (PCC), Fig. 6b shows the root-mean-square (RMS) value of the AC current at PCC1, Fig. 6c shows the RMS value of the AC side current of MMC-1, Fig. 6d shows the DC voltage of each terminal and Fig. 6e shows the DC current of each terminal.

In Fig. 6a, the increase of the active power at T1 is observed due to the short-circuit fault. Since MMC-1 is successfully blocked 2 ms later, isolating the fault from the DC grid, the active power at T3 and T4 is maintained stable at the nominal value and the loss of power at T1 is balanced by MMC-2. The AC current of PCC1 surges to a peak of 0.84 kA following the fault and reduces to zero when CB1 is tripped, as shown in Fig. 6b. Due to the successful fast blocking of MMC-1, no overcurrent is observed at the MMC AC side, as shown in Fig. 6c. The DC voltages of the other three terminals decrease initially and resume to the nominal value smoothly, as shown in Fig. 6d, since MMC-1 has been operated as a rectifier to

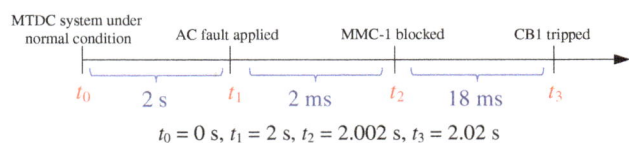

Fig. 5 Control and protection sequence of MTDC system

(a) Active powers of MTDC

(b) RMS value of AC current at PCC1

(c) RMS value of MMC-1 AC side current

(d) MMC DC side voltages

(e) MMC DC side currents

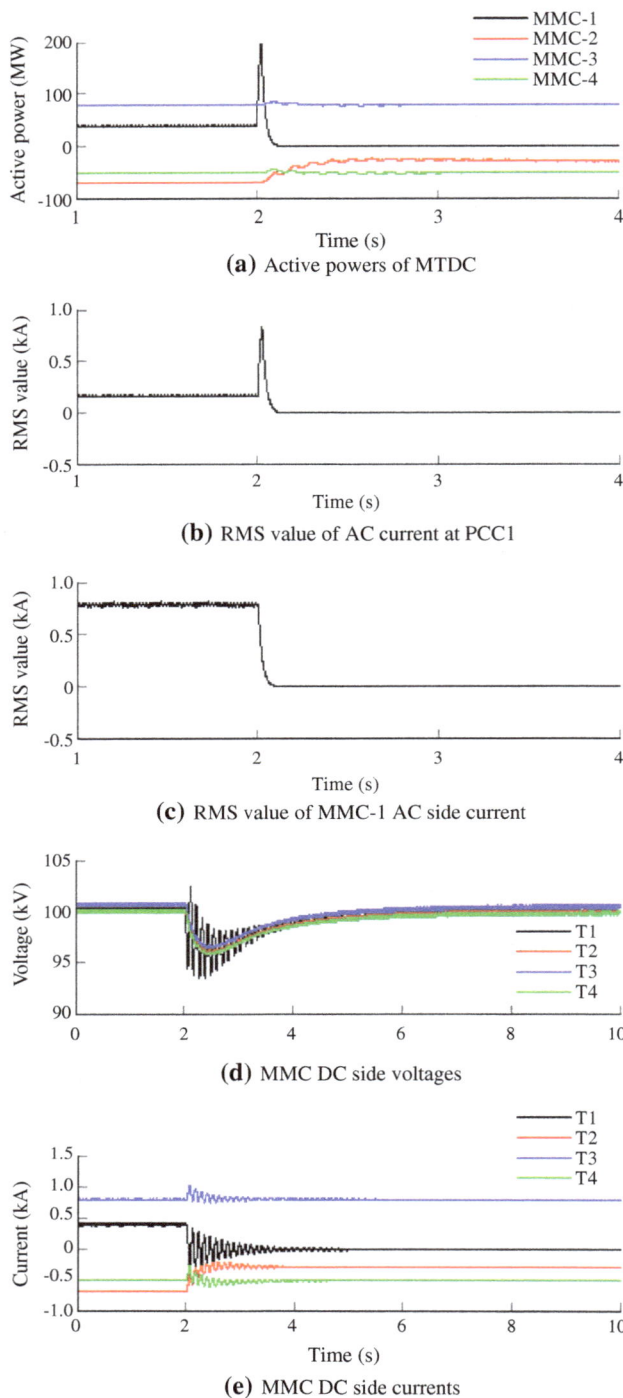

Fig. 6 MTDC dynamic performance with normal converter blocking

inject the power from the AC system to the DC grid. The DC current at T1 gradually decreases to zero, while the DC current of the others is not significantly influenced, as shown in Fig. 6e. Therefore, the simulation results verify that with normal blocking operation of the converter, the proposed control and protection strategy is effective against the fault at the MMC AC side.

4 Control and protection strategy against converter-side AC fault during converter blocking failure

Previous analysis and simulation were made with normal converter blocking operation. In this section, the same three-phase short-circuit fault is applied; however, the MTDC system is under different conditions of blocking failure of MMC-1.

4.1 Whole MMC blocking failure

If MMC-1 is completely not blocked, i.e., all SMs on six arms fail to be blocked, and only CB1 is tripped after the fault, the DC current of T2 for power balancing together with the current of T3 will flow into the faulted point at T1 through MMC-1. This process is illustrated by the red arrow and the power flow direction of each terminal is shown by the blue arrow before the occurrence of the fault, as shown in Fig. 7.

During the transient period when the fault is applied, the voltage drop at the faulted point will lead to a fault current surge, which comes from the AC utility side and MMC-1 AC side, and may potentially damage the system equipment. Because of the tripping of CB1, the fault current from the utility side is interrupted and then contributed by the current from MMC-1 AC side. The fault characteristics with complete blocking failure of MMC-1 are shown in Fig. 8. Figure 8a shows RMS value of MMC-1 AC side current, which presents the fault current from the converter side due to the blocking failure. The MMC DC side current of each terminal is shown in Fig. 8b, which presents the impact of the blocking failure on the adjacent terminals.

In Fig. 8a, the MMC-1 AC side current increases and stays at a peak of 5.6 kA, which is 7 times of the nominal value and may damage the system equipment, particularly the IGBTs of MMC-1 and MMC-2. Large DC over currents can be observed in both T1 and T2, and the current reversal of T2 can be seen in Fig. 8b. Therefore, it is indicated that the fault is not fully isolated, necessitating special control and protection to isolate the fault, preventing the power from further injecting into the faulted point.

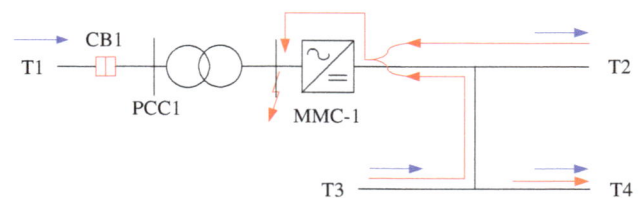

Fig. 7 Schematic diagram of MMC-1 blocking failure and tripping of CB1 following fault

(a) RMS value of MMC-1 AC side current

(b) MMC DC side current of each terminal

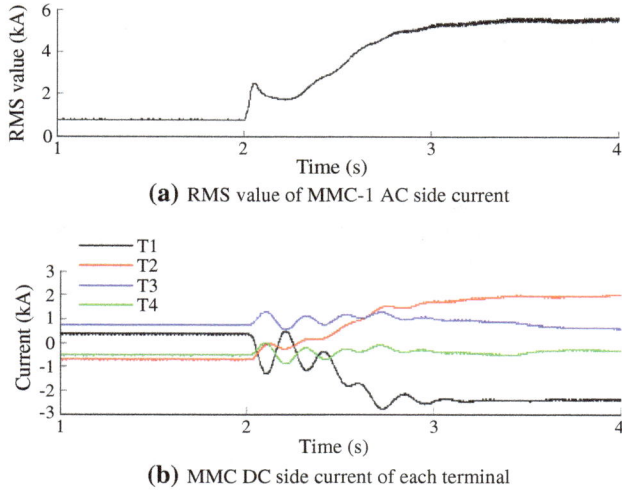

Fig. 8 MTDC system fault characteristics with blocking failure of whole MMC-1 converter

Firstly, the IGBTs of the other MMCs need to be protected immediately by blocking the adjacent MMCs. Nevertheless, the freewheeling diode cannot prevent the AC currents flowing into the DC grid, so the AC circuit breakers of the interconnected terminals need to be tripped. Figure 9 shows the proposed special control and protection strategy of the MTDC system.

After tripping all the AC circuit breakers, the MTDC system has no power sources and the current of the DC grid gradually decreases. When the DC current at T1 reduces to zero, the DC ISO1 is tripped to realize the isolation of the fault section at T1. Then, T2, T3 and T4 could be restored. Figure 10 shows the MTDC system with the fault section isolation and restoring of the other three terminals.

With the blocking failure of whole MMC-1, special control and protection is activated where the control and protection sequence is shown in Fig. 11. The fault is applied at 2 s. MMC-1 completely fails to be blocked and only CB1 is tripped at 20 ms later after the fault. When the

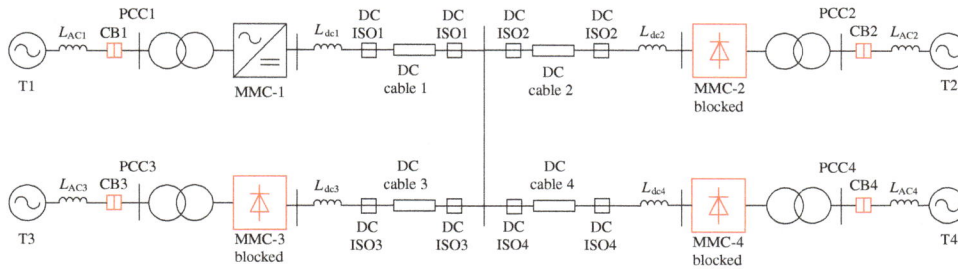

Fig. 9 Special control and protection strategy against whole MMC-1 blocking failure

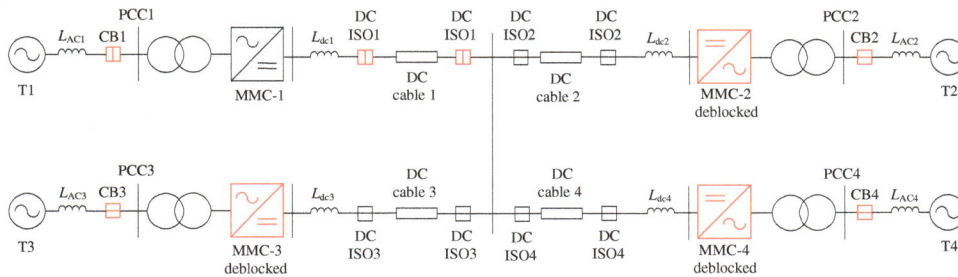

Fig. 10 MTDC system recovery after the faulted section isolation

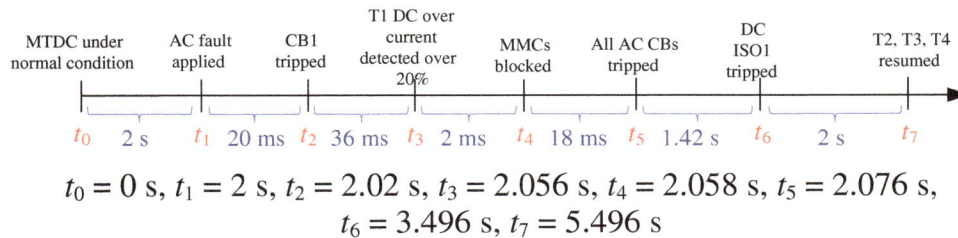

$t_0 = 0$ s, $t_1 = 2$ s, $t_2 = 2.02$ s, $t_3 = 2.056$ s, $t_4 = 2.058$ s, $t_5 = 2.076$ s, $t_6 = 3.496$ s, $t_7 = 5.496$ s

Fig. 11 Control and protection sequence of MTDC system under AC fault during blocking failure of whole MMC-1

DC current of any terminal is detected over 20% of its nominal value, the MMCs of the other three terminals will be blocked within 2 ms with tripping the AC circuit breakers within 20 ms. In the simulation, the DC current of T1 is the first to be detected over 20% of its nominal value. DC ISO1 is tripped when the DC current of T1 reduces to zero. In the case study, it takes approximately 1.42 s for the DC current of T1 to decay to zero after tripping all the AC circuit breakers. T2, T3 and T4 can be resumed after the tripping of DC ISO1.

The control and protection characteristics of the MTDC system are shown in Fig. 12. Figure 12a shows the active power of the MTDC system, Fig. 12b shows the RMS value of MMC-1 AC side current, Fig. 12c shows the MMC DC side voltages and Fig. 12d shows the MMC DC side currents.

The active power of the MTDC system decreases to zero within 0.15 s by applying the proposed control and protection strategy, as shown in Fig. 12a. The RMS value of MMC-1 AC side current increases to a peak of 2.5 kA when MMC-1 completely fails to be blocked, as shown in

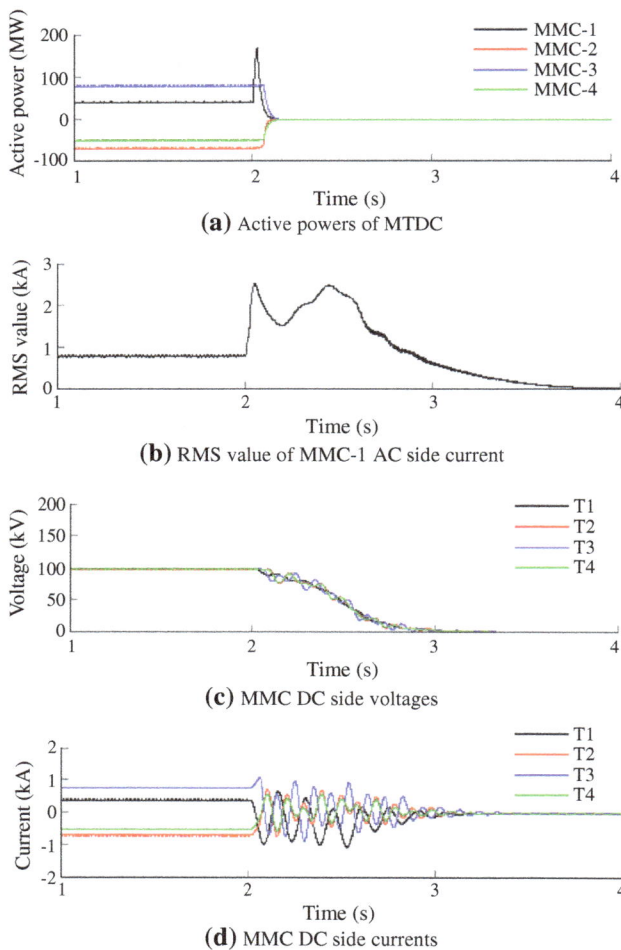

(a) Active powers of MTDC

(b) RMS value of MMC-1 AC side current

(c) MMC DC side voltages

(d) MMC DC side currents

Fig. 12 Control and protection characteristics with whole MMC-1 blocking failure

Fig. 12b. The RMS value of MMC-1 AC side current in this figure is much lower than that of the previous case without special protection and reduces to zero within 2 s. The DC voltage of the MTDC system gradually reduces to zero after tripping all the AC circuit breakers, as shown in Fig. 12c. The DC current at T1 is monitored over 0.48 kA at 2.056 s, as shown in Fig. 12d, leading to the blocking of the adjacent MMCs and tripping of all the AC circuit breakers. The DC current of the other terminals reduces to zero with small oscillations. It can be seen that the oscillations and over current have been reduced, and the fault is fully isolated using the proposed control and protection strategy when the complete blocking failure of MMC-1 appears.

4.2 One converter arm complete blocking failure

A higher possibility is considered to have the blocking failure occurred on part of MMC-1 instead of the whole converter. The impact of such a condition on the MTDC system should be less significant and will determine the necessity of utilizing the special control and protection.

The condition discussed in this section is that one of the six converter arms has blocking failure, i.e., all the SMs on one arm fail to be blocked, while the other five arms are blocked normally. The upper arm of phase A of MMC-1 is selected to have blocking failure in the simulation, the simulation results are shown in Fig. 13. Figure 13a shows the current in phase A of MMC-1 AC side, Fig. 13b shows the MMC DC side current of each terminal. The control and protection sequence is the same as that in Fig. 5.

The current in phase A of MMC-1 AC side increases to a peak of 6.2 kA, as shown in Fig. 13a. MMC-1 DC side current at T1 reverses and eventually stabilizes at −1.1 kA. Therefore,

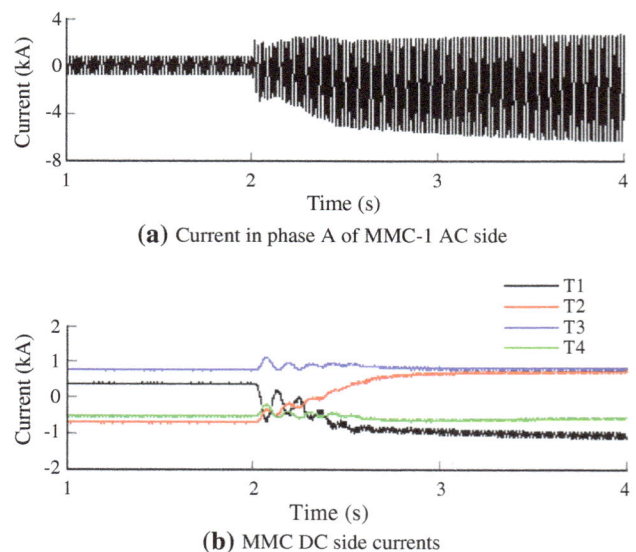

(a) Current in phase A of MMC-1 AC side

(b) MMC DC side currents

Fig. 13 Blocking failure of one converter arm of MMC-1

it is indicated that, due to the blocking failure of the upper arm of phase A, the AC and DC side of MMC-1 are not fully isolated. This leads to the result that the MMC DC side current of T1 does not reduce to zero after the protection operation, but flows through MMC-1 and injects into the faulted point. Therefore, the blocking failure of one arm of the converter and the whole converter blocking failure are both considered as emergency condition, necessitating the fault isolation with the special control and protection strategy proposed in Section 4.1.

4.3 One converter arm partly blocking failure

In comparison with the complete blocking failure of one converter arm, the possibility of having the blocking failure on part of a converter arm would be higher in practical operations. Three different conditions of blocking failure on one arm are considered: ① all SMs except one; ② half of the SMs; ③ only one SM.

4.3.1 All SMs but one blocking failure

Despite the fact that all SMs fail to be blocked except one, the one that is blocked successfully provides the possibility of interrupting the DC current flowing to the MMC AC side. This is because that the arm current must flow through the diode D1, as shown in Fig. 2 and charge the blocked SM capacitor before flowing to the MMC AC side. When the voltage of the SM capacitor is charged higher than the positive DC voltage of T1, D1 will work in the reverse direction to block the current.

Five SMs of the upper arm of phase A are simulated to have blocking failure. The simulation results are shown in Fig. 14. Figure 14a shows the current on phase A of MMC-1 AC side, Fig. 14b shows the MMC DC side current of each terminal and Fig. 14c shows the voltages of the SM capacitors on the upper arm of phase A.

Figure 14a shows that the current in phase A of MMC-1 AC side is largely limited with a peak value of 1.2 kA and reduces to zero within 1.5 s, due to the existence of the blocked SM. In Fig. 14b, the DC current of T1 decreases to zero indicating the isolation of AC and DC side of MMC-1. However, the SM, which has been blocked successfully, is charged until its voltage increases to the nominal positive voltage of T1 (50 kV), so as to interrupt the arm current and achieve the isolation. Therefore, even the overcurrent is largely restricted and the isolation of the fault side is achieved, the blocked SM capacitor is identified to have a potential risk of being overcharged and breakdown.

4.3.2 Half converter arm blocking failure

In comparison with the condition that all SMs have blocking failure except one, it is more likely to have the

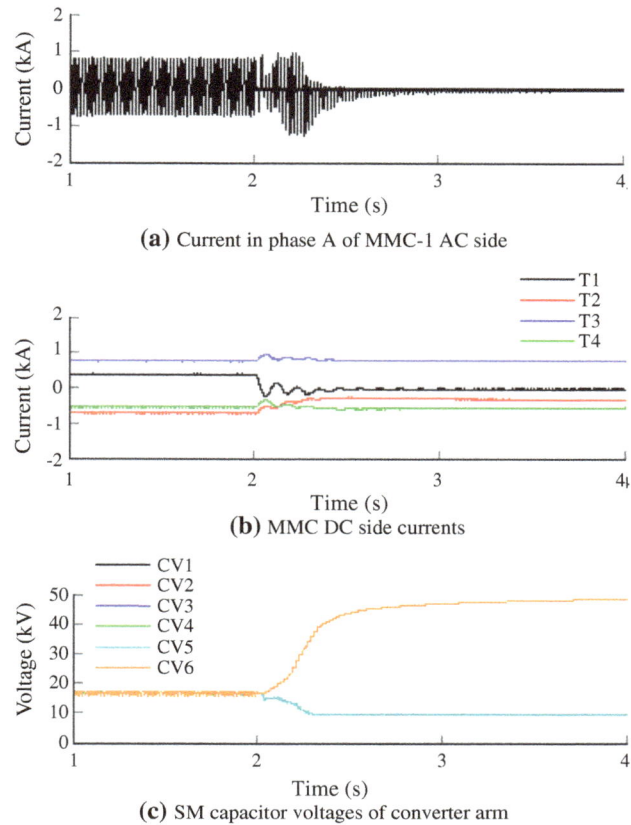

(a) Current in phase A of MMC-1 AC side

(b) MMC DC side currents

(c) SM capacitor voltages of converter arm

Fig. 14 Blocking failure of five SMs of one converter arm of MMC-1

blocking failure on half of the SMs. When half of the SMs are blocked successfully, the blocked SM will equally divide the positive DC voltage when positive DC voltage is larger than their sum. According to the applied strategy of the SM capacitor balancing, the voltages of the blocked SM capacitors will eventually stabilize at the nominal value. Three SMs of the upper arm of phase A are simulated to have blocking failure with the simulation results, as shown in Fig. 15. Figure 15a shows the current in phase A of MMC-1 AC side, Fig. 15b shows the MMC DC side current of each terminal and Fig. 15c shows the voltages of SM capacitors on the upper arm of phase A.

The peak of the current on phase A of MMC-1 AC side further reduces to 1 kA when half of the SMs have been blocked successfully and the current decreases to zero within 1 s, as shown in Fig. 15a. In addition, there is no overvoltage on the capacitors of the blocked SMs, as shown in Fig. 15c.

4.3.3 One SM blocking failure

In practical application, there are tens or hundreds of SMs on each arm, so the blocking failure of one SM is

(a) Current in phase A of MMC-1 AC side

(b) MMC DC side currents

(c) SM capacitor voltages of converter arm

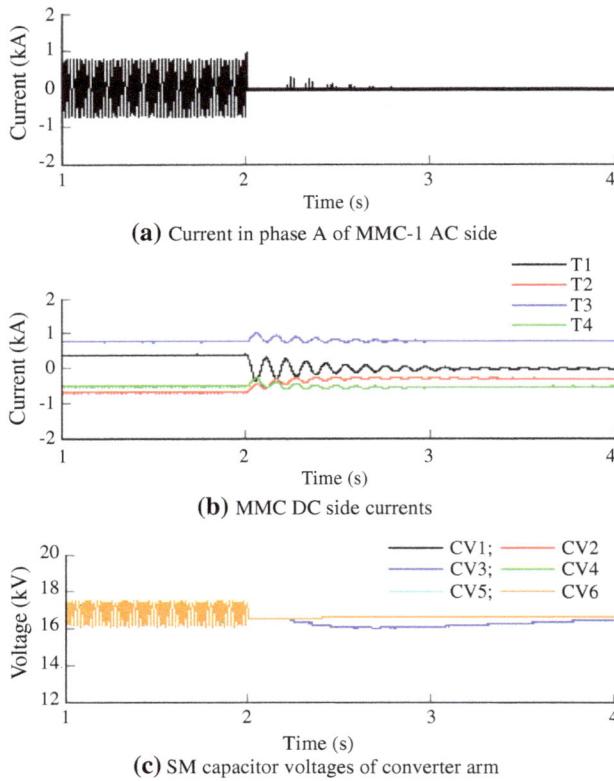

Fig. 15 Blocking failure of three SMs of one converter arm of MMC-1

(a) Current in phase A of MMC-1 AC side

(b) MMC DC side currents

(c) SM capacitor voltages of converter arm

Fig. 16 Blocking failure of one SM of one converter arm of MMC-1

considered to be more probable than the others. In the previous section, there was little overcurrent on the AC side of MMC-1 and no overvoltage on the capacitors of the blocked SMs when half of the SMs on one arm failed to be blocked following the fault. Therefore, it can be predicted that the impact of one SM blocking failure on the MTDC system should be even smaller. Figure 16a shows the current in phase A of MMC-1 AC side, Fig. 16b shows the MMC DC side current of each terminal and Fig. 16c shows the voltages of the SM capacitors on the upper arm of phase A.

When only one SM has blocking failure after the fault, there is no overcurrent observed in Fig. 16a. Since the positive DC voltage of T1 is smaller than the sum of the capacitor voltages of the blocked SMs, there would be no charging current on the converter arm and thus no overvoltage on the capacitors of the blocked SMs.

5 Discussion

The investigated blocking conditions are: ① the converter is normally blocked; ② the converter has complete blocking failure; ③ one converter arm has complete blocking failure; ④ one converter arm has partly blocking

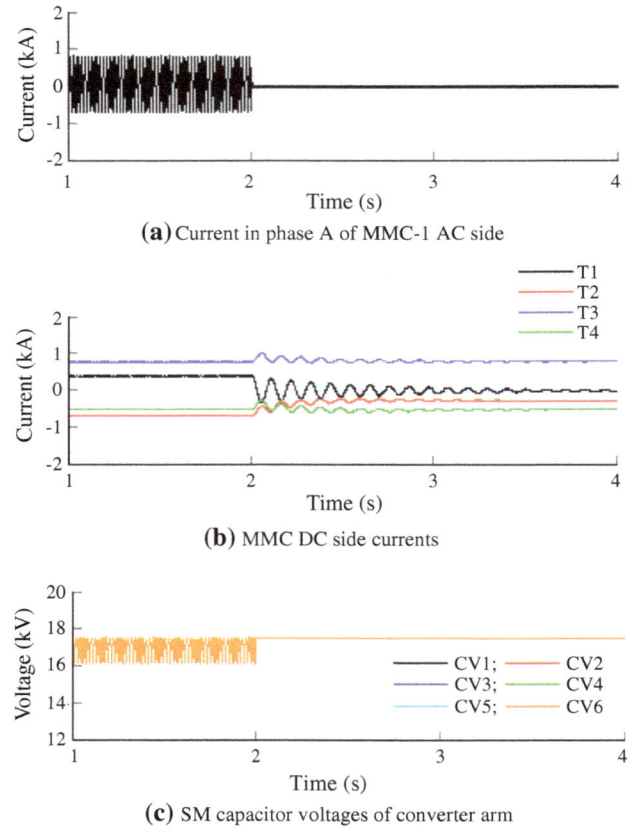

failure, including all SMs except one have blocking failure; half of SMs has blocking failure; only one SM has blocking failure.

According to the analytical study and simulation results, condition ② is considered to be the most severe condition with least occurring possibility, but has significant impact on the MTDC system, necessitating the special control and protection. As for condition ③ and the conditions between ② and ③, since the isolation of the faulted AC and DC sides is not achieved eventually, they will cause large influence on the MTDC system and are regarded as emergency conditions which require the special control and protection.

In condition ④, the isolation of the faulted AC side can be achieved without the special control and protection, so the basic control and protection strategy under condition ① is sufficient. The fault analyzed and simulated in this paper is the most severe AC fault, three-phase short-circuit fault. Therefore, the other types of faults can be effectively isolated in the MTDC system by applying the special control and protection strategy proposed.

During the protection period of condition described in Section 4.3.1, the capacitor of the blocked SM is identified to have the potential risk of being overcharged and breakdown. Furthermore, another potential problem with low probability, but may exist in condition described in Section 4.3.2, is found in the simulation. That is, at the instant of the blocking operation of the converter, if SMs that have blocking failure are under the switch-on condition, i.e. the capacitors are not bypassed; an impulse of the charging current will emerge on the converter arm leading to the charging process on the blocked SMs. Hence, attention needs to be paid on such conditions that there may be potential risk of overvoltage on the blocked SM capacitors, although the possibility is small.

The fault is assumed to occur at T1 in this paper. Due to the same control strategy of the converter station of T1, T3 and T4, which is the constant active power control, the control and protection strategy will be applicable for the fault at the other two terminals. But for the fault that occurs at T2, the constant DC voltage control is used, the control and protection operation will block the control system leading to the voltage of the DC grid out of control. Under this condition, one of the other three terminals needs to take over the control of the DC voltage, which can be achieved by the voltage margin control [20] or voltage droop control [18] instead of constant real power control.

6 Conclusions

Since there are a large number of series-connected SMs on each arm of MMC HVDC converters, the possibility of contingent failure on one or more SMs cannot be neglected. This paper proposes the issue of potential blocking failure of the MMC at the fault side after an AC fault. The blocking failure may occur on one or several SMs, or even the whole converter.

In this paper, the impact of different blocking failures in an MTDC system after a converter-side AC fault is analyzed with simulation studies. According to the performance and impact on the MTDC, they have been divided into the emergency condition and non-emergency condition. Under the emergency condition, the isolation between the faulted AC and DC sides is not achieved using the normal control and protection strategy. A special control and protection strategy, which achieves both the isolation of the fault terminal with reducing the fault current and recovery of the other terminals, is proposed. Under the non-emergency condition, a potential risk of overcharging the blocked SMs has been identified. The simulation is conducted using the RTDS. Simulation results prove the effectiveness of the proposed special control and protection strategy under the emergency condition.

Acknowledgement This work is supported by UK EPSRC and UK National Grid.

References

[1] Kundur P (1994) Power system stability and control. McGraw-Hill, New York

[2] Ibrahim M (2012) Disturbance analysis for power systems. Wiley, New York

[3] Tang L, Ooi BT (2007) Locating and isolating DC faults in multiterminal DC systems. IEEE Trans Power Deliv 22(3):1877–1884

[4] Lu W, Ooi BT (2003) DC overvoltage control during loss of converter in multiterminal voltage-source converter-based HVDC (M-VSC-HVDC). IEEE Trans Power Deliv 18(3):915–920

[5] Jovcic D, Ooi BT (2010) Developing DC transmission networks using DC transformers. IEEE Trans Power Deliv 25(4):2535–2543

[6] Callavik M, Blomberg A, Häfner J et al (2012) The hybrid HVDC breaker. In: ABB Grid Syst, Technical Paper

[7] Adam GP, Ahmed KH, Finney SJ et al (2010) AC fault ride-through capability of a VSC-HVDC transmission systems. In: Proceedings of energy conversion congress and exposition (ECCE), Atlanta, US, 12–16 Sept 2010, pp 3739–3745

[8] Yue W, Zhao C, Lu Y et al (2010) Study and simulation of VSC-HVDC under AC power system faults. In: Proceedings of 5th international conference on critical infrastructure (CRIS), Beijing, China, 20–22 Sept 2010, pp 1–6

[9] Suul JA, Luna A, Rodríguez P et al (2012) Power control of VSC HVDC converters for limiting the influence of AC unbalanced faults on multi-terminal DC grids. In: Proceedings of 9th IET international conference on AC and DC power transmission (ACDC 2012), Birmingham, UK, 4–5 Dec 2012, pp 1–7

[10] Chen X, Sun H, Wen J et al (2011) Integrating wind farm to the grid using hybrid multiterminal HVDC technology. IEEE Trans Ind Appl 47(2):965–972

[11] Davies M, Dommaschk M, Dorn J et al (2008) HVDC PLUS—basics and principle of operation. In: Siemens Tech Rep, 2008

[12] Saeedifard M, Iravani R (2010) Dynamic performance of a modular multilevel back-to-back HVDC system. IEEE Trans Power Deliv 25(4):2903–2912

[13] Marquardt R, (2001) Stromrichterschaltungen Mit Verteilten Energiespeichern. German Patent DE 10103031A1, 24 Jan 2001

[14] Huang H (2009) Multilevel voltage-sourced converters for HVDC and FACTS applications. In: Cigré session, B4-401, 2009

[15] Li K, Zhao C (2010) New technologies of modular multilevel converter for VSC-HVDC application. In: Proceedings of power and energy engineering conference (APPEEC), Chengdu, China, 28–31 Mar 2010, pp 1–4

[16] Schauder C, Mehta H (1993) Vector analysis and control of advanced static VAR compensators. IEE Proc Gener Transm Distrib 140(4):299–306

[17] Papic I, Zunko P, Povh D et al (1997) Basic control of unified power flow controller. IEEE Trans Power Syst 12(4):1734–1739

[18] Dierckxsens C, Srivastava K, Reza M et al (2012) A distributed DC voltage control method for VSC MTDC systems. Electr Power Syst Res 82(1):54–58

[19] Saad H, Peralta J, Dennetiere S et al (2013) Dynamic averaged and simplified models for MMC-based HVDC transmission systems. IEEE Trans Power Deliv 28(3):1723–1730

[20] Nakajima T, Irokawa S (1999) A control system for HVDC transmission by voltage sourced converters. In: Proceedings of Power Engineering Society Summer Meeting, Edmonton, Alta, 18–22 Jul 1999, pp 1113–1119

Puyu WANG received the B.Eng. degree from University of Birmingham, UK and Huazhong University of Science and Technology (HUST), China, in 2011, both in electrical engineering. He is now pursuing the Ph.D. degree and also a research fellow in electrical power systems at the University of Birmingham, UK. His research interest includes HVDC technology, power electronics, DC-DC converters, and integration of renewable energy.

Xiao-Ping ZHANG received B.Eng., M.Sc. and Ph.D. degrees in electric engineering from Southeast University, China in 1988, 1990, 1993, respectively. He is currently a Professor in electric power systems at the University of Birmingham, UK, and he is also the Director of the University Institute for Energy Research and Policy. Before joining the University of Birmingham, he was an Associate Professor in the School of Engineering at the University of Warwick, UK. From 1998 to 1999, he was visiting UMIST. From 1999 to 2000, he was an Alexander-von-Humboldt Research Fellow with the University of Dortmund, Germany. He worked at China State Grid EPRI on EMS/DMS advanced application software research and development between 1993 and 1998. He is the co-author of the monograph *Flexible AC Transmission Systems: Modeling and Control* (New York: Springer, 2006 and 2012). Prof Zhang is an Editor of the IEEE Transactions on Smart Grid and IEEE Transactions on Power Systems, and he has also been serving on the editorial board of Journal of Modern Power Systems and Clean Energy.

Paul F. COVENTRY is a technical leader in HVDC technologies at National Grid.

Zhou LI was a Ph.D. student in the School of Electrical Engineering at Southeast University in China, as well as the School of Electronic, Electrical and Computer Engineering School of the Birmingham University, UK. He is currently a lecturer at Southeast University. His research interest is electric power system operation & control with HVDC link.

22

Active stabilization methods of electric power systems with constant power loads: a review

Mingfei WU (✉), Dylan Dah-Chuan LU

Abstract Modern electric power systems have increased the usage of switching power converters. These tightly regulated switching power converters behave as constant power loads (CPLs). They exhibit a negative incremental impedance in small signal analysis. This negative impedance degrades the stability margin of the interaction between CPLs and their feeders, which is known as the negative impedance instability problem. The feeder can be an LC input filter or an upstream switching converter. Active damping methods are preferred for the stabilization of the system. This is due to their higher power efficiency over passive damping methods. Based on different sources of damping effect, this paper summarizes and classifies existing active damping methods into three categories. The paper further analyzes and compares the advantages and disadvantages of each active damping method.

Keywords Stabilization, LC filters, Constant power loads

1 Introduction

In some electric power systems, if the outputs of switching power converters are tightly regulated, the instantaneous input power of the converter is constant within the regulation bandwidth. Therefore, from the view

M. WU, D. D.-C. LU, School of Electrical and Information Engineering, The University of Sydney, Darlington, NSW, Australia
(✉) e-mail: miwu2816@uni.sydney.edu.au;
mingfei.wu@sydney.edu.au

of their feeders, these tightly regulated converters behave as constant power loads (CPLs) [1–4]. CPLs have an inverse proportional v-i characteristic and exhibit negative incremental impedance in small signal analysis. This negative incremental impedance can degrade the stability margin of system interaction between CPLs and their feeder system, and is known as negative impedance instability problem [5–7]. This can be illustrated by a cascaded system as shown in Fig. 1.

The upstream circuit as shown in Fig. 1 can be a passive circuit, for example, an LC filter. It can also be another power converter. The downstream circuit is a tightly regulated converter. Both upstream and downstream circuits are designed to be stable individually. i.e.; the transfer functions of the upstream circuit

$$G_A(s) = v_{A\_o}(s)/v_{A\_in}(s) \tag{1}$$

and the downstream circuit

$$G_B(s) = v_{B\_o}(s)/v_{B\_in}(s) \tag{2}$$

are stable. However, the stability of the cascaded system with a transfer function of

$$G_{AB}(s) = G_A(s)G_B(s)/(1 + Z_{in}(s)/Z_o(s)) \tag{3}$$

depends on the stability of $Z_o(s)/(Z_{in}(s) + Z_o(s))$. The negative incremental impedance characteristic of $Z_i(s)$ can degrade the stability margin of $Z_o(s)/(Z_i(s) + Z_o(s))$ and consequently destabilize the operation of the cascaded system. The sufficient and necessary condition of the cascaded system stability is that the Nyquist contour of $T(s)$, as shown in (4), does not encircle the point $(-1, 0)$ [8].

$$T(s) = Z_i(s)/Z_o(s) \tag{4}$$

where $T(s)$ denotes the minor loop gain of the cascaded system.

However, how to use this stability criterion for active damping design is not straightforward. Middlebrook's

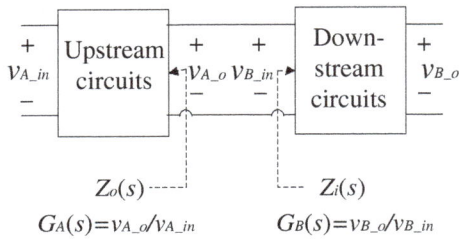

Fig. 1 The cascaded system for illustration of system stability

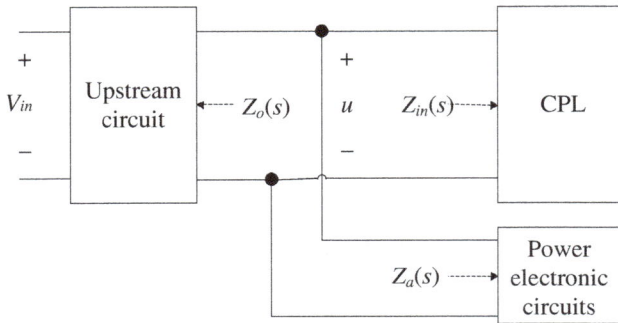

Fig. 2 Three main approaches of active damping methods

stability criterion, which is an illustrative stability criterion, has been proposed. A sufficient condition is proposed to guarantee the system stability which is shown in (5) [9].

$$\|Z_o(j\omega)\| \ll \|Z_{in}(j\omega)\| \tag{5}$$

Several similar stability criteria have also been proposed [10–12]. This paper focuses on the active methods that stabilize the cascaded system with CPL as shown in Fig. 1. The Middlebrook's stability criterion is adopted in this paper to mathematically show the effectiveness of these methods, because most of these methods are designed based on this stability criterion.

In order to stabilize the unstable system due to CPLs, several active stabilizing methods have been proposed. However, these active methods are based on different electric power system architectures and applications. In order to clarify the differences between these methods and make it easier for engineers and researchers to find a suitable method for a given system with CPLs, this paper classifies the existing methods into three categories according to different sources of stabilizing effect. Then, each method is analyzed to show its advantages and disadvantages.

The general approach of active stabilization methods is to modify $Z_o(s)$ or/and $Z_{in}(s)$ to fulfill the Middlebrook's stability criterion and consequently stabilize the cascaded system. These methods can be classified into three categories according to different sources of stabilizing effect as shown in Fig. 2.

1) Modify the output impedance of the upstream circuit $Z_o(s)$ by increasing the bandwidth of the control loop

of the upstream converter [13], or by adding an extra damping loop [14–21].

2) Modify the input impedance of the CPL, $Z_{in}(s)$ by injecting a stabilizing current into the CPL [2, 22–28].

3) Add an auxiliary power electronic circuit between the upstream and downstream circuits and control the input impedance of the auxiliary power electronic circuit, $Z_a(s)$. $Z_a(s)$ can be used to modify either $Z_o(s)$ or $Z_{in}(s)$ [29, 30].

In Sections 2, 3 and 4, these three main approaches of active damping methods are analyzed respectively. In Section 5, comparisons of these active damping methods are made and conclusion is provided.

2 Active damping method 1: modifying the output impedance $Z_o(s)$ of the upstream converter

In some electric power system configurations, the upstream circuit is a switching power converter. For example, in electric vehicles (EVs) onboard DC power systems, the feeder of CPLs is another stage of converter. This upstream converter could be an intermediate bus converter or a source converter of a distributed generator. The upstream converter could be either a DC/DC converter or an AC/DC converter. In signal analysis point of view, if the converter works in continuous conduction mode (CCM) and is in open loop control, inductor and capacitor in these source converters serve as an LC filter. This LC filter can result in a peak value around its resonant frequency in the output impedance of the source converters. Therefore, the Middlebrook's stability criterion may be violated and the system can be unstable. In addition, if the converter is in closed loop control and its bandwidth is lower than the resonant frequency, the LC filter resonant characteristic in the output impedance still exist beyond the bandwidth of the closed loop control as illustrated in Fig. 3. The relationship between the bandwidth of the closed loop control and the output impedance of the converter will be illustrated and explained in Section 2.1. Meanwhile, if the converter works in discontinuous conduction mode (DCM), zero inductor current operation changes the small signal operation of the inductor. Consequently, the output impedance of the converter may change and the cascaded system is stable. The stability analysis of DC/DC converter in DCM is presented in [31, 32].

This paper focuses on the stabilizing method of switching converters in CCM. A general approach to stabilizing this type of systems is to modify the output impedance of the source converter. One approach is to increase the closed loop control bandwidth of the source

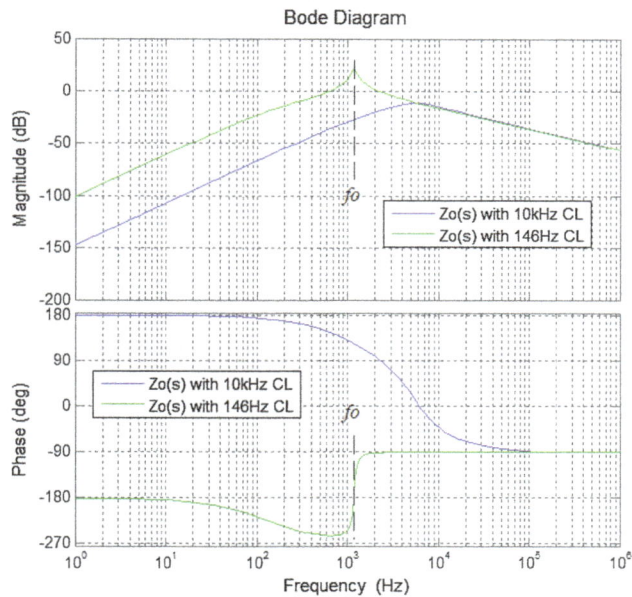

Fig. 3 The output impedance of the buck converter with different closed loop control bandwidth (CL in labels denotes the closed loop control)

converter. Another approach is to add an extra stabilizing loop.

2.1 Increasing the closed loop control bandwidth of the upstream converter

The upstream converter is represented by a buck converter as an example. The output impedance of the buck converter is

$$Z_o(s) = Ls/(LCs^2 + V_{in}F(s) + 1) \qquad (6)$$

where L and C are inductor and capacitor of the buck converter respectively, V_{in} is the input voltage and

$$F(s) = d(s)/v_o(s) \qquad (7)$$

is the transfer function of the voltage controller. The bode diagram of the output impedance of the buck converter with different closed loop control bandwidth is shown in Fig. 3. In this example, $L = 200$ μH and $C = 100$ μF. The resonant frequency, f_o, of the LC filter is 1.13 kHz.

From Fig. 3, it can be found that, when the bandwidth of the closed loop control is lower than the resonant frequency of the LC filter, the output impedance of the buck converter is close to the output impedance of an LC filter. The magnitude around the resonant frequency is high, possibly causing instability operation. In contrast, when the bandwidth of the closed control loop is higher than the resonant frequency, the magnitude response is much lower. This phenomenon can be briefly explained by (6). If the cut off frequency, f_c, of closed control loop is lower than the

resonance, f_o, i.e. $f_c < f_o$, where $f_o = 1/(2\pi\sqrt{LC})$, the magnitude of $F(2\pi f_o)$ will be much less than 1. Thus Z_o $(2\pi f_o)$ will become large. In contrast, if $f_c > f_o$, the magnitude of $F(2\pi f_o)$ will be much larger than 1, and Z_o $(2\pi f_o)$ will become small and it is easier to fulfill (5).

In other type of DC/DC and AC/DC converters, the relationship between the output impedance and the bandwidth of the closed loop control is similar to the buck converters. The peak value of the converter output impedance is large and is around the resonant frequency of the LC filter inside the converter if a low bandwidth closed loop control is applied. In this case, output impedance of the DC/DC converter can be reduced by increasing the closed loop control bandwidth to be larger than the resonance of LC filter. However, this is usually not available for AC/DC converter. Because, in conventional control of the AC/DC converter, the control bandwidth should be designed to be lower than the line frequency, i.e. 50 or 60 Hz, due to the need for power factor correction. In [33–37], several fast controllers for AC/DC converters have been proposed. However, the bandwidth of the closed loop control can only be increased up to around 200 Hz which is normally lower than the resonant frequency of the LC filter inside the converter. Consequently, the LC resonance peak in the output impedance is almost not reduced. Therefore, the stabilizing method through increasing the control bandwidth may not applicable for AC/DC converter.

Based on the above analysis, when the upstream converter is a DC/DC converter, one simple method to reduce the output impedance is to have a high closed loop control bandwidth. In [13], Du, et al have given the criteria for closed loop controller design of upstream DC/DC converter for the stability of the cascaded system with CPL.

2.2 Adding an extra stabilizing loop

In some applications, high closed loop control bandwidth is not available or the upstream converter works in open loop control. In this situation, an extra stabilizing loop is required for stabilization of the cascaded system. For small signal stability, linear methods are proposed. However, in some applications, large signal stability is required, or the CPL works in several operating points. In these conditions, several nonlinear methods are proposed.

2.2.1 DC/DC converter

Several active stabilizing methods have been proposed for DC/DC converter with CPL. One linear and effective method is to build a virtual resistor in series with the inductor in the DC/DC converter [14]. This virtual resistor

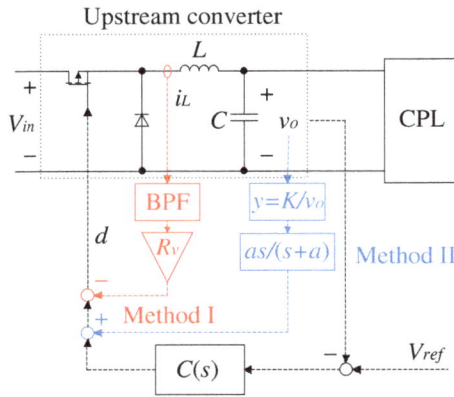

Fig. 4 The configuration of the two extra stabilizing loops for a buck converter loaded by CPL

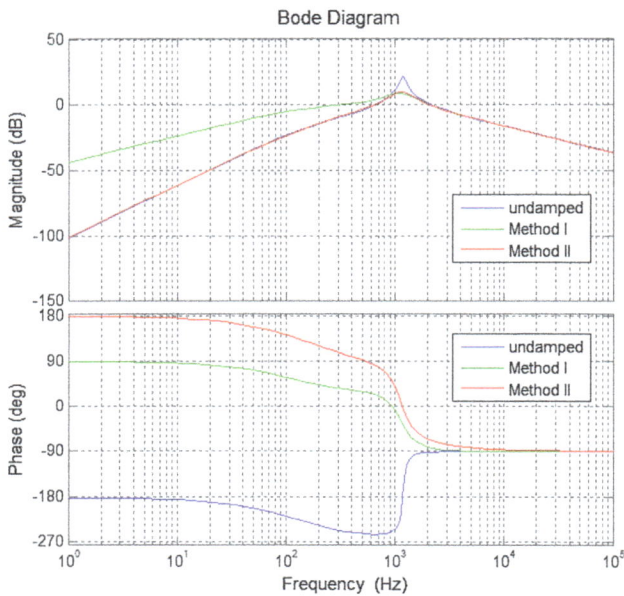

Fig. 5 Bode diagrams of the output impedances of the buck converter with the two active damping methods

reduces the peak value of the output impedance. The configuration of this stabilizing loop is shown in Fig. 4 and is named as Method I. In Fig. 4, a buck converter is used as an example. Another method is to add a nonlinear feedback loop to cancel the nonlinearity of the CPL and thus stabilize the system [15]. In small signal analysis, this method builds a virtual resistor in parallel with the CPL. Thus, the output impedance of the DC/DC converter can be reduced. The configuration of this method is illustrated in Fig. 4 and named as Method II. The bode diagram of the output impedance of the buck converter with these two methods is shown in Fig. 5.

From Fig. 5, it can be observed that both Methods I and II can effectively reduce the output impedance of the DC/

DC converter around the resonant frequency of LC filter. In Method I, the virtual resistor R_v, has to be within the small or large signal stability criteria as shown in (8) and (9) respectively [38].

$$L/CR < R_v < R \tag{8}$$

$$\sqrt{L/C} < R_v < R \tag{9}$$

where L and C are the inductor and capacitor of the buck converter and R is the resistor of the CPL. However, if the ratio of L/C is large and R is small, R_v may not exist. In contrast, in Method II, given any power level of CPL, there always exist a constant K to stabilize the cascaded system. In addition, in Method I, a current sensor is used meanwhile a voltage sensor is used in Method II.

In addition, there are other nonlinear active stabilization methods. Sliding mode control is proposed for stabilization of a DC/DC buck converter with CPL. The benefits of this method is that it can stabilize the cascaded system in several operating points [18]. A passivity based control method is also proposed for a DC/DC boost converter. The main advantage of this method is that it can obtain a fast response when load condition changes [19]. A boundary control method for DC/DC converter with CPL is proposed in [16, 17]. This method can provide faster transient, more robust operation compared with conventional PID controller. However, all these three methods require both voltage and current sensors.

2.2.2 AC/DC converter

The upstream converter can also be an AC/DC converter. In [20], a linear active stabilization method by modifying the output impedance of AC/DC converter is presented. Three kinds of stabilizing loop are proposed and compared. The configuration of the most effective stabilizing loop is shown in Fig. 6. The output impedance of the AC/DC converter with this stabilizing loop is shown in Fig. 7. With the stabilizing method, the output impedance of the AC/DC converter can be reduced and the Middlebrook's stability criterion can be fulfilled.

In addition, sliding mode control method is also proposed for AC/DC converters with CPLs. Sliding mode control guarantees the stability in large range of operating points [21]. However, this method requires both voltage and current sensors.

3 Active damping method 2: modifying the input impedance $Z_{in}(s)$ of CPLs

In some DC power electric systems, the feeder of a CPL is an LC input filter. In addition, in AC power systems, a

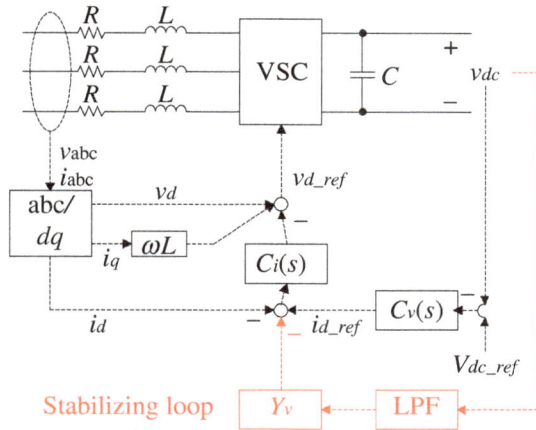

Fig. 6 The configuration of the stabilizing loop in the AC/DC converter

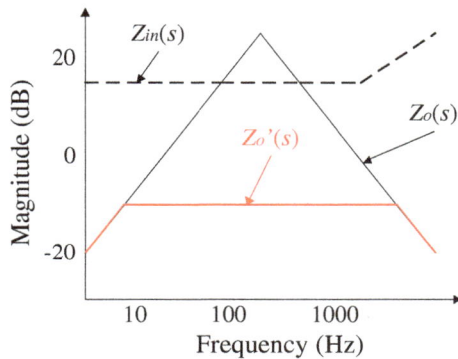

Fig. 7 The magnitude response of the output impedance of the AC/DC converter without stabilizing method (in solid black) and with stabilizing method (in solid red)

diode type rectifier is equivalent to an LC filter as shown in Fig. 8. In Fig. 8, an AC load is used as an example but it can also be a DC load.

In these configurations, the upstream circuit is a passive LC filter. High bandwidth control of the upstream circuit is not available. Therefore, the damping effort can be only from CPLs themselves. In order to stabilize such cascaded system, some active damping methods have been proposed. These methods can be classified into linear methods and nonlinear methods. In linear methods, compensating current are injected into the CPL to modify the input impedance of CPL, $Z_{in}(s)$, such that Middlebrook's stability criterion is fulfilled.

3.1 Linear methods

In linear methods, a stabilizing power is injected into the CPL to modify its input impedance. The configuration of the linear methods is shown in Fig. 9.

AC power supplies a CPL

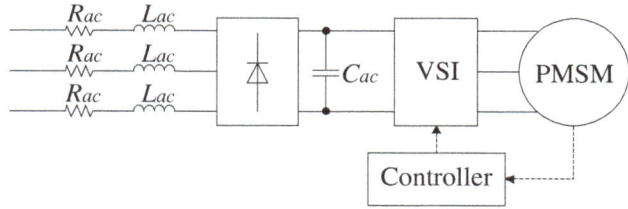

DC power supplies a CPL

LC input filter

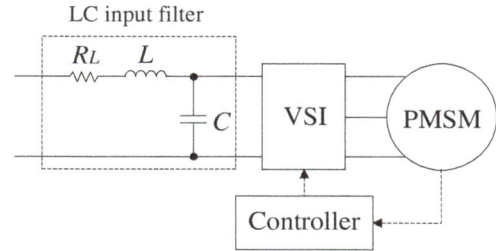

Equivalent circuit

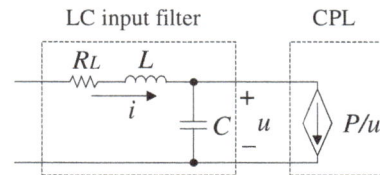

Fig. 8 A CPL in AC power system and DC power system and their equivalent circuit

The operation of the cascaded system with linear methods can be described in (10).

$$\begin{cases} \dfrac{\mathrm{d}i}{\mathrm{d}t} = -\dfrac{R_L}{L}i - \dfrac{1}{L}u + \dfrac{1}{L}V_{in} \\ \dfrac{\mathrm{d}u}{\mathrm{d}t} = \dfrac{1}{C}i - \dfrac{P + P_{stab}}{Cu} \end{cases} \tag{10}$$

where, i and u are inductor current and capacitor voltage respectively. L and C are the inductor and capacitor of the LC filter respectively. R_L is the equivalent resistor of the power cable and physical resistor of the inductor. V_{in} is the input voltage of the LC filter. P is the power of CPL and P_{stab} is the power used for stabilize the system. P_{stab} is injected into the CPL to build virtual resistor or virtual capacitor parallel connected with LC filter capacitor depending on different linear methods as illustrated in Fig. 9.

There are two requirements for P_{stab} [39].

1) With P_{stab}, the input impedance of the CPL can be modified to fulfill the Middlebrook's stability criterion.
2) In steady state operation, $P_{stab} = 0$.

In the linear methods, the stabilizing power P_{stab} is represented by stabilizing current, i_{stab} and

Equivalent circuit

Function of these linear methods

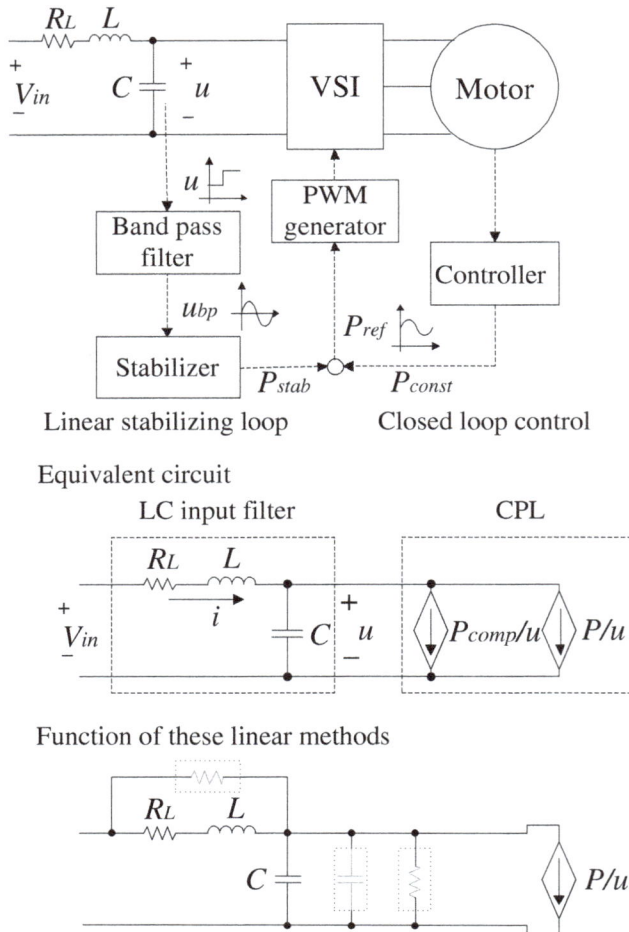

Fig. 9 The configuration of the cascaded system with linear methods

$$i_{stab} = \frac{P_{stab}}{Cu} \qquad (11)$$

After linearization around the operation point, (10) can be rewritten as

$$\begin{cases} \dfrac{di}{dt} = -\dfrac{R_L}{L}i - \dfrac{1}{L}u + \dfrac{1}{L}V_{in} \\[2mm] \dfrac{du}{dt} = \dfrac{1}{C}i + \dfrac{u}{CR} + i_{stab} \end{cases} \qquad (12)$$

where R is the resistance of the CPL.

In order to see the effectiveness of active stabilization methods, the performances of the DC brushless motor without damping methods are shown in Fig. 10. From Fig. 10, it can be found that, during the start of the motor, when the rotating speed ω_r is small, P is small and R is large and is larger than $\|Z_o(s)\|$. Hence, the system is stable. However, when ω_r increases larger, then R becomes small and is smaller than $\|Z_o(s)\|$, the system becomes unstable.

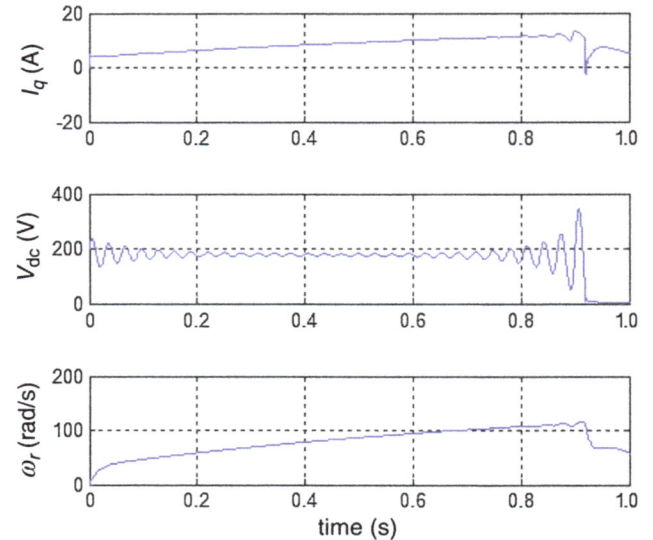

Fig. 10 The performances of the DC brushless motor without damping methods

3.1.1 Virtual resistor R_v [2, 23–25]

Modifying (11) so that

$$i_{stab} = \frac{C_b(s)u}{CR_v} \qquad (13)$$

where, $C_b(s)$ is the transfer function of a band pass filter with unity gain, it represents a virtual resistor R_v that is being built in parallel with the LC filter capacitor. The function of the band pass filter is to pick the oscillation component which is around the resonant frequency of the LC filter and attenuate the steady state value and switching noises. In order to stabilize the system, R_v needs to be chosen such that Middlebrook's stability criterion is fulfilled as shown in Fig. 11.

3.1.2 Virtual capacitor C_v [22, 23]

When (11) is modified so that,

$$i_{stab} = \frac{sC_vC_l(s)u}{C} \qquad (14)$$

where $C_l(s)$ is a low pass filter with unity gain, a virtual capacitor C_v is built in parallel with the LC filter capacitor. Thus the overall capacitance of the LC filter is increased. Also, the value of C_v needs to be chosen to make the Middlebrook's stability criterion fulfilled as shown in Fig. 12.

3.1.3 Comparison

From Figs. 13 and 14, it can be found that, the performance of active damping method by building R_v is better

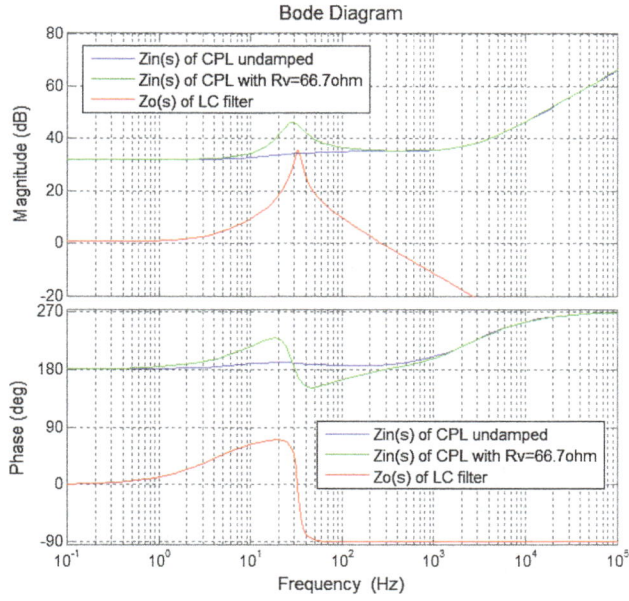

Fig. 11 The bode diagram of $Z_o(s)$ of LC filter and $Z_{in}(s)$ of CPL without damping method and with virtual resistor active damping method

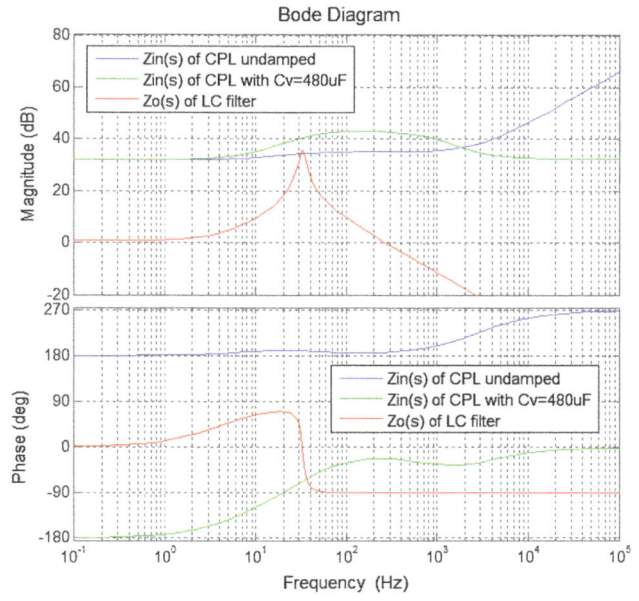

Fig. 13 The performances of a DC brushless motor as a CPL with active damping method by building virtual resistor $R_v = 10\,\Omega$

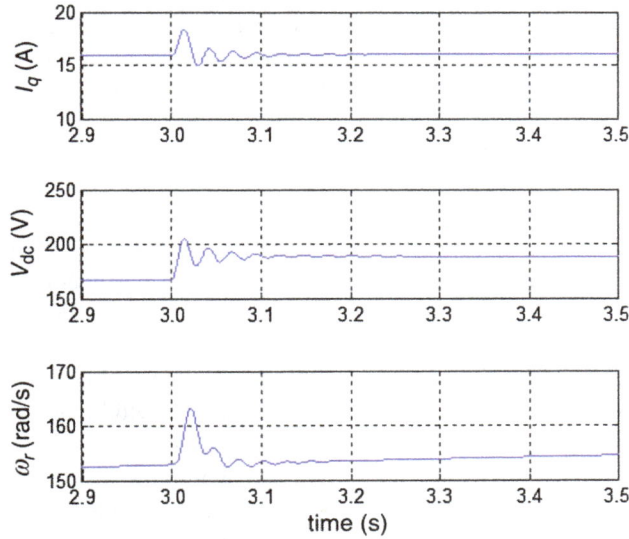

Fig. 12 The bode diagram of $Z_o(s)$ of LC filter and $Z_{in}(s)$ of CPL without damping method and with virtual capacitor active damping method

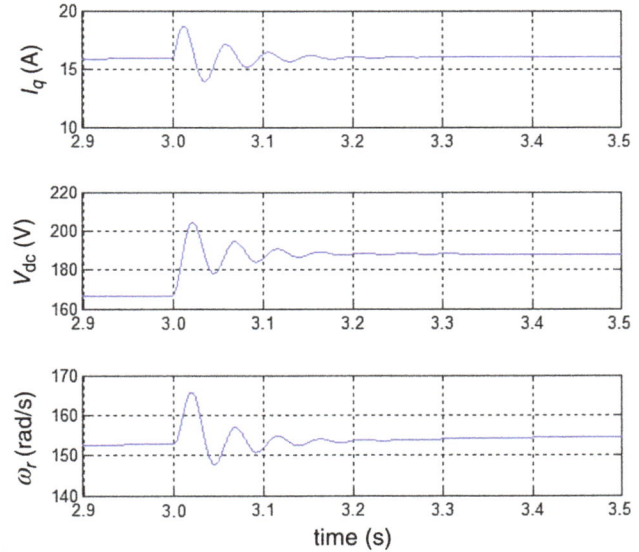

Fig. 14 The performances of a DC brushless motor as a CPL with active damping method by building virtual capacitor $C_v = 1,200\,\mu F$

than the active damping method by building C_v, because the former method can achieve almost the same damping effect with relatively smaller undesirable oscillation in rotating speed.

Mathematically, the comparison can be made based on root locus method. Assume that these two methods are applied in an unstable cascaded system separately. Due to

that both $C_b(s)$ and $C_l(s)$ have unity gain, the amplitude of i_{stab} only depends on $1/R_v$ in (13) or ωC_v in (14). We can compare these two methods by examining the damping effort with the same amplitude of i_{stab}. Thus, we assume that

$$\frac{1}{R_v} = \omega C_v \qquad (15)$$

Therefore, the amplitude of i_{stab} in these two methods are the same. In another words, the undesirable effects on

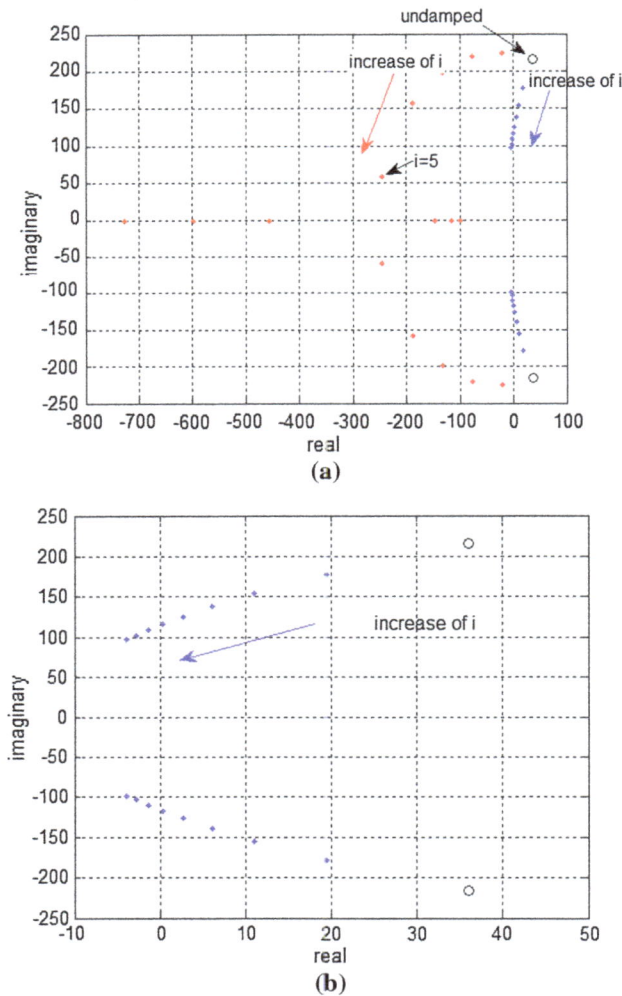

Fig. 15 a The root loci of the cascaded system with method of building virtual resistor R_v (in red) and with method of building virtual capacitor C_v (in blue) and **b** the details around zero

the load performance are the same. A series of R_v and C_v are selected as shown in (16) and (17).

$$C_v = 0.5Ci \tag{16}$$

$$R_v = \frac{1}{0.5\omega Ci} \tag{17}$$

where $i = 1, 2, \ldots, 8$. The root loci of the cascaded system with two active methods are shown separately in Fig. 15.

In Fig. 15, it can be observed that when i increases, the poles of the system with the method of building virtual resistor can approach to the negative real part much quicker comparing with the method of building virtual capacitor. Therefore, with the same amplitude of i_{stab}, i.e. the same undesirable effect on the load performance, the damping effort of the method of building virtual resistor is better than the method of building virtual capacitor.

3.2 Nonlinear methods

Sudhoff S. D. et al proposed a nonlinear method [28]. However, this method has limited damping effect compared with the linear method. This has been presented in [39]. A passivity based control method is proposed in [39]. This passivity based control can provide better damper without large undesirable load performance. The passivity based control algorithm is shown in (13) and (14).

$$\theta = -R_L \frac{P}{V_{dc}} + R_1 \left(i - \frac{P}{V_{dc}} \right) + V_{in} \tag{18}$$

$$P_{stab} = u \left(\frac{P}{V_{dc}} - C \frac{d\theta}{dt} \right) - P + \frac{(u - \theta)u}{R_2} \tag{19}$$

where R_1 and R_2 are two coefficients. θ and V_{dc} are reference value of i and u. However, complicated calculation with measurements i, u and P as shown (18) and (19) can result in large error and noises. As a result, this method is difficult to be implemented.

3.3 Sensitivity and availability

As shown in Fig. 9, a stabilizing power is injected into the CPL to stabilize the cascaded system. This stabilizing power is a transient oscillating component. It can result in undesirable load performances, such as the oscillation in the rotating speed of the motors. Therefore, there is always a compromise between the damping of the oscillation in LC input filter and the load performances. Therefore, sensitivity from input voltage of CPL to the rotating speed is important in the design of active stabilization method.

As the injected power is realized by the downstream converters, a large stabilizing power injected into the CPL is required to achieve a greater damping effect. This implies a wide range of duty cycle values in the downstream converters. However, the duty cycles are within the range of (0, 1). Therefore, the stabilizing effect is limited by available duty cycle range, e.g. high step-up converters usually operate at duty cycle between 0.5 and 1. Thus the availability of the stabilizing effect of the active stabilization method is another aspect to be considered.

4 Active damping method 3: adding a shunt impedance $Z_a(s)$

Owing to the sensitivity and availability problems in the active stabilization in Section 3, a new type of methods are proposed. In this type of methods, an auxiliary DC/DC converter is added across the input voltage of the CPL. This DC/DC converter works as a power buffer and also

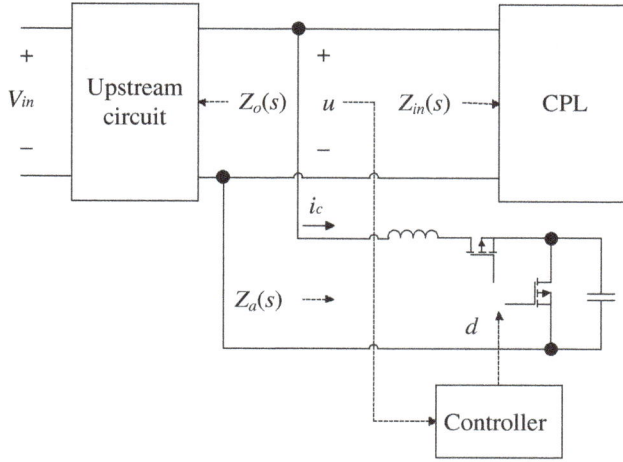

Fig. 16 The schematic diagram of the extra power electronic circuit

can be designed to decouple the interaction between the CPL and its LC input filter.

As shown in Fig. 16, an auxiliary power electronic circuit is added to stabilize the system. The input current i_c is controlled depending on the input voltage u to control its input impedance $Z_a(s)$. In order to fulfill the Middlebrook's stability, $Z_a(s)$ should fulfill

$$\|Z_o(s)\| < \left\| \frac{Z_a(s)Z_{in}(s)}{Z_a(s) + Z_{in}(s)} \right\| \tag{20}$$

Or

$$\left\| \frac{Z_a(s)Z_o(s)}{Z_a(s) + Z_o(s)} \right\| < \|Z_{in}(s)\| \tag{21}$$

In addition, $\|Z_a(s)\|$ should approach infinity when frequency is zero.

Theoretically, this auxiliary DC/DC converter can stabilize the system as the CPL does as described in Section 3 and it does not inherit the sensitivity problem. It can work as a virtual resistor or a virtual capacitor or use to realize nonlinear methods.

In [30], an auxiliary DC/DC converter is presented to mimic a virtual capacitor. This is realized by the control of the input current of the auxiliary DC/DC converter, i_c, according to the capacitor voltage u. i.e.

$$i_c = C_v \frac{\mathrm{d}u}{\mathrm{d}t} \tag{22}$$

where C_v is the virtual capacitor.

In [29], an auxiliary DC/DC converter is used to realize a pole placement method. In this method, the duty cycle of the auxiliary converter is controlled to replace the poles of the whole dynamic system. The capacitor voltage, u, and the input current of the auxiliary converter, i_c, are measured and second order observer is used. Thus, all four

system states, inductor current of LC filter, i, capacitor voltage of the auxiliary converter, v_c, i_c and u are available for controller design. Therefore, the poles of the closed loop control system can be replaced to obtain a good damping performance.

5 Comparison and conclusion

This paper analyzes several existing active stabilization methods. The methods have been evaluated through Middlebrook's stability criterion. According to the source of the stabilizing effect, these methods are classified into three categories. The advantages and disadvantages of these three types of methods are summarized as follows:

The active damping method which reduces the output impedance of the upstream converter is discussed in Section 2. The main advantage of this type of methods is that it can stabilize the cascaded system without affecting the operation and performance of CPL. The disadvantage of these methods is that the upstream circuit needs to be controllable, e.g. a switching regulator. These methods are suitable on the systems in which the feeder of CPLs is another stage converter.

The active damping method which increases the input impedance of the CPL is discussed in Section 3. The main merit is that the CPL can overcome the negative impedance instability by itself. These methods are suitable for the systems in which the feeder of CPL is an uncontrollable LC filter. The demerit of these methods is that the stabilizing power injected into the CPL can result in some undesirable load performances. There is always a compromise between the damping effect and the load performances.

The active damping methods by adding a shunt impedance through an extra power electronics circuit is discussed in Section 4. The strength of this method is that the extra power electronic converter can stabilize the system without the undesirable effect on the load performances. This type of methods has a high potential to achieve similar function as the CPL does to stabilize the system, but without the compromise. The weak side is that extra circuit is required resulting in additional cost and power losses.

References

[1] Rahimin AM, Khaligh A, Emadi A (2006) Design and implementation of an analog constant power load for studying cascaded converters. In: Proceedings of the 32nd annual conference

on IEEE industrial electronics society (IECON'06), Paris, 6–10 Nov 2006, pp 1709–1714

[2] Liu XY, Forsyth AJ, Cross AM (2007) Negative input-resistance compensator for a constant power load. IEEE Trans Ind Electron 54(6):3188–3196

[3] Kwasinski A, Krein PT (2007) Passivity-based control of buck converters with constant-power loads. In:Proceedings of the 2007 IEEE power electronics specialists conference (PESC'07), Orlando, 17–21 Jun 2007, pp 259–265

[4] Emadi A, Khaligh A, Rivetta CH et al (2006) Constant power loads and negative impedance instability in automotive systems: Definition, modeling, stability, and control of power electronic converters and motor drives. IEEE Trans Veh Technol 55(4):1112–1125

[5] Rivetta C, Williamson GA (2004) Global behaviour analysis of a DC-DC boost power converter operating with constant power load. In: Proceedings of the 2004 international symposium on circuits and systems (ISCAS'04), vol 5. Vancouver, Canada, 23–26 May 2004, pp 956–959

[6] Rivetta CH, Emadi A, Williamson GA et al (2006) Analysis and control of a buck DC-DC converter operating with constant power load in sea and undersea vehicles. IEEE Trans Ind Appl 42(2):559–572

[7] Rahimi AM, Emadi A (2009) An analytical investigation of DC/DC power electronic converters with constant power loads in vehicular power systems. IEEE Trans Veh Technol 58(6):2689–2702

[8] Riccobono A, Santi E (2012) Comprehensive review of stability criteria for DC power distribution systems. In: Proceedings of the 2012 IEEE energy conversion congress and exposition (ECCE'12), Raleigh, 15–20 Sept 2012, pp 3917–3925

[9] Middlebrook RD (1976) Input filter considerations in design and application of switching regulators. In: Proceedings of the IEEE industry applications society annul meeting (IAS'76), Chicago, 11–14 Oct 1976, pp 366–382

[10] Wildrick CM, Lee FC, Cho BH et al (1995) A method of defining the load impedance specification for a stable distributed power system. IEEE Trans Power Electron 10(3):280–285

[11] Sudhoff SD, Glover SF, Lamm PT et al (2000) Admittance space stability analysis of power electronic systems. IEEE trans aero electron syst 36(3-Part 1):965–973

[12] Feng XG, Liu JJ, Lee FC (2002) Impedance specifications for stable DC distributed power systems. IEEE Trans Power Electron 17(2):157–162

[13] Du W, Zhang J, Zhang Y et al (2013) Stability criterion for cascaded system with constant power load. IEEE Trans Power Electron 28(4):1843–1851

[14] Rahimi AM, Emadi A (2009) Active damping in DC/DC power electronic converters: A novel method to overcome the problems of constant power loads. IEEE Trans Ind Electron 56(5):1428–1439

[15] Rahimi AM, Williamson GA, Emadi A (2010) Loop-cancellation technique: A novel nonlinear feedback to overcome the destabilizing effect of constant-power loads. IEEE Trans Veh Technol 59(2):650–661

[16] Onwuchekwa CN, Kwasinski A (2009) Boundary control of buck converters with constant-power loads. In: Proceedings of the 31st international telecommunications energy conference (INTELEC'09), Incheon, 18–22 Oct 2009, 6 pp

[17] Onwuchekwa CN, Kwasinski A (2011) Analysis of boundary control for boost and buck-boost converters in distributed power architectures with constant-power loads. In: Proceedings of the 26th annual IEEE applied power electronics conference and exposition (APEC'11), Fort Worth, 6–11 Mar 2011, pp 1816–1823

[18] Zhao Y, Qiao W, Ha D (2014) A sliding-mode duty-ratio controller for DC/DC buck converters with constant power loads. IEEE Trans Ind Appl 50(2):1448–1458

[19] Zeng J, Zhang Z, Qiao W (2014) An interconnection and damping assignment passivity-based controller for a DC-DC boost converter with a constant power load. IEEE Trans Ind Appl 50(4):2314–2322

[20] Radwan AAA, Mohamed YARI (2012) Linear active stabilization of converter-dominated DC microgrids. IEEE Trans Smart Grid 3(1):203–216

[21] Zhang XN, Vilathgamuwa DM, Foo G, et al (2013) Cascaded sliding mode control for global stability of three phase AC/DC PWM rectifier with rapidly varying power electronic loads. In: Proceedings of the 39th annual conference of the IEEE industrial electronics society, (IECON'13), Vienna, 10–13 Nov 2013, pp 4580–4587

22. Magne P, Marx D, Nahid-Mobarakeh B et al (2012) Large-signal stabilization of a DC-link supplying a constant power load using a virtual capacitor: Impact on the domain of attraction. IEEE Trans Ind Appl 48(3):878–887

[23] Mohamed YARI, Radwan AAA, Lee TK (2012) Decoupled reference-voltage-based active DC-link stabilization for PMSM drives with tight-speed regulation. IEEE Trans Ind Electron 59(12):4523–4536

[24] Lee WJ, Sul SK (2014) DC-link voltage stabilization for reduced DC-link capacitor inverter. IEEE Trans Ind Appl 50(1):404–414

[25] Liutanakul P, Awan AB, Pierfederici S et al (2010) Linear stabilization of a DC bus supplying a constant power load: A general design approach. IEEE Trans Power Electron 25(2): 475–488

[26] Glover SF, Sudhoff SD (1998) An experimental validated nonlinear stabilizing control for power electronics based power systems. In: Proceedings of the SAE aerospace power system conference (APSC'98), Detroit, 21–23 Apr 1998, pp 71–88

[27] Mosskull H, Galic J, Wahlberg B (2007) Stabilization of induction motor drives with poorly damped input filters. IEEE Trans Ind Electron 54(5):2724–2734

[28] Sudhoff SD, Corzine KA, Glover SF et al (1998) DC link stabilized field oriented control of electric propulsion system. IEEE Trans Energ Conver 13(1):27–33

[29] Inoue K, Kato T, Inoue M, et al (2012) An oscillation suppression method of a DC power supply system with a constant power load and a LC filter. In: Proceedings of the IEEE 13th workshop on control and modeling for power electronics (COMPEL'12), Kyoto, 10–13 Jun 2012, 4 pp

[30] Zhang X, Ruan X, Kim H et al (2013) Adaptive active capacitor converter for improving stability of cascaded DC power supply system. IEEE Trans Power Electron 28(4):1807–1816

[31] Grigore V, Hatonen J, Kyyra J et al (1998) Dynamics of a buck converter with a constant power load. In: Proceedings of the 29th annual IEEE power electronics specialists conference (PESC'98), vol 1. Fukuoka, 17–22 May 1998, pp 72–78

[32] Rahimi AM, Emadi A (2010) Discontinuous-conduction mode DC/DC converters feeding constant-power loads. IEEE Trans Ind Electron 57(4):1318–1329

[33] Spiazzi G, Mattavelli P, Rossetto L (1997) Power factor pre-regulators with improved dynamic response. IEEE Trans Power Electron 12(2):343–349

[34] Prodic A, Chen JQ, Maksimovic D, et al (2003) Self-tuning digitally controlled low-harmonic rectifier having fast dynamic response. IEEE trans power electron 18(1-Part 2):420–428

[35] Lamar DG, Fernandez A, Arias M et al (2008) A unity power factor correction preregulator with fast dynamic response based on a low-cost microcontroller. IEEE Trans Power Electron 23(2):635–642

[36] Wall S, Jackson R (1997) Fast controller design for single-phase power-factor correction systems. IEEE Trans Ind Electron 44(5):654–660

[37] Rezaei K, Golbon N, Moschopoulos G (2014) A new control scheme for an AC-DC single-stage buck-boost PFC converter

with improved output ripple reduction and transient response. In: Proceedings of the 29th annual IEEE applied power electronics conference and exposition (APEC'14), Fort Worth, 16-20 Mar 2014, pp 1866–1873

[38] Belkhayat M, Cooley R, Witulski A (1995) Large signal stability criteria for distributed systems with constant power loads. In: Proceedings of the 26th annual IEEE power electronics specialists conference (PESC '95), vol 2. Atlanta, 18–22 Jun 1995, pp 1333–1338

[39] Liu XY, Forsyth AJ (2005) Comparative study of stabilizing controllers for brushless DC motor drive systems. In: Proceedings of the 2005 IEEE international conference on electric machines and drives, San Antonio, 15–18 May 2005, pp 1725–1731

Mingfei WU received the B.Eng. degree in Electrical and Electronic Engineering from Northumbria University, UK, and M.Sc. degree in Control Systems from Imperial College London, UK, in 2009 and 2010, respectively. From 2012, he started to pursue his Ph.D. degree in the University of Sydney, Australia. His current research focuses on the stability of DC microgrids.

Dylan DAH-CHUAN LU received his Ph.D. degree in Electronic and Information Engineering from The Hong Kong Polytechnic University, Hong Kong, China, in 2004. In 2003, he joined PowereLab Ltd. as a Senior Engineer. His major responsibilities include project development and management, circuit design, and contribution of research in the area of power electronics. In 2006, he joined the School of Electrical and Information Engineering, The University of Sydney, Australia, where he is currently a Senior Lecturer. He presently serves as an Associate Editor of the International Journal of Electronics, and the Australian Journal of Electrical and Electronics Engineering. He is an author/co-author of over 90 technical articles in the areas of power electronics and engineering education. He has two patents on efficient power conversion. His current research interests include power electronics circuits and control for efficient power conversion, lighting, renewable electrical energy systems, microgrid and power quality improvement, and engineering education. Dr. LU received the Dean's Research Award in 2011. He is a senior member of the IEEE and a member of the IEAust.

23

Dynamic frequency response from electric vehicles in the Great Britain power system

Jian MENG, Yunfei MU (✉), Jianzhong WU,
Hongjie JIA, Qian DAI, Xiaodan YU

Abstract With the large penetration of renewable energy, fulfilling the balance between electricity demand and supply is a challenge to the modern power system. According to the UK government, the wind power penetration will reach 30% by the year 2020. The role of electric vehicles (EVs) contributing to frequency response was investigated. A dynamic frequency control strategy which considers the comfort level of vehicle owners was developed for EVs to regulate their power consumption according to the deviation of system frequency. A simulation model of a population of EVs equipped with such control was implemented in Matlab/Simulink platform. In this paper, a simplified Great Britain power system model is used to study the contribution of EVs to dynamic frequency control. The case study showed that using EVs as a demand response resource can greatly reduce the frequency deviations. And the rapid response from EVs can help reduce the operation cost of conventional generators.

Keywords Electric vehicles (EVs), Dynamic frequency response, Vehicle to grid (V2G), State of charge (SOC)

J. MENG, Y. MU, H. JIA, X. YU, Key Laboratory of Smart Grid of Ministry of Education, Tianjin University, Tianjin 300072, China
(✉) e-mail: yunfeimu@tju.edu.cn
J. WU, School of Engineering, Institute of Energy, Cardiff University, Cardiff, UK
Q. DAI, China Electric Power Research Institute, Haidian District, Beijing 100192, China

1 Introduction

With the increasing concerns over the environmental and energy problems, the concept 'low-carbon economy' has been more and more popular. Consequently, several countries around the world have taken appropriate means in order to de-carbonize the power system and transport sector. In the UK, a large scale of renewable energy will come from wind turbines in the following years. It is estimated that the wind generation capacity will reach 30 GW within a generation capacity of 100 GW which serves a load of around 60 GW by the year of 2020. To realize the target of decarbonizing the domestic transport sector, the number of EVs will be greatly increased, which will play an important role in the future [1].

However, a large scale integration of wind generation may bring inevitable concerns over the power system operation due to the variability of wind generation, especially the active power balance between the generation and demand. As a consequence, a high penetration of wind energy will definitely increase the difficulty in frequency control. In order to keep the system frequency within an acceptable range, several researches focused on the frequency stability of power system in the presence of high renewable energy penetration. Oudalov and Pascal used battery energy storage system (BESS) for frequency control [2, 3]. BESS can substantially reduce the frequency deviations caused by sudden supply/demand variations. However the cost of BESS is very high and it is difficult to bring profit by now. The impact of wind turbine on the system frequency stability in the UK was investigated by Pearmine [4]. Ramtharan studied the capability of doubly fed induction generator wind turbines to serve the frequency control [5]. A frequency droop controller was developed to realize that function by adjusting the generator torque set point. Short investigated the possibility of utilizing refrigerators as frequency response resource [6].

The temperature set points which adjust according to the system frequency were used to decide the on-off state.

Recently, EV technology is one of the distributed energy technologies that have been increasingly deployed. The vehicle-to-grid (V2G) concept allows EV to serve as a mobile battery storage unit. With a bidirectional power interface, EV battery could either charge or discharge. The capability of storing energy and the instantaneous active power control of fast-switching converters of EVs are two attractive features that enable EVs to provide various ancillary services, e.g., the provision of primary frequency control in island and grid connected systems [7–9]. A droop control allows EVs to adjust their charging power in response to the system frequency signal. The controlling effect of EVs for frequency control was investigated in [10], which controlled the system frequency by a sudden disconnection of all EVs from the power system following a large disturbance in the system. However, this control approach can result in undesired condition over frequency responses when the response power of EVs is more than the power imbalance in the system. Also, the effect of EVs constraints on the provision of dynamic frequency response in the Great Britain (GB) power system has not been discussed comprehensively yet.

In this paper, the effect of utilizing EVs for dynamic frequency response in the GB power system at the year of 2020 was studied. A dynamic frequency control (DFC) strategy which considers the comfort level of EV owners was developed for EVs to regulate their power consumption dynamically according to the deviation of system frequency along a typical day. A participation factor k with battery state of charge (SOC) was introduced to evaluate the comfort levels of EV owners. Based on a simplified GB power system model, the effect of EVs in dynamic frequency response was studied considering the V2G characteristic. Results of two EV charging scenarios: "dumb" charging, and "smart" charging, were compared.

2 Frequency control in the GB system

2.1 Frequency control

To maintain the frequency stability, the upper/lower frequency limit of the GB power system is 49.5/50.5 Hz [11]. In the UK, the power system is designed to accept a loss of 1320 MW generation. On this occasion, the frequency deviation will be less than 0.8 Hz and the frequency will be restored to 49.5 Hz within 60 seconds. Figure 1 shows a typical frequency transient for a generation loss of 1320 MW based on a severe frequency event [12, 13]. To reach the target of 'low-carbon economy', more wind energy will be integrated into the power system through power electronic interfaces. Considering the uncertainties

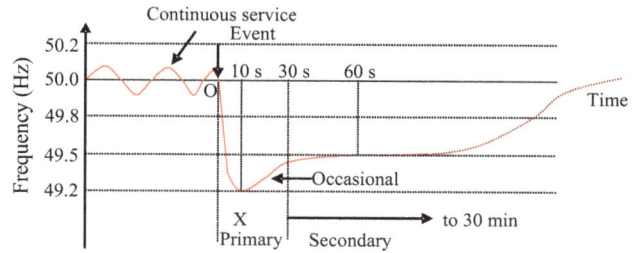

Fig. 1 Frequency curve with a loss of 1320 MW generation

brought by wind power, the GB power system needs to be able to accept up to a new maximum of 1800 MW generation [14]. Therefore larger frequency response capacity is likely to be required.

2.2 Simplified GB power system model

In this paper, a simplified system model was utilized to represent the GB system, as shown in Fig. 2. This model considers the characteristics of generators and frequency dependent loads. The power plants were represented in the model by blocks which consist of governors and turbines. The power plants adjust their power output according to the frequency deviation. Relevant parameters are shown in Table 1 [15, 20]. R_{eq} represents the speed control of turbines and operates on the speed deviation which is formed between the actual speed and the reference speed. T_G is a typical time constant related to the governor actuator. A transient droop compensation was set between the governor and the turbine. T_1 and T_2 are time constants of lead-lag transfer function. The turbine model is characterized by a time constant T_T. It represents the mechanical power output following the governor action. H_{eq} is set in consideration of the operational generation capacity in the 2020 system. D is a single damping constant which represents the damping of frequency dependent loads. T_{EV} is the time constant which represents EV charger time delay. These parameters capture the characteristics of different governors and turbines. The values were set according to a

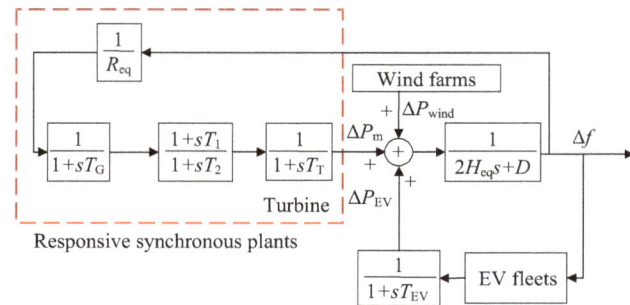

Fig. 2 Simplified GB power system model

Table 1 Parameters of the simplified GB system

Parameters	Value
R_{eq} (p.u.)	−0.09
T_G (sec)	0.2
T_1 (sec)	2
T_2 (sec)	12
T_T (sec)	0.3
H_{eq} (sec)	4.44
D (p.u.)	1

serious frequency event happened in the GB power system in 2008 [13].

Most of the responsive synchronous plants in service at present will remain in service by 2020. New plants to be built will share similar characteristics. Thus the characteristics of responsive synchronous plants will be changeless by 2020. Only the operational capacity of plants will vary because some of them will be replaced by wind plants.

Because of the intermittency of wind energy, wind farms will play a significant role in the system. The power disturbance from the wind farms is represented by ΔP_{wind}.

3 Dynamic frequency control strategy

3.1 A generic EV charging model

A generic EV charging model (GECM) is used to simulate the charging and discharging procedure of EV batteries. As depicted in Fig. 3, the nonlinear GECM is composed of a generic EV charger model and a generic battery model.

3.1.1 A generic EV charger model

The model is composed of a converter and an inverter. The voltage is adjusted to an appropriate level by the converter for EV battery. The reactor L connects the inverter and the power system. The resistance of the reactor is represented by R_C. The inverter operation is based on a pulse-width modulation (PWM) switching technique. The active/reactive power controls were employed through vector control and were used to generate modulating signals for PWM [16]. The current i is calculated as follow [17, 18]:

$$L\frac{di}{dt} + R_c i - V_{grid} + V_{EV} = 0 \tag{1}$$

The voltage of EV battery (V_{EV}) is calculated by [17, 18]

$$V_{EV} = \alpha M V_{DC} \sin(\omega t + \delta) \tag{2}$$

where M is the PWM modulation index used for inverter; δ is the angle between V_{EV} and V_{grid}; α is a constant (0.5 or 1) which depends on the exact topology of the inverter.

Equation (1) indicates that the current (or power) exchanged between the battery and the system is governed by a first order differential equation. Thus a first order lag with a time constant $T_{EV} = L/R_c$ was used for frequency studies.

3.1.2 A generic battery model

As shown in Fig. 4, the generic battery model is composed of a voltage source and a constant resistance. The model that only uses battery SOC as a state variable is chosen to accurately reproduce the manufacturer's curves for four major types of battery chemistries.

The battery voltage V_{batt} is described by

$$V_{batt} = E - R_b i \tag{3}$$

$$E = E_0 - \frac{K}{SOC} + A \exp\left(-B \int_0^t i\,dt\right) \tag{4}$$

$$SOC = 1 - \frac{\int_0^t i(t)\,dt}{Q} \tag{5}$$

where E_0 is the battery constant voltage; K is the polarisation voltage; Q is the battery capacity; i is the charging current; A is the exponential zone amplitude; B is

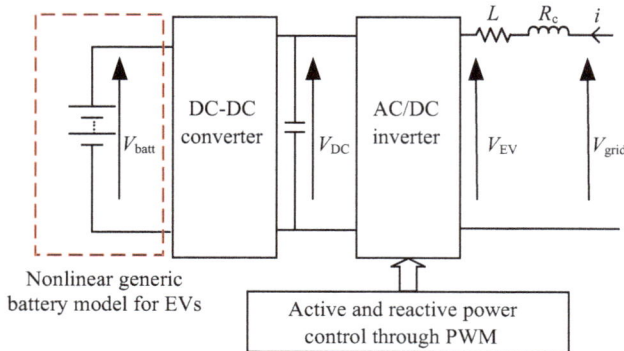

Fig. 3 A generic EV charger model of EV

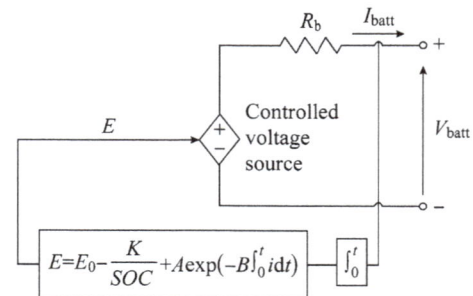

Fig. 4 A generic battery model for EVs

the exponential zone time constant inverse; R_b is the internal resistance.

To simplify the integral part, the charging power P is calculated as

$$P = E_0 i - \frac{Ki}{SOC} + Ai \exp(-Bit) - R_b i^2 \qquad (6)$$

where i is a constant charging current.

There are only a few parameters in this proposed generic battery model. Based on the battery type and capacity, the SOC can be calculated using this model.

3.2 Dynamic frequency control strategy

A dynamic frequency control strategy which considers the comfort level of the EV owners is developed for the EVs to regulate their power consumption dynamically according to the system frequency deviation. In this strategy, the comfort level is evaluated by the participation factor k. Instead of tripping EVs directly, the SOC set point is varied dynamically according to the deviations of frequency and k. EVs can serve the frequency regulation in two ways. One approach would be to switch off (on) EVs that are being charged (idle). The other approach is supporting the grid through discharging. There are two control modes in the developed dynamic frequency control strategy.

(1) V1G mode

In V1G mode, EVs participate in frequency response through adjusting their power consumption. The SOC set point of EVs is controlled as

$$\begin{aligned} SOC_{max}^{V1G} &= SOC_{max}^{normal} + k\Delta f \\ \Delta f &= f - f_0 \\ SOC_{min} &\leq SOC_{max}^{V1G} \leq 1 \\ \Delta f &\geq \Delta f_{cr} \end{aligned} \qquad (7)$$

where SOC_{max}^{normal} is nominal high SOC set point of EV battery; SOC_{max}^{V1G} is the modified high SOC set point under V1G mode; SOC_{min} is the minimum SOC value for satisfying the travelling requirement of EV owners; f_0 is the system nominal frequency e.g. 50 Hz for the GB system; f is the actual system frequency; Δf_{cr} is the value of frequency decrease when $SOC_{max}^{V1G} = SOC_{min}$; k $(k > 0)$ is the participation factor of EVs. The larger value of k, the larger variation of SOC_{max}^{V1G}. As a result, the frequency response capability is larger, and the travelling comfort level is lower, and vice versa.

When system frequency increases ($\Delta f > 0$), the EVs with SOC between SOC_{max}^{normal} and SOC_{max}^{V1G} were changed from idle status to charging status to increase their power consumption.

When system frequency decreases ($\Delta f_{cr} < \Delta f < 0$), the EVs with SOC between SOC_{max}^{normal} and SOC_{max}^{V1G} were

changed from charging status to idle status to decrease their power consumption. When $\Delta f = \Delta f_{cr}$, the EV resources for V1G mode is used up. The EVs turn to V2G mode which starts from $SOC_{max}^{V2G} = 1$.

(2) V2G mode

When the system frequency falls lower than Δf_{cr} ($\Delta f < \Delta f_{cr} < 0$), the V2G mode is used for frequency control. In V2G mode, EVs participate in frequency response through adjusting their discharging power. The EVs with SOC between 1 and SOC_{max}^{V2G} discharge their stored battery energy back to the system. The SOC set point of EVs is controlled according to

$$\begin{aligned} SOC_{max}^{V2G} &= 1 + k\Delta f + (SOC_{max}^{normal} - SOC_{min}) \\ SOC_{min} &\leq SOC_{max}^{V2G} \leq 1 \\ \Delta f &\leq \Delta f_{cr} \end{aligned} \qquad (8)$$

where SOC_{max}^{V2G} is the modified high SOC set point under V2G mode.

Figure 5 is used to illustrate the dynamic frequency control strategy. EVs connecting to system are divided into charging group, discharging group and idle group.

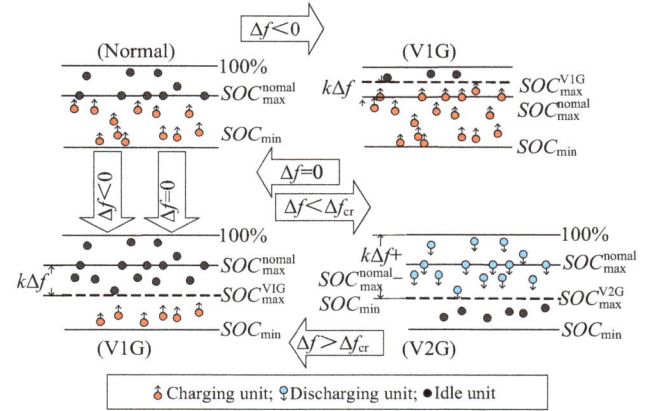

Fig. 5 Dynamic frequency control strategy

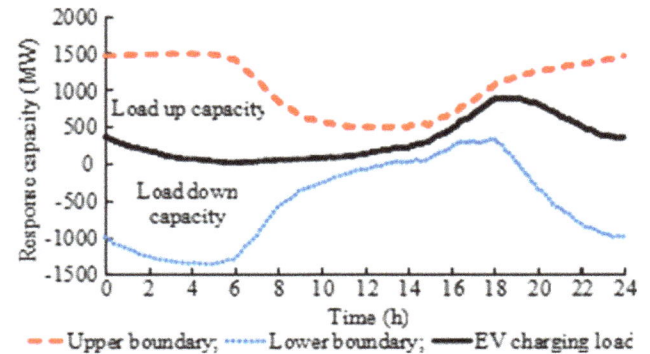

Fig. 6 Response capacity (dumb charging)

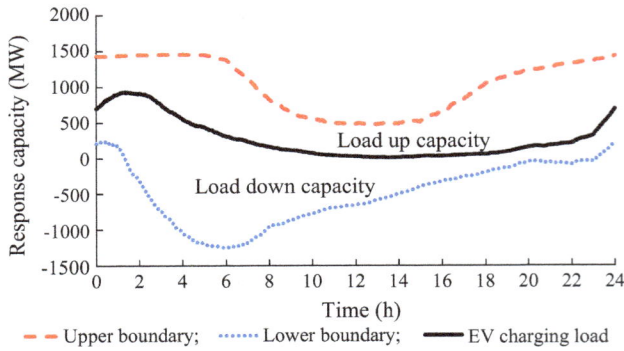

Fig. 7 Response capacity (smart charging)

The two control modes switch smoothly following the frequency deviation, which can serve the frequency regulation of power system. At the same time, the comfort level is considered in the strategy.

4 Case study

The GB power system model discussed above was utilized to study EVs' ability to serve frequency response. Two charging scenarios proposed in [19] (dumb charging & smart charging) were used in this section to consider the travel behaviors and comforts of EV owners. The number

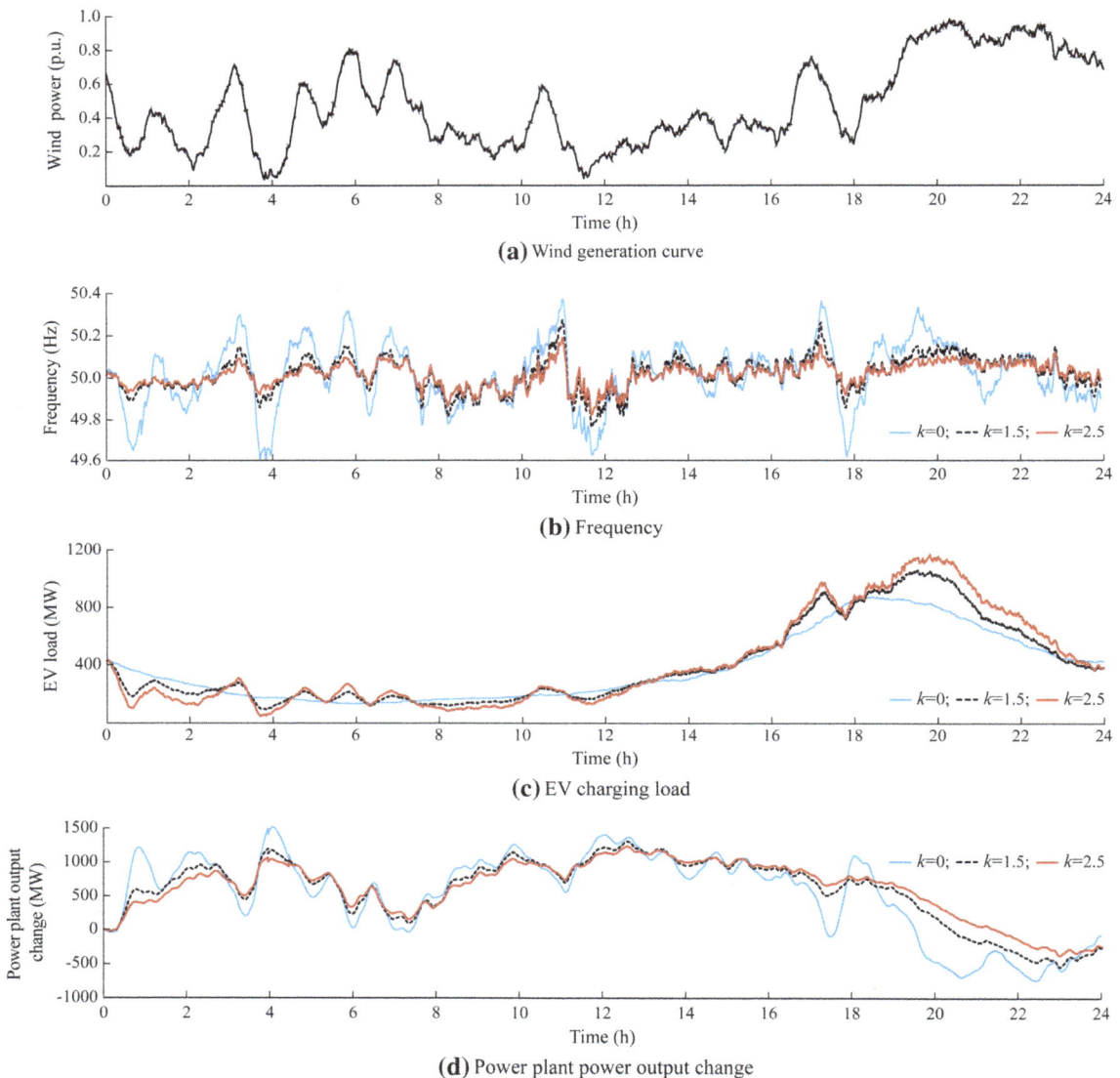

(a) Wind generation curve

(b) Frequency

(c) EV charging load

(d) Power plant power output change

Fig. 8 Simulation results (dumb charging)

of EVs which are anticipated in 2020 is the same as [20]. They constitute 1.74% of vehicles in GB in the year of 2011 [21]. The charger parameter L is set as 7 mH and R_c as 0.2 Ω for the EV charging model ($T_{EV} = 35$ ms) [20]. The SOC lower limit SoC_{min} is 30%. The normal SOC upper limit SOC_{max}^{normal} is 80%.

4.1 Response capacity of EVs

In this section, the EV aggregation model is built to investigate the capacity during a whole day considering EV characteristics and travel behaviors. It is built with the consideration of battery type, battery capacity, energy consumption per kilometer, daily travel distance, daily travelling time and demand SOC levels for travelling [22]. In dumb charging, all of the EVs are considered to start charging soon after their daily trip. Smart charging is used to shift the charging load from peak load time to valley hours.

Figures 6 and 7 show the EV charging power with its upper/lower boundary, which show the available response capacity (load up/ load down) of EVs along a whole day.

The upper boundary refers to the charging power of EVs that with SOC lower than 100% all turn to charging status. The lower boundary refers to the charging power of EVs that with SOC higher than 30% all turn to discharging status. Due to the transportation behaviors of EVs, the available response capacity changes over time. The response capacity is relatively small during the day time (from 6:00 to 18:00) [25].

In dumb charging, as shown in Fig. 6, larger response capacity is obtained at night (from 18:00 to 6:00). Smart charging fails to charge immediately after EVs come back from daily trips. Thus the load down capacity is smaller from 18:00 to 2:00, as shown in Fig. 7. The capacity has a strong impact on the effect of frequency response.

4.2 Simulation results for frequency response

The simulation results under dumb charging are shown in Fig. 8. Fig. 8a shows the wind generation curve (P_{wind}) along a day. The intermittence of wind power will increase the difficulty in frequency control. As depicted in Fig. 8b, the frequency fluctuations with EVs' response ($k > 0$) are

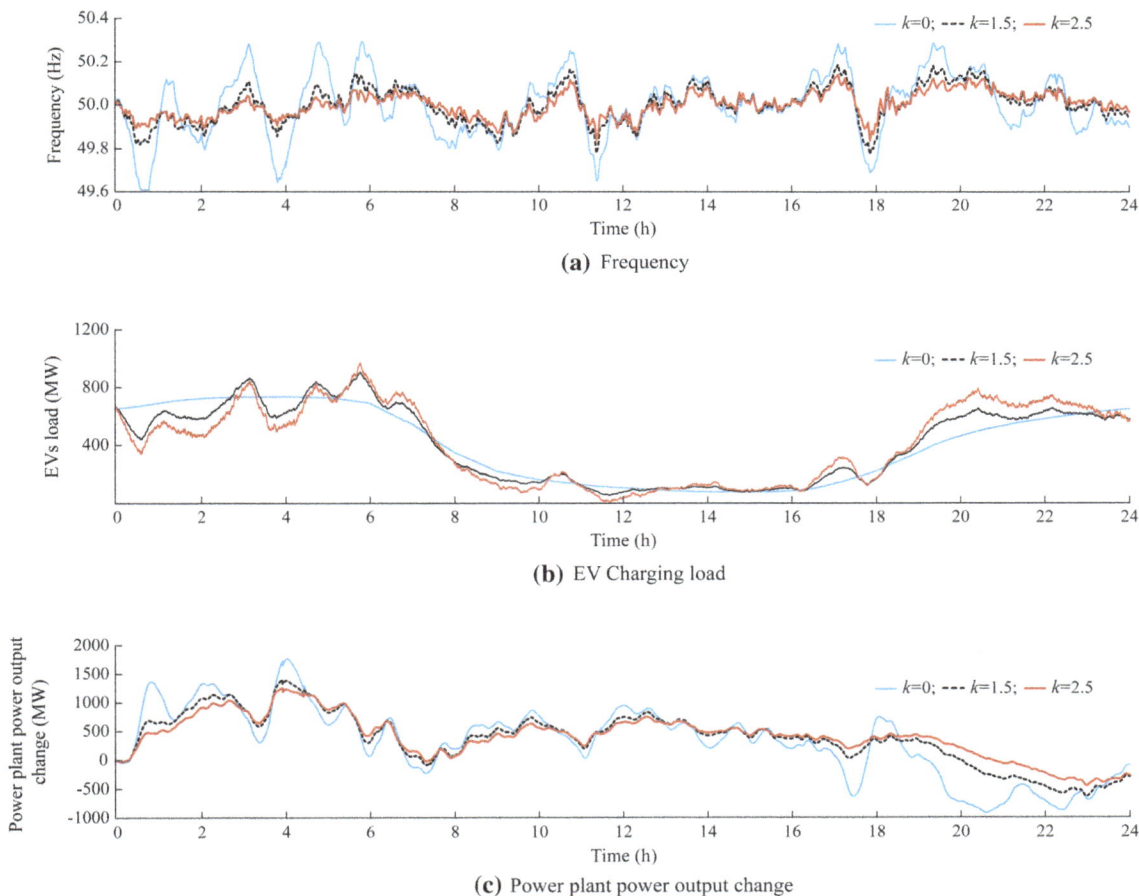

(a) Frequency

(b) EV Charging load

(c) Power plant power output change

Fig. 9 Simulation results (smart charging)

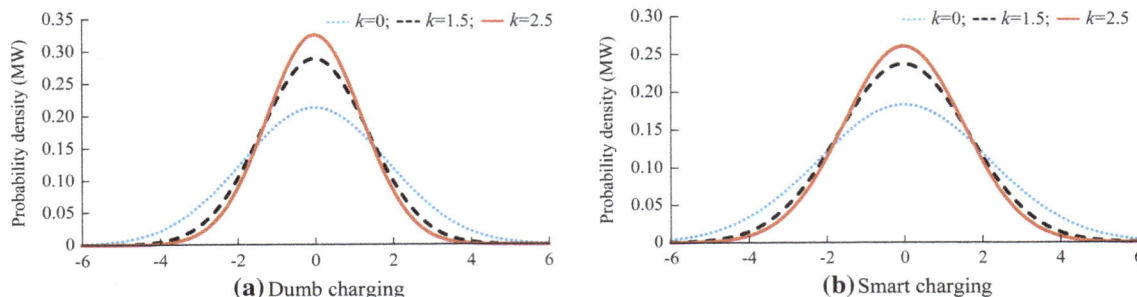

Fig. 10 Probability density of power plant output variation on minute-scale

much smaller than that without EVs' response ($k = 0$). A larger k means a better frequency control effect. When $k = 0$, the frequency deviation is within [−0.4, 0.4 Hz]. With the increasing value of k, the participation factor of EVs increases. The frequency deviation is mainly within [−0.2, 0.2 Hz] when k equals to 1.5 and 2.5. The effect of frequency response is also constrained by the available response capacity of EVs, which can be seen in Fig. 8b. In the day time (6:00–18:00), the frequency deviation is mainly within [−0.2, 0.2 Hz]. The range is narrowed to [−0.1, 0.1 Hz] at night (18:00–6:00). This can also be verified by Fig. 6.

Figure 8c shows the EV charging load along the day. The curve with $k = 0$ shows the EV charging load without participating in frequency response. EV charging load will either decrease when $\Delta f < 0$ or increase when $\Delta f > 0$. The load variation will increase with the growth of k. EV charging profiles are integrated into the 24-hour load demand [23, 24].

Figure 8d shows the power output variations from conventional power plant along a whole day. With EVs participating in frequency response, the power output fluctuation of power plant is reduced. A larger k brings smaller power fluctuation range.

The simulation results under smart charging are shown in Fig. 9. Due to the lack of 'load down' capacity at night, the frequency response from 17:30 to 18:30 is worse than that with dumb charging, as shown in Fig. 9a. Smart charging can also cause a shift of EV charging load, as shown in Fig. 9b–c. The comparison results between dumb and smart charging illustrate that the ability of EVs to provide dynamic frequency service has a close relationship with the EV charging scenarios.

The benefits from EV frequency response can also be illustrated by the probability density distribution of power plant output in Fig. 10. With the growth of k, the fluctuation of the power output variation is reduced with both dumb and smart charging scenarios, which can significantly reduce the operational cost of power plants.

5 Conclusions

In this paper, we studied the potential of EVs for dynamic frequency regulation in the UK in 2020. A dynamic frequency control strategy from EVs considering the owner comfort level was developed. The strategy was embedded to the simplified GB system with a GECM for the dynamic frequency response study. The charging load under two charging scenarios (dumb charging & smart charging) with two control modes were considered. The following conclusions can be drawn:

1) The EV aggregation have great potential for providing system dynamic frequency response. Using EVs to maintain the frequency stability can significantly reduce the frequency deviations along the whole day. At the same time, the power fluctuations of conventional generators are decreased to some extent. However, the regulation capacity of EVs has a distribution along 24 hours. Thus the frequency response from traditional power plants should be scheduled dynamically along the whole day.

2) The regulation capacity of the EV aggregation to serve dynamic frequency regulation is closely related to the charging scenarios. For the EVs with dumb charging, large regulation capacity can only be reached at peak load hour of the charging loads. In contrast, smart charging is able to distribute the regulation capability to most of the time along the whole day.

3) The impact of the time delay caused by the chargers is deemed negligible as shown in the simulation results. EVs' ability to provide frequency response is closely related to the response capacity and participation factor k. Two charging scenarios (dumb/smart charging) with different available response capacity show different frequency response characteristics. In this paper, the whole EV aggregation shares the same k during the whole day. How to determine the optimum k under different conditions considering personalized demand and travelling comfort requirements of different EV owners should be studied in future research.

Acknowledgements This work was supported in part by UK-China NSFC/EPSRC EV (Nos. 51361130152 & EP/L001039/1), the project National Natural Science Foundation of China (Nos. 51307115, 51377117, and 51277128), the National High Technology R&D Program (863 Program) of China (No. 2015AA050403), the Special funding for "Thousands Plan" of SGCC (No. XT71-12-028), Top & Tail Transformation Program (No. EP/I031707/1), and Tianjin Municipal Science and Technology Development Program of China (No. 13TXSYJC40400).

References

[1] Department for Environment Food and Rural Affairs (2008) Local authority CO_2 emissions estimates 2006—statistical summary. DEFRA, London

[2] Oudalov A, Chartouni D, Ohler C (2007) Optimizing a battery energy storage system for primary frequency control. IEEE Trans Power Syst 22(3):1259–1266

[3] Mercier P, Cherkaoui R, Oudalov A (2009) Optimizing a battery energy storage system for frequency control application in an isolated power system. IEEE Trans Power Syst 24(3):1469–1477

[4] Pearmine R, Song YH, Chebbo A (2007) Influence of wind turbine behavior on the primary frequency control. IET Renew Power Gener 1(2):142–150

[5] Ramtharan G, Ekanayake JB, Jenkins N (2007) Frequency support from doubly fed induction generator wind turbines. IET Renew Power Gener 1(1):3–9

[6] Short JA, Infield DG, Ferris LL (2007) Stabilization of grid frequency through dynamic demand control. IEEE Trans Power Syst 22(3):1284–1293

[7] Pillai JR, Bak-Jensen B (2010) Vehicle-to-grid systems for frequency regulation in an Islanded Danish distribution network. In: Proceedings of the 2010 IEEE vehicle power and propulsion conference (VPPC'10), Lille, France, 1–6 Sept 2010, 6 pp

[8] Teimourzadeh Baboli P, Parsa Moghaddam M, Fallahi F (2011) Utilizing electric vehicles on primary frequency control in smart power grids. In: International proceedings of chemical, biological and environmental engineering (IPCBEE), vol 26: Proceedings of the 2011 international conference on petroleum and sustainable development (ICPSD'11), Dubai, United Arab Emirates, 28–30 Dec 2011, pp 6–10

[9] Ersal T, Ahn C, Hiskens IA et al (2011) Impact of controlled plug-in EVs on microgrids: A military microgrid example. In: Proceedings of the 2011 IEEE Power and Energy Society general meeting, San Diego, CA, USA, 24–29 July 2011, 7 pp

[10] Peças Lopes JA, Soares FJ, Rocha Almeida PM (2011) Integration of electric vehicles in the electric power system. Proc IEEE 99(1):168–183

[11] National Electricity Transmission System Security and Quality of Supply Standard (2009). NETS SQSS, Version 2.0, National Grid, London, UK

[12] Erinmez IA, Bickers DO, Wood GF et al (1999) NGC experience with frequency control in England and Wales—provision of frequency response by generators. In: Proceedings of the 1999

IEEE Power and Energy Society winter meeting, vol 1. New York, NY, USA, 31 Jan–4 Feb 1999, pp 590–596

[13] National Grid (2009) Report of the investigation into the automatic demand disconnection following multiple generation losses and the demand control response, National Grid, London, UK, 27 May 2008

[14] National Grid (2009) Future balancing services requirements: reserve. National Grid, London

[15] Ekanayake JB, Jenkins N, Strbac G (2008) Frequency response from wind turbines. Wind Eng 32(6):573–586

[16] Peças Lopes JA, Rocha Almeida PM, Soares FJ (2009) Using vehicle-to-grid to maximize the integration of intermittent renewable energy resources in islanded electric grids. In: Proceedings of the international conference on clean electrical power (ICCEP'09), Capri, Italy, 9–11 June 2009, pp 290–295

[17] Zhao Y, Hu XH, He ZY et al (2008) A study of mathematic modeling of VSC for electromechanical transient analysis. In: Proceedings of the China international conference on electricity distribution (CICED'08), Guangzhou, China, 10–13 Dec 2008, 6 pp

[18] Yin M, Li GY, Li GK, et al (2004) Modeling of VSC-HVDC and its active power control scheme. In: Proceedings of the 2004 international conference on power system technology (POWERCON'04), vol 2. Singapore, 21–24 Nov 2004, pp 1351–1355

[19] Qian KJ, Zhou CZ, Allan M et al (2011) Modeling of load demand due to EV battery charging in distribution systems. IEEE Trans Syst 26(2):802–810

[20] Mu YF, Wu JZ, Ekanayake J et al (2013) Primary frequency response from electric vehicles in the Great Britain power system. IEEE Trans Smart Grid 4(2):1142–1150

[21] Department for Transport (2012) Vehicle licensing statistics Great Britain 2011. Department for Transport, London

[22] Wang MS, Zeng PL, MU YF, et al (2014) An efficient power plant model of electric vehicles considering the travel behaviors of EV users. In: Proceedings of the 2014 IEEE international conference on the power system technology (POWERCON'14), Chengdu, China, 20–22 Oct 2014, pp 3322–3327

[23] Yang ZL, Li K, Niu Q et al (2014) A self-learning TLBO based dynamic economic/environmental dispatch considering multiple plug-in electric vehicle loads. J Mod Power Syst Clean Energy 2(4):298–307

[24] Smith R, Meng K, Dong ZY et al (2013) Demand response: a strategy to address residential air-conditioning peak load in Australia. J Mod Power Syst Clean Energy 1(3):223–230

[25] Wang MS, Mu YF, Jia HJ et al (2015) A preventive control strategy for static voltage stability based on an efficient power plant model of electric vehicles. J Mod Power Syst Clean Energy 3(1):103–113

Jian MENG is a master-degree student majoring in Power System and its Automation in Tianjin University, China. His main research interests include power system stability and control, and integration of electric vehicles.

Yunfei MU is a lecturer at Tianjin University. His main research interests include power system security analysis, electric vehicles and smart grids.

Jianzhong WU is currently a reader in Institute of Energy, Cardiff School of Engineering, Cardiff University, U.K. His research activities are focused on energy infrastructure and smart grids.

Hongjie JIA is a professor of Tianjin University. His research interests include power system stability analysis and control, distribution network planning, renewable energy integration, and smart grids.

Qian DAI is an engineer of China Electric Power Research Institute. Her main research interest includes power system planning under uncertainty, power system reliability, distributed generation and energy management of smart grid.

Xiaodan YU is an associate professor of Tianjin University. Her research interests include power system stability analysis and control, and smart grids.

Permissions

The contributors of this book come from diverse backgrounds, making this book a truly international effort. This book will bring forth new frontiers with its revolutionizing research information and detailed analysis of the nascent developments around the world.

We would like to thank all the contributing authors for lending their expertise to make the book truly unique. They have played a crucial role in the development of this book. Without their invaluable contributions this book wouldn't have been possible. They have made vital efforts to compile up to date information on the varied aspects of this subject to make this book a valuable addition to the collection of many professionals and students.

This book was conceptualized with the vision of imparting up-to-date information and advanced data in this field. To ensure the same, a matchless editorial board was set up. Every individual on the board went through rigorous rounds of assessment to prove their worth. After which they invested a large part of their time researching and compiling the most relevant data for our readers.

The editorial board has been involved in producing this book since its inception. They have spent rigorous hours researching and exploring the diverse topics which have resulted in the successful publishing of this book. They have passed on their knowledge of decades through this book. To expedite this challenging task, the publisher supported the team at every step. A small team of assistant editors was also appointed to further simplify the editing procedure and attain best results for the readers.

Apart from the editorial board, the designing team has also invested a significant amount of their time in understanding the subject and creating the most relevant covers. They scrutinized every image to scout for the most suitable representation of the subject and create an appropriate cover for the book.

The publishing team has been an ardent support to the editorial, designing and production team. Their endless efforts to recruit the best for this project, has resulted in the accomplishment of this book. They are a veteran in the field of academics and their pool of knowledge is as vast as their experience in printing. Their expertise and guidance has proved useful at every step. Their uncompromising quality standards have made this book an exceptional effort. Their encouragement from time to time has been an inspiration for everyone.

The publisher and the editorial board hope that this book will prove to be a valuable piece of knowledge for researchers, students, practitioners and scholars across the globe.

List of Contributors

Fei YANG and Liang GUO
NARI Technology Development Co., Ltd., Nanjing 210061, China

Robert SMITH and Robert SIMPSON
Ausgrid, Sydney, NSW 2000, Australia

Ke MENG and Zhaoyang DONG
Center for Intelligent Electricity Networks (CIEN), The University of Newcastle, Newcastle, NSW, Australia

Lei GUO
North Subsection of State Grid Corporation of China, Beijing 100053, China

Qiwei QIU
Shanghai Electric Power Company, Shanghai 201400, China

Jian LIU and Yu ZHOU
North Subsection of State Grid Corporation of China, Beijing 100053, China

Linglei JIANG
Maintenance Company of Shanghai Electric Power Company, Shanghai 200000, China

Guangfu TANG, Zhiyuan HE and Hui PANG
State Grid Smart Grid Research Institute, Beijing 102200, China

Chengxin LI and Junyong LIU
The School of Electrical Engineering and Information, Sichuan University, Chengdu 610065, China

Guo CHEN
The School of Electrical and Information Engineering, The University of Sydney, Camperdown, NSW 2006, Australia

Dong YANG
State Grid Shandong Electric Power Research Institute, Jinan 250002, China

Kang ZHAO and Yutian LIU
Key Laboratory of Power System Intelligent Dispatch and Control of Ministry of Education, Shandong University, Jinan 250061, China

Jianmin ZHANG
School of Automation, Hangzhou Dianzi University, Hangzhou 310017, China

Qianzhi ZHANG
School of Electrical, Computer, and Energy Engineering, Ira A. Fulton Schools of Engineering, Arizona State University, Tempe, AZ 85287-5706, USA

Chunyu ZHANG, Yi DING, Pierre PINSON and Jacob ØSTERGAARD
Center for Electric Power and Energy, Technical University of Denmark, 2800 Kgs. Lyngby, Denmark

Niels Christian NORDENTOFT
Danish Energy Association, 1970 Frederiksberg, Denmark

Zhile YANG and Kang LI
School of Electrical, Electronics and Computer Science, Queen's University Belfast, Belfast BT9 5AH, UK

Qun NIU
Shanghai Key Laboratory of Power Station Automation Technology, School of Mechatronic Engineering and Automation, Shanghai University, Shanghai 200072, China

Yusheng XUE
NARI Group Corporation (State Grid Electric Power Research Institute), Nanjing 211106, China

Aoife FOLEY
School of Mechanical and Aerospace Engineering, Queen's University, Belfast BT9 5AH, UK

Ming NIU, Can WAN and Zhao XU
The Hong Kong Polytechnic University, Hong Kong, China

Wenming GONG
University of Chinese Academy of Sciences, Beijing 100190, China

Shuju HU and Honghua XU
Institute of Electrical Engineering, Chinese Academy of Sciences, Beijing 100190, China

Martin SHAN
Control Engineering and Energy Storage Systems, Fraunhofer IWES, 34119 Kassel, Germany

Francesco Paolo DEFLORIO, Luca CASTELLO, Ivano PINNA and Paolo GUGLIELMI
Politecnico di Torino, 10129 Turin, Italy

Fengji LUO, Junhua ZHAO, Haiming WANG and Yingying CHEN
Centre for Intelligent Electricity Networks (CIEN), The University of Newcastle, Callaghan, NSW 2308, Australia

Xiaojiao TONG
The Hunan First Normal University, Changsha, China

Zhao Yang DONG
The University of Sydney, Sydney, NSW 2006, Australia

Zhiguo LEI
School of Mechanical and Electrical Engineering, Fujian Agriculture and Forestry University, Fuzhou 350002, Fujian, China

Chengning ZHANG, Junqiu LI and Guangchong FAN
National Engineering Laboratory for Electric Vehicle, Beijing Institute of Technology, Beijing 100081, China

Zhewei LIN
Ningde Entry-Exit Inspection and Quarantine Bureau, Fujian 355017, China

Hong LIU, Jianyi GUO and Shaoyun GE
Key Laboratory of Smart Grid of Ministry of Education, Tianjin University, Tianjin 300072, China

Pingliang ZENG
Electric Power Research Institute of China, Beijing 100192, China

Huiyu WU
University of Wisconsin Madison, Madison, WI 53706, USA

Wuhua LI, Chi XU, Hongbin YU, Yunjie GU and Xiangning HE
Zhejiang University, Hangzhou 310027, China

Jiuqing CAI, Changsong CHEN, Peng LIU and Shanxu DUAN
State Key Laboratory of Advanced Electromagnetic Engineering and Technology, School of Electrical and Electronic Engineering, Huazhong University of Science and Technology, Wuhan, China

Jian ZHANG, Xiaodong YUAN and Yubo YUAN
Jiangsu Electric Power Company Research Institute, Nanjing 211103, China

Yongjie FANG
NARI Group Corporation, Nanjing 211106, China

Mingwei LI, Xu WANG, Yu GONG, Yangyang LIU and Chuanwen JIANG
Department of Electrical Engineering, Shanghai Jiao Tong University, Shanghai 200240, China

Puyu WANG, Xiao-Ping ZHANG
University of Birmingham, Birmingham B15 2TT, UK

Paul F. COVENTRY
COVENTRY, National Grid, Warwick CV34 6DA, UK

Zhou LI
Southeast University, Nanjing 210096, China

Mingfei WU and Dylan Dah-Chuan LU
School of Electrical and Information Engineering, The University of Sydney, Darlington, NSW, Australia

Jian MENG, Yunfei MU, Hongjie JIA and Xiaodan YU
Key Laboratory of Smart Grid of Ministry of Education, Tianjin University, Tianjin 300072, China

Jianzhong WU
School of Engineering, Institute of Energy, Cardiff University, Cardiff, UK

Qian DAI
China Electric Power Research Institute, Haidian District, Beijing 100192, China